Rock Reinforcement and Rock Support

ISRM Book Series
Series editor: Xia-Ting Feng
Institute of Rock and Soil Mechanics, Chinese Academy of Sciences, Wuhan, China

ISSN : 2326-6872
eISSN: 2326-778X

Volume 6

International Society for Rock Mechanics

ISRM

Rock Reinforcement and Rock Support

Ömer Aydan

*Department of Civil Engineering, University of the Ryukyus,
Nishihara, Okinawa, Japan*

CRC Press
Taylor & Francis Group
Boca Raton London New York

CRC Press is an imprint of the
Taylor & Francis Group, an **informa** business

A BALKEMA BOOK

Cover illustration: Okumino Underground Power House. Courtesy of Chubu Electric
Power Company.

Published by:
CRC Press/Balkema
P.O. Box 447, 2300 AK Leiden, The Netherlands
e-mail: Pub.NL@taylorandfrancis.com
www.crcpress.com – www.taylorandfrancis.com

First issued in paperback 2021

ISBN-13: 978-1-03-209587-5 (pbk)
ISBN-13: 978-1-138-09583-0 (hbk)

Typeset by Apex CoVantage, LLC

Library of Congress Cataloging-in-Publication Data
Names: Aydan, Ömer, author.
Title: Rock reinforcement and rock support / Ömer Aydan, University of Ryukyus Dept.
    of Civil Engineering.
Description: Boca Raton : CRC Press/Balkema, 2017. | Series: ISRM book series,
    ISSN 2326-6872 ; volume 6 | Includes bibliographical references and index.
Identifiers: LCCN 2017045063 (print) | LCCN 2017045840 (ebook) |
    ISBN 9781315391304 (ebook) | ISBN 9781138095830 (hardcover : alk. paper)
Subjects: LCSH: Rock mechanics. | Engineering geology.
Classification: LCC TA706 (ebook) | LCC TA706 .A9295 2017 (print) |
    DDC 624.1/5132—dc23
LC record available at https://lccn.loc.gov/2017045063

Contents

About the author   xi
Acknowledgments   xiii

1 Introduction   1

**2 Mechanism of failure in rock engineering structures
and its influencing factors**   7
  2.1   Rock, discontinuities, and rock mass   7
        2.1.1   Rocks   7
        2.1.2   Origin of discontinuities in rock and their mechanical
                behavior   9
        2.1.3   Rock mass and its mechanical behavior   13
  2.2   Modes of instability about underground openings   16
  2.3   Modes of instability of slopes   23
  2.4   Modes of instability of foundations   27

**3 Design philosophy of rock support and rock
reinforcement**   31
  3.1   Introduction   31
  3.2   Empirical design methods   33
        3.2.1   Rock Quality Designation (RQD) method   35
        3.2.2   Rock Mass Rating (RMR)   35
        3.2.3   Q-system (rock tunneling quality index)   37
        3.2.4   Rock Mass Quality Rating (RMQR)   39
  3.3   Analytical approach   40
        3.3.1   Hydrostatic in situ stress state   40
        3.3.2   Non-hydrostatic in situ stress state   41
  3.4   Numerical methods   42
  3.5   Methods for stabilization against local instabilities   46
        3.5.1   Estimation of suspension loads   46
        3.5.2   Sliding loads   48
        3.5.3   Loads due to flexural toppling   49
  3.6   Integrated and unified method of design   51

3.7   Considerations on the philosophy of support and reinforcement
       design of rock slopes   56
       3.7.1   Empirical design systems   56
       3.7.2   Kinematic approach   58
       3.7.3   Integrated stability assessment and design system
                for rock slopes   58
3.8   Considerations on philosophy of support design of pylons   62
       3.8.1   Geological, geophysical, and mechanical investigations   64
       3.8.2   Specification of material properties   64
3.9   Considerations on the philosophy of foundation design of dams
       and bridges   65

4 Rockbolts (rockanchors)   67
4.1   Introduction   67
4.2   Rockbolt/rockanchor materials and their mechanical behaviors   69
       4.2.1   Yield/failure criteria of rockbolts   70
       4.2.2   Constitutive modeling of rockbolt material   72
4.3   Characteristics and material behavior of bonding annulus   76
       4.3.1   Push-out/pull-out tests   76
       4.3.2   Shear tests   83
4.4   Axial and shear reinforcement effects of bolts in continuum   94
       4.4.1   Contribution to the deformational moduli of the medium   94
       4.4.2   Contribution to the strength of the medium   94
       4.4.3   Improvement of apparent mechanical properties of rock
                and confining pressure effect   97
4.5   Axial and shear reinforcement effects of bolts in medium with
       discontinuities   99
       4.5.1   Increment of the tensile resistance of a discontinuity
                plane by a rockbolt   101
       4.5.2   Increment of the shear resistance of a discontinuity
                plane by a rockbolt   101
       4.5.3   Response of rockbolts to movements at/along
                discontinuities   106
4.6   Estimation of the cyclic yield strength of interfaces for
       pull-out capacity   110
4.7   Estimation of the yield strength of interfaces in boreholes   111
4.8   Pull-out capacity   113
       4.8.1   Constitutive equations   114
       4.8.2   Governing equations   116
4.9   Simulation of pull-out tests   127
4.10   Mesh bolting   132
       4.10.1   Evaluation of elastic modulus of reinforced medium   132
       4.10.2   Evaluation of tensile strength of reinforced medium   134

5 Support members   137
5.1   Introduction   137
5.2   Shotcrete   137
     5.2.1   Historical background   137
     5.2.2   Experiments on shotcrete 138
     5.2.3   Constitutive modeling   146
     5.2.4   Structural modeling of shotcrete 147
5.3   Concrete liners   148
     5.3.1   Historical background   148
     5.3.2   Mechanical behavior of concrete 148
     5.3.3   Constitutive modeling of concrete 150
     5.3.4   Structural modeling   150
5.4   Steel liners and steel ribs/sets   151
     5.4.1   Steel liners   151
     5.4.2   Steel ribs/sets   151
     5.4.3   Constitutive modeling   151
     5.4.4   Structural modeling   151

6 Finite element modeling of reinforcement/support system   155
6.1   Introduction   155
6.2   Modeling reinforcement systems: rockbolts   155
     6.2.1   Mechanical modeling of steel bar 156
     6.2.2   Mechanical modeling of grout annulus 156
     6.2.3   Finite element formulation of rockbolt element 159
6.3   Finite element modeling of shotcrete   166
6.4   Finite element modeling of steel ribs/sets or shields 168
6.5   Finite element analysis of support and reinforcement systems 170
6.6   Discrete finite element method (DFEM-BOLT) for the analysis of support and reinforcement systems 171
     6.6.1   Mechanical modeling   171
     6.6.2   Finite element modeling 172
     6.6.3   Finite element modeling of block contacts 173
     6.6.4   Considerations of support and reinforcement system 175

7 Applications to underground structures   177
7.1   Introduction   177
7.2   Analytical approach   177
     7.2.1   Solutions for hydrostatic in situ stress state for support system and fully grouted rockbolts 177
     7.2.2   Solutions for hydrostatic in situ stress state for pre-stressed rockanchors 199
     7.2.3   Analytical solutions for non-hydrostatic in situ stress state 207

7.3		Numerical analyses on the reinforcement and support effects in continuum	217
	7.3.1	Effect of bolt spacing	217
	7.3.2	Effect of the magnitude of the allowed displacement before the installation of the bolts	217
	7.3.3	Effect of elastic modulus of the surrounding rock	218
	7.3.4	Effect of equipping rockbolts with bearing plates	219
	7.3.5	Effect of bolting pattern	221
	7.3.6	Applications to actual tunnel excavations	223
	7.3.7	Comparison of reinforcement effects of rockbolts and shotcrete	227
	7.3.8	Application to Tawarazaka Tunnel	232
7.4		Mesh bolting in compressed air energy storage schemes	235
	7.4.1	Analytical solution	235
	7.4.2	Applications	238
7.5		Reinforcement effects of rockbolts in discontinuum	242
	7.5.1	Reinforcement against separation: suspension effect	242
	7.5.2	Pillars: shear reinforcement of a discontinuity by a rockbolt	244
	7.5.3	Shear reinforcement against bending and beam building effect	246
	7.5.4	Reinforcement against flexural and columnar toppling failure	248
	7.5.5	Reinforcement against sliding	254
	7.5.6	Arch formation effect	256
7.6		Support of subsea tunnels	267
7.7		Reinforcement and support of shafts	268
7.8		Special form of rock support: backfilling of abandoned room and pillar mines	271
	7.8.1	Short-term experiments	272
	7.8.2	Long-term experiments	279
	7.8.3	Verification of the effect of backfilling through in situ monitoring	281
	7.8.4	Analysis of backfilling of abandoned mines	283
8 Reinforcement and support of rock slopes			**289**
8.1		Introduction	289
8.2		Reinforcement against planar sliding	289
	8.2.1	Finite element analysis	290
	8.2.2	Physical model experiments	292
	8.2.3	Discrete finite element analyses	293
8.3		Reinforcement against flexural toppling failure	296
	8.3.1	Limit equilibrium method	296

		8.3.2	Finite element method	297
		8.3.3	Discrete finite element analyses	298
	8.4	Reinforcement against columnar toppling failure		300
		8.4.1	Physical model experiments	301
		8.4.2	Discrete finite element analyses	302
	8.5	Reinforcement against combined sliding and shearing		304
		8.5.1	Formulation	304
		8.5.2	Stabilization	309
		8.5.3	Applications	309
	8.6	Physical model tests on the stabilization effect of rockbolts and shotcrete on discontinuous rock slopes using tilting frame apparatus		312
		8.6.1	Model materials and their properties	312
		8.6.2	Apparatuses and testing procedure	312
		8.6.3	Test cases	313
		8.6.4	Results and discussions	313
	8.7	Stabilization of slope against buckling failure		316

9 Foundations **317**

	9.1	Introduction		317
	9.2	Foundations under tension		317
		9.2.1	Pylons	317
		9.2.2	Design of anchorages	338
		9.2.3	Suspension bridges	353
	9.3	Foundations under compressions		365
		9.3.1	Base foundations	365
		9.3.2	Cylindrical sockets (piles)	370

10 Dynamics of rock reinforcement and rock support **375**

	10.1	Introduction		375
	10.2	Dynamic response of point-anchored rockbolt model under impulsive load		376
	10.3	Dynamic response of yielding rockbolts under impulsive load		378
	10.4	Turbine induced vibrations in an underground powerhouse		381
	10.5	Dynamic behavior of rockbolts and rockanchors subjected to shaking		383
		10.5.1	Model tests on rockanchors restraining potentially unstable rock block at sidewall of underground openings	383
		10.5.2	Model tests on rockanchors restraining potentially unstable rock block in roof of underground openings	386
	10.6	Planar sliding of rock slope models		387
	10.7	A theoretical approach for evaluating axial forces in rockanchors subjected to shaking and its applications to model tests		394
	10.8	Application of the theoretical approach to rockanchors of an underground powerhouse subjected to turbine-induced shaking		395

10.9	Model tests on fully grouted rockbolts restraining a potentially unstable rock block against sliding	397
10.10	Excavations	404
	10.10.1 Unbolted circular openings	405
	10.10.2 Bolted circular openings	406
10.11	Dynamic response of rockbolts and steel ribs during blasting	407

11 Corrosion, degradation, and nondestructive testing **409**

11.1	Introduction	409
11.2	Corrosion and its assessment	409
	11.2.1 The principle of iron corrosion	409
	11.2.2 Factors controlling corrosion rate	410
	11.2.3 Experiments on corrosion rate of rockbolts	411
	11.2.4 Observations of iron bolts at Koseto hot spring discharge site	415
	11.2.5 Corrosion of iron at Ikejima Seashore	419
	11.2.6 Corrosion of deformed bar at Tekkehamam hot spring site	420
	11.2.7 Corrosion of an iron bar at Moyeuvre abandoned iron mine and its investigation by X-ray CT scanning technique	421
	11.2.8 Simulation of corrosion	423
	11.2.9 Effect of corrosion on the physico-mechanical properties of tendon	424
	11.2.10 Estimation of failure time of tendons	426
11.3	Effect of degradation of support system	428
11.4	Nondestructive testing for soundness evaluation	429
	11.4.1 Impact waves for nondestructive testing of rockbolts and rockanchors	430
	11.4.2 Guided ultrasonic wave method	451
	11.4.3 Magneto-elastic sensor method	452
	11.4.4 Lift-off testing technique	452
11.5	Conclusions	452

12 Conclusions **455**

Bibliography	463
Subject index	481

About the author

Born in 1955, Professor Aydan studied Mining Engineering at the Technical University of Istanbul, Turkey (B.Sc., 1979), Rock Mechanics and Excavation Engineering at the University of Newcastle upon Tyne, UK (M.Sc., 1982), and finally received his Ph.D. in Geotechnical Engineering from Nagoya University, Japan, in 1989. Prof. Aydan worked at Nagoya University as a research associate (1987–1991), and then at the Department of Marine Civil Engineering at Tokai University, first as Assistant Professor (1991–1993), then as Associate Professor (1993–2001), and finally as Professor (2001–2010). He then became Professor of the Institute of Oceanic Research and Development at Tokai University and is currently Professor at the University of Ryukyus, Department of Civil Engineering & Architecture, Nishihara, Okinawa, Japan. He has furthermore played an active role on numerous ISRM, JSCE, JGS, SRI, and Rock Mech. National Group of Japan committees and has organized several national and international symposia and conferences. Professor Aydan has received the 1998 Matsumae Scientific Contribution Award, the 2007 Erguvanlı Engineering Geology Best Paper Award, the 2011 Excellent Contributions Award from the International Association for Computer Methods in Geomechanics and Advances, and the 2011 Best Paper Award from the Indian Society for Rock Mechanics and Tunneling Technology, and he was awarded the 2013 Best Paper Award at the 13th Japan Symposium on Rock Mechanics and the 6th Japan-Korea Joint Symposium on Rock Engineering. He was also made Honorary Professor in Earth Science by Pamukkale University in 2008 and received the 2005 Technology Award, the 2012 Frontier Award, and the 2015 Best Paper Award from the Japan National Group for Rock Mechanics.

Acknowledgments

The author sincerely acknowledges Prof. Xia-Ting Feng for inviting the author to contribute to the ISRM Book Series on "Rock Reinforcement and Rock Support". The content of this book is an outcome of the studies carried out by the author at Nagoya University, Tokai University, and University of the Ryukyus in Japan over more than three decades. The author would like to thank Emeritus Prof. Dr Toshikazu Kawamoto of Nagoya University, Prof. Dr Yasuaki Ichikawa of Okayama University, Prof. Dr Yuzo Obara of Kumamoto University, Prof. Dr Takashi Kyoya of Tohoku University, Prof. Dr Tomoyuki Akagi and Prof. Dr Takashi Ito of Toyota National College of Technology, Assoc. Prof. Dr Naohiko Tokashiki of University of the Ryukyus, Dr Seiji Ebisu of Okumura Corporation, and his deceased friends Dr Mitsuhiro Sezaki of Miyazaki University and Mr Jun Itoh of Oriental Consultants Global Co. Ltd for their guidance, help, and suggestions at various stages of his studies quoted in this book.

The author also acknowledges particularly Dr Y. Miyaike, Mr S. Tsuchiyama, Mr H. Okuda, Dr S. Komura and Mr F. Uehara of Chubu Electric Power Company (Nagoya, Japan) for collaboration with the author in accessing many sites and providing financial funds to develop a dynamic direct shear testing device used in conventional, cyclic, creep, and dynamic behavior of interfaces of rockbolts and rockanchors and non-destructive equipment for the inspection of rockbolts/rockanchors utilized in the studies quoted in this book. He also acknowledges his former students, Mr Mitsuo Daido and Dr Yoshimi Ohta of Tokai University, for their help during experiments and computations reported in this study.

The author would also like to thank Alistair Bright, acquisitions editor at CRC Press/Balkema, for his patience and collaboration during the preparations for this book and Ms José van der Veer, the production editor, for her great efforts in producing this book.

Finally, I want to thank my wife, Reiko, my daughter, Ay, my son, Turan Miray, and my parents for their continuous help and understanding, without which this book could not have been completed.

Chapter 1

Introduction

The stability of underground and surface geotechnical structures during and after excavation is of great concern to designers, as any kind of instability may result in damage to the environment, as well as high repair costs and time consumption (Figs. 1.1–1.4). The rock in nature is not always continuous and may have numerous discontinuities that vary in scale. As a result, the safety evaluation of a structure under consideration is a highly complex problem and requires very careful investigation. Accordingly, it is always necessary to examine the most likely forms of instability in relation to the physical nature of the rock mass and the geometry of the structure and its site, as well as the pre-existing state of stress. The forms of instability and their mechanism and the factors and conditions associated with them must be clearly understood to correctly stabilize the structure.

Figure 1.1 Various underground structures in rock.

Figure 1.2 Tunnels in rock.

Figure 1.3 Foundations on rock.

Figure 1.4 Rock slopes.

In addition to the stability problems, the environmental requirements and functional duties of structures may need to be carefully evaluated. All these factors together with those related to the stabilization procedure will result in setting the conditions for the selection of support members that satisfy mechanical as well as environmental and functional requirements.

The design of support members and the evaluation of the stability of structures are not possible unless one understands what rock mass really is. Most of the available approaches are either mechanically orientated without proper consideration of rock mass or geologically orientated without paying proper attention to the mechanics. In this respect, the present volume attempts to bridge the two approaches and bring a unified approach for the design of support and reinforcement systems for rock engineering structures, from not only the mechanical engineering but also the geological engineering point of view.

Rockbolts of various types (i.e. mechanically anchored, grouted, etc.) have recently become one of the principal support members in the civil and mining engineering fields. This probably results from the ease of their transportation, storage, and installation and their rapidly developing reinforcement effects as compared with other support members, such as steel sets and concrete liners. Their superior reinforcement effects in securing the stability of geotechnical engineering structures excavated in various types of ground and states of stress are very well known qualitatively in engineering practice. However, the first fundamental study for quantifying the reinforcement effects of rockbolts has been carried out by

Aydan (1989) in his doctorate study. Subsequent studies by Pellet (1994); Moosavi (1997); Marence and Swoboda (1995) and Ebisu et al. (1994a, 1994b) have made further contributions on the behavior of rockbolts under different conditions. The studies on rockbolts, cable rockbolts, and rockanchors are now orientated towards their response under dynamic conditions (e.g. Aydan et al., 2012; Owada et al., 2004; Owada and Aydan, 2005; Li, 2010).

In the last decade, the use of shotcrete has rapidly increased, particularly in tunnel construction, and shotcrete has become an important element of modern tunnel-support techniques. The development of the early age strength of shotcrete is a decisive factor, because the excavation cycle and attainable excavation speeds are significantly influenced by it. The first fundamental study on the characteristics of shotcrete and its representation in numerical simulations was undertaken by Sezaki (1990) and his colleagues (Sezaki et al., 1989, 1992; Aydan et al., 1992).

Steel ribs or steel sets have long been used in many rock excavations. Their design concept is based as a moment-resisting structure under uniform or concentrated loads, and their load-bearing capacity is evaluated by assuming moment resistance capacity or buckling failure.

Despite decades of use of concrete liners in rock excavations, the supporting effects of concrete liners is not well understood. This is due to a poor understanding of how they interact with the surrounding rock mass, together with the incorporation of other support and reinforcement members and in relation to the installation stage in the overall construction scheme. The concrete liners are auxiliary support members rather than main load-bearing structures. Therefore, there is a strong debate whether they are necessary support members. In this book, various aspects of concrete liners are also presented and discussed.

The present book has been undertaken to highlight the reinforcement functions of rockbolts/rockanchors and support systems consisting of shotcrete, steel ribs, and concrete liners under various conditions and to evaluate their reinforcement and supporting effects, both qualitatively and quantitatively.

The book consists of 12 chapters. The contents of 10 chapters out of 12 are described briefly as follows:

Chapter 2 is devoted to the mechanism and influencing factors of failure phenomena in rock engineering structures. The rock and types of discontinuities encountered in natural rock are briefly described, and their combined effects on the mechanical response of rock mass as a structure are discussed together with the implications on real rock structures. Then, classifications on the forms of instability in underground openings, slopes, and foundations, under both compressive and tensile stress fields, are described in relation with the structure of rock mass.

Chapter 3 is concerned with the present design philosophy of support and reinforcement for rock engineering structures. A brief description of available design approaches, such as empirical, analytical, and numerical methods, are given and discussed. The approaches, which are used independently of each other, are presented in a unified manner. The presently available support members and their functions are briefly described and discussed, with an emphasis on rockbolts and rockbolting.

Chapter 4 describes experimental studies undertaken on the mechanical behavior of the rockbolt system. First, the behavior of the bolt material used in practice is given, then the experimental study undertaken for the anchorage performance of rockbolts in push-out and pull-out tests and subsequent shear tests on the mechanical behavior of interfaces within the system and grouting material are described. In this chapter, the constitutive laws for the rockbolt system are described. A constitutive law for the bar is derived based on the classical incremental elasto-plasticity theory, as bar materials such as steel exhibit a non-dilatant plastic behavior. On the other hand, the constitutive law for the grout annulus and interfaces is derived based on the multi-response theory proposed by Ichikawa (Ichikawa, 1985; Ichikawa et al., 1988), as the grout annulus and interfaces exhibit a dilatant plastic behavior. Then, procedures to determine the parameters for the constitutive laws from the experimental data are described and several examples are given. Evaluation of the contribution of rockbolts/rockanchors for improving the properties of rock mass is described and the shear reinforcement effect of rockbolts on rock discontinuities is presented in view of some theoretical and experimental findings. A detailed presentation of estimation of pull-out capacity of rockbolts/rockanchors under various conditions are described. Furthermore, the evaluation of reinforcement effect of mesh bolting on rock masses subjected to tensile stresses are presented.

Chapter 5 describes the characteristics of various support elements, such as shotcrete, concrete liner, and steel ribs/sets. The constitutive laws of each support member and various experimental studies on their characteristics are presented. Furthermore, the concepts for their mechanical modeling are also explained.

Chapter 6 describes the models representing reinforcement and support systems in numerical analyses, particularly in finite element studies. Details of rockbolt elements, shotcrete, and beam elements are presented.

Chapter 7 is concerned with the analytical and numerical methods for evaluating support and reinforcement systems and their effects in underground excavations. Analytical methods for evaluating the ground-response-support reaction, which incorporates various support members, rockbolts, and rockanchors, and the face effect are presented, and several examples of applications are given. Furthermore, a theoretical formulation of the effect that mesh bolting has for compressed air energy storage schemes is given, and several examples of excavations are presented. A series of finite element simulations are presented to show the effects of various conditions for the effective utilization of reinforcement and support systems for underground structures. The effect of rockbolting with other support members is investigated in relation to some practical situations. Several examples are analyzed on the response of rockbolts in discontinuum, and their implications for interpreting field measurements of rockbolt performances are discussed. Furthermore, the presently available proposals on the suspension effect, the beam building effect, and the arch formation effect of rockbolts are re-examined and more generalized solutions are presented. In addition to covering the reinforcement effect of rockbolts against the sliding type of failure, solutions for the reinforcement effect of bolts against the flexural and columnar type of toppling failure are given.

Chapter 8 describes the effect of support and reinforcement systems for the stabilization of rock slopes. Procedures for stabilizing the rock slopes against some typical failure modes are presented, along with several examples of applications. Furthermore, the chapter presents applications of the discrete finite element method, incorporating the effect of rockbolts to rock slope stability problems. In addition, model experiments on the effect of rockbolting against planar sliding and block-toppling modes are given and compared with estimations from the limit equilibrium technique.

Chapter 9 is concerned with the stabilization of the foundations of bridges, pylons, and dams subjected to tension or compressive forces. Examples of applications include the potential use of rockanchors as foundations of pylons and of tunnel-type anchorage for suspension bridges. The use of rockanchors for the stabilization of bridge and dam foundations under compression is also presented and discussed.

Chapter 10 deals with dynamic issues such as rockburst, earthquakes, and blasting, which cause dynamic loads on rock support and rock reinforcement. Theoretical, numerical, and experimental studies on rockbolts and rockanchors under shaking are presented, along with several examples of applications.

Chapter 11 describes the mechanisms and techniques for evaluating corrosion in steel and iron materials in relation to the long-term performance and degradation of reinforcement and support systems and provides site examples. Furthermore, some procedures are presented for non-destructive evaluation of support and reinforcement systems.

Chapter 2

Mechanism of failure in rock engineering structures and its influencing factors

This chapter deals with natural rock, the types of discontinuities encountered in it, rock mass, and the mechanism of the modes of instability in underground and surface structures and associated factors and conditions.

The first part of this chapter is devoted to the geological description of rocks and of the formation and types of discontinuities in rocks and rock mass. Then, the mechanical behavior of rock mass is discussed, considering the behaviors of intact rock, discontinuities, and the structure of the rock mass.

In the second part of the chapter, the discussion of various modes of instability of rock engineering structures and the factors associated with the modes of instability are presented. Then, Aydan's classifications for the modes of instability in rock engineering structures are presented in relation to the elements associated with the modes of instability (Aydan, 1989).

2.1 ROCK, DISCONTINUITIES, AND ROCK MASS

2.1.1 Rocks

Rocks in nature can be geologically classified into three main groups: igneous, sedimentary, and metamorphic, and each of these groups may be further subdivided into several classes. For example, igneous rocks are subdivided into three classes: extrusive, intrusive, and semi-intrusive, although the chemical composition of the three types may be same (Fig. 2.1). The order of minerals and the internal structure of rocks is a result of the chemical composition of rising magma, its velocity, and the environmental conditions during the cooling process, which greatly affects the discontinuity formation in such rocks.

Sedimentary rocks, on the other hand, result from the accumulation of particles differing in size, shape, and chemical composition in some certain geographical locations and a rebonding through certain physical or chemical agents or processes under various thermo-hydro environmental physical conditions (Fig. 2.2). The rocks belonging to this group are usually found in the form of layers, and the orientation of grains or minerals have some regularity in relation to the sedimentation process.

Metamorphic rocks are the result of the restructuring of existing rocks, which may be sedimentary, igneous, or even metamorphic under high pressures and/or high temperatures (Fig. 2.3). Because of high pressures and temperatures, the internal structure of rocks becomes highly anisotropic.

Figure 2.1 Views of some igneous rocks.

Figure 2.2 Views of some metamorphic rocks.

Figure 2.3 Views of some sedimentary rocks.

All rocks are an assemblage of a single mineral or several minerals of regular or irregular shapes differing in size and arranged in certain patterns, depending on the chemical and thermal phase changes and physical conditions at the time of their occurrence. The mechanical behavior of rocks is an apparent behavior of the mechanical response of minerals or grains and the interaction taking place among the grains due their shape and spatial distributions in relation to the applied constraint and force conditions.

2.1.2   Origin of discontinuities in rock and their mechanical behavior

Discontinuities in rocks are termed cracks, fractures, joints, bedding planes, schistosity, or foliation planes and faults. Discontinuities are products of certain phenomena the rocks were exposed to in their geological past and are expected to be regularly distributed within a rock mass. They can be classified into the four groups outlined below according to the mechanical or environmental process they underwent (Erguvanlı, 1973; Yüzer and Vardar, 1983; Miki, 1986; Ramsay and Huber, 1987; Aydan *et al.*, 1988b, etc.) (Fig. 2.4).

i)   Tension discontinuities due to

- Cooling
- Drying
- Freezing
- Bending
- Flexural slip
- Uplifting
- Faulting
- Stress relaxation due to erosion, glacier retreat, or human-made excavation

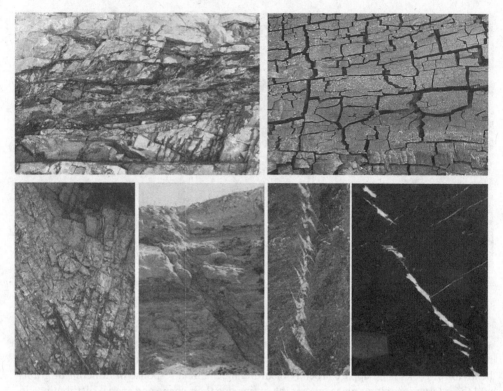

Figure 2.4 Views of discontinuities *in situ.*

ii) Shear discontinuities due to

- Folding
- Faulting

iii) Discontinuities due to periodic sedimentation
iv) Discontinuities due to metamorphism

Because of the discontinuities resulting from one or more of the combined actions of the abovementioned processes, the structure of rock mass in nature may look like an assemblage of blocks of typical shapes (Figs. 2.5 and 2.6). The most common block shapes are rectangular, rhombohedral, hexagonal, or pentagonal prisms. While hexagonal and/or pentagonal prismatic blocks are commonly observed in extrusive basic igneous rocks, such as andesite or basalt, and some fine-grained sedimentary rocks underwent cooling or drying processes, the most common block shapes are between a rectangular prism and a rhombohedral prism. The lower and upper bases of the blocks are usually limited by planes called flow planes, bedding planes, and schistosity or foliation planes in igneous, sedimentary, and metamorphic rocks, respectively. These discontinuities can be regarded very continuous for most of the rock structures concerned. Other discontinuities are usually found in, at least, two or three sets, crossing these planes orthogonally or obliquely. These secondary sets, if present, may

Figure 2.5 Views of rock mass in nature.

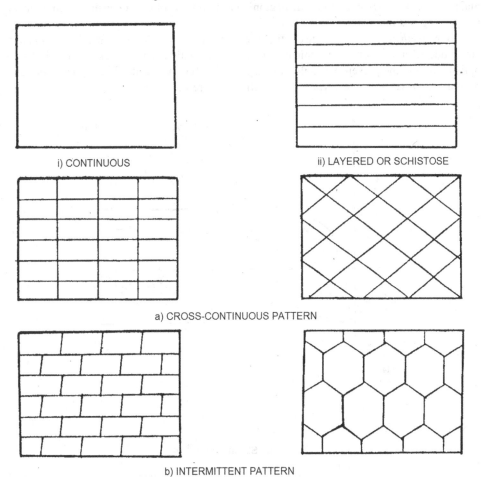

i) CONTINUOUS

ii) LAYERED OR SCHISTOSE

a) CROSS-CONTINUOUS PATTERN

b) INTERMITTENT PATTERN

iii) BLOCKY

Figure 2.6 Geometrical modeling of rock mass.

be very continuous or intermittent. As a result, the rock mass may be viewed as (Fig. 2.6) (Goodman, 1976; Aydan *et al.*, 1988b):

- Continuous medium
- Tabular (layered) medium
- Blocky medium

Blocky medium can be further subdivided into two groups, depending upon the continuity of secondary sets as follows (Aydan and Kawamoto, 1987; Shimizu *et al.*, 1988):

- Cross-continuously arranged blocky medium
- Intermittently arranged blocky medium

Discontinuities, although they may be viewed as planes in large scale, have undulating surfaces varying in irregularity. As a result, they may be regarded as bands with a certain thickness associated with the amplitude of the undulations. The discontinuities may be filled with material, such as calcite, quartzite, or weathering products of host rock or transported materials, or they may exist from the beginning as thin films of clay deposits in sedimentary rocks along bedding planes.

The mechanical behavior of discontinuities is mostly associated with the inclination and amplitude of undulations, mechanical response of discontinuity wall rock, the level of normal stress, and the presence and the thickness of infilling materials. The typical shear and normal responses of various types of discontinuities are illustrated in Figure 2.7.

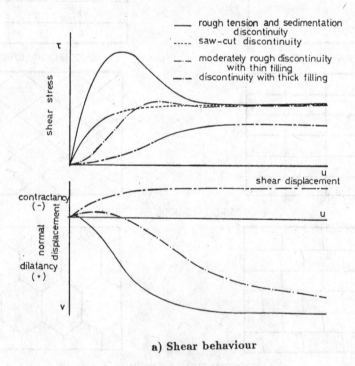

a) Shear behaviour

Figure 2.7 Mechanical behavior of discontinuities.

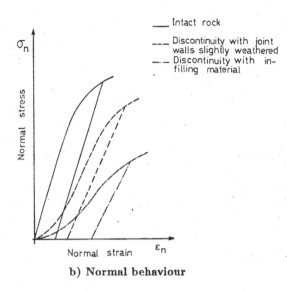

b) **Normal behaviour**

Figure 2.7 (Continued)

2.1.3   Rock mass and its mechanical behavior

Rock mass generally consists of blocks or layers of rock bounded by discontinuities, which look like a masonry wall with or without cementation (Figs. 2.4 and 2.5). As a result, its mechanical behavior depends on the mechanical behaviors of the rock element and of discontinuities and their orientations with respect to the applied load and constraint conditions. Although rock mass is modeled as an equivalent continuum in many studies and projects, the rock mass should be regarded as a structure and its mechanical response as a structural response rather than a material response. It is always pointed out that the strength of rock samples and discontinuities measured in the laboratory are not of much use for evaluating the stability of rock engineering structures. Let us consider a sample with a continuous discontinuity set subjected to a triaxial state of stress and assume that the failure is only governed by shearing. The triaxial strength of such a sample can be shown to be (Jaeger, 1962; Aydan *et al.*, 1987b):

$$\sigma_1^d = \frac{2c_d + \sigma_3(1+\cos 2\alpha)\tan\phi}{\sin 2\alpha - (1-\cos 2\alpha)\tan\phi} \qquad (2.1)$$

where

c_d = cohesion of discontinuity set
α = inclination of discontinuity set from horizontal
ϕ_d = friction angle of discontinuity
σ_3 = least lateral principal stress
σ_1^d = strength of rock mass involving only the failure at a discontinuity plane)

When $\sigma_1^d \leq \sigma_1^i$, ($\sigma_1^i$ is strength of rock mass involving only the failure of intact rock), the strength of the mass is equal to the strength offered by the discontinuity set. On the other hand, if $\sigma_1^d \geq \sigma_1^i$, the strength of the mass is governed by the intact rock element (Fig. 2.8), except at some transition zones where the failure by tensile splitting, bending, or buckling

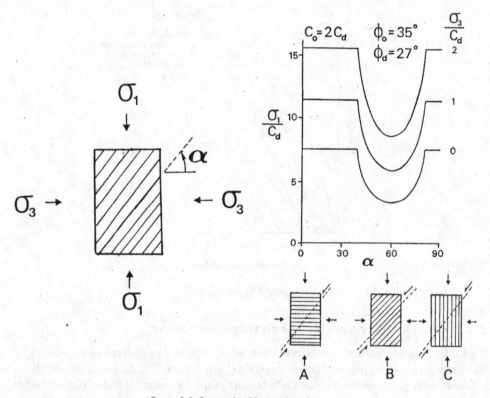

Figure 2.8 Strength of layered rock mass.

may be prevailing. The next problem is what the relation between the behavior of such samples with the situations in actual rock engineering structures is. Let us consider three specific cases in which rock mass is layered (Fig. 2.9):

- Slope
- Foundation (a dam abutment)
- Underground opening

  - Shallow underground opening
  - Deep underground opening

and assume that failure takes place by shearing. The states corresponding to the states denoted by A, B, and C in Figure 2.8 for the sample are indicated in each Figure for three specific cases in Figure 2.9. These simple illustrations clearly show that the important elements are the strength of rock elements and discontinuities in association with the specific loading condition and the geometry of the structure. Therefore, the stability of any rock engineering structure in a rock mass should be evaluated in terms of the mechanical response of the rock element and the discontinuity sets and the structure of rock mass, although it may be quite cumbersome due to the input of geometrical and material parameters in analyses.

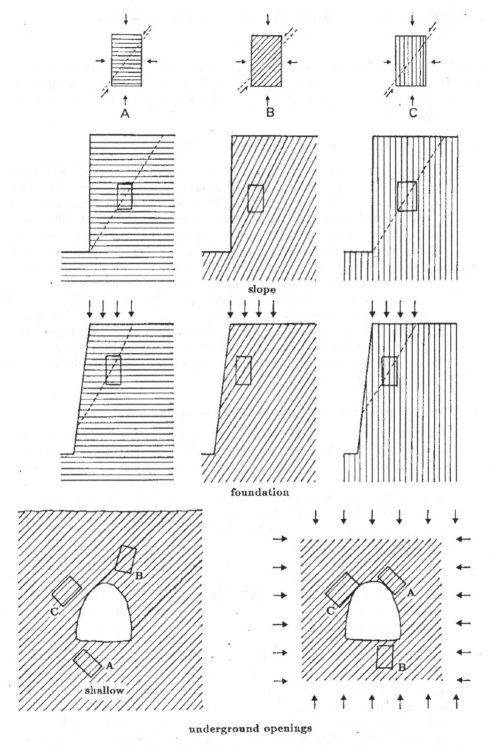

Figure 2.9 Situations in structures in layered rock mass corresponding to the situations in laboratory tests.

2.2   MODES OF INSTABILITY ABOUT UNDERGROUND OPENINGS

In the light of previous discussion on rock mass, the modes of instability likely to take place in the vicinity of underground openings may be classified as below, depending upon the structure of rock mass as shown in Figures 2.10, 2.11, and 2.12 (Aydan *et al.*, 1987c; Kawamoto and Aydan, 1988):

i)   Failure modes involving only intact rock

Rockbursting: This type of instability results from the combined action of initial shearing and the subsequent splitting, resulting in sudden detachment of rock slabs with a high velocity. This type of failure is usually observed in brittle hard rocks, such as unweathered igneous rocks and siliceous sedimentary rocks (Panet, 1969; Bieniawski and van Tonder, 1969; Hoek and Brown, 1980; Aydan, 1989, etc.). As the rock becomes less brittle, the rockbursts become less severe. Figure 2.13 shows plots of some compiled data on underground excavations in which rockbursts were observed.

Squeezing: This type of instability is results from the complete shearing of rock surrounding an excavation. This type of failure can be observed in ductile materials, such as rock salt, thickly bedded mudstone, halite, chalk, etc. (Terzaghi, 1946; Sperry and Heuer, 1979, etc.). It should be noted that σ_c denotes the uniaxial strength of the rock element, not that of the rock mass in Figure 2.13. These plots confirm that the critical parameter controlling the stability in rockburst and squeezing phenomena is the strength of rock element.

ii)   Failure modes involving discontinuities and intact rock

Bending: This type of instability is usually observed in sedimentary rocks due to gravitational forces, when layers are generally parallel to the roof and *in situ* stresses parallel to layering is relatively low. Figure 2.11 shows a typical example of a bending failure observed in a model test. This type of failure is associated with the tensile strength of layers at the early stages of failure (Birön and Arıoğlu, 1983; Hoek and Brown, 1980; Whittaker and Reddish, 1989, etc.). This is confirmed by the plots of some failed excavations due to bending (Fig. 2.14).

Buckling: Contrary to bending failure, this type of instability is observed when high *in situ* stresses parallel to layering are present and the thickness of layers in comparison with the span is relatively small. Figures 2.10 and 2.11 show some field examples and examples of model openings failed through buckling (Everling, 1964; Detzlhofer, 1970; Amberg, 1983, etc.). It is usually observed in metamorphic rocks and thinly layered sedimentary rocks. The plots of some data on excavations where buckling was observed confirm this conclusion (Fig. 2.14).

Punching and sliding: This highly localized form of instability is observed when the rock is relatively thinly layered. Some field examples are reported by Arnold *et al.* (1972).

Flexural toppling: This type of failure is also a localized form of instability, and it can be observed particularly in roofs and sidewalls of openings excavated in sedimentary and metamorphic rocks. Some examples of this type of instability are shown in Figures 2.10 and 2.11 (Goodman, 1977; Aydan *et al.*, 1988c). Layers of rock bend and fail like interacting cantilevers that fail in flexure.

Bursting & Spalling (Granite)
(after Panet 1969)

Buckling (Shale)[3]

Bending (Whittaker & Reddish 1989)

Flexural Toppling

Shearing and Sliding (Hill & Bauer 1984)

Block Falls & Block sliding

Figure 2.10 Pictures of failures observed in underground openings in the field.

Spalling (Ewy et al. 1988)

Shearing & Sliding (Kaiser 1979)

Buckling (Everling 1964)

Bending

Block Falls

Flexural Toppling

Figure 2.11 Pictures of failures observed in underground openings in model tests.

FAILURES INVOLVING ONLY INTACT ROCK

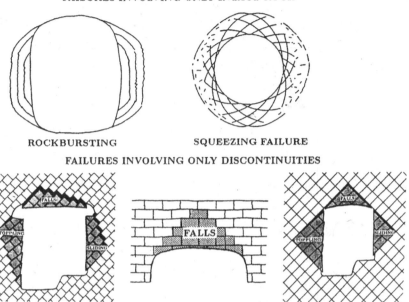

ROCKBURSTING   SQUEEZING FAILURE

FAILURES INVOLVING ONLY DISCONTINUITIES

FALLS   SLIDING   TOPPLING

FAILURES INVOLVING INTACT ROCK AND DISCONTINUITIES

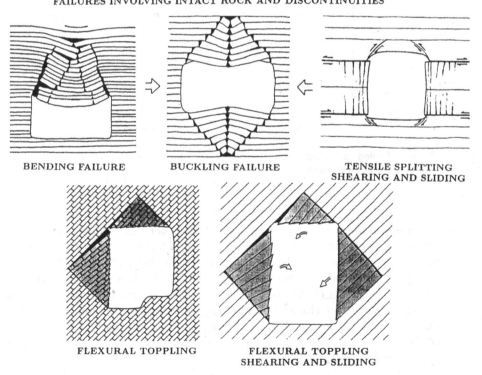

BENDING FAILURE   BUCKLING FAILURE   TENSILE SPLITTING
SHEARING AND SLIDING

FLEXURAL TOPPLING   FLEXURAL TOPPLING
SHEARING AND SLIDING

Figure 2.12 Classifications of modes of instability in underground openings.

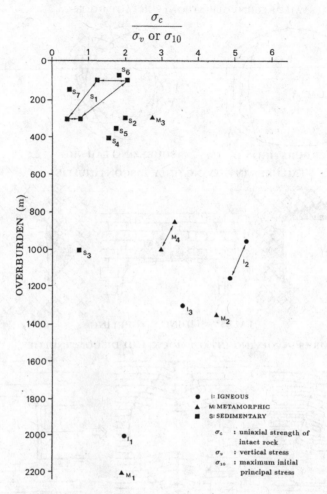

Figure 2.13 Plots of failed case studies involving only intact rock. Note that the strength σ_c is the strength of rock element.

Shearing and sliding: This type of failure involves combined sliding of unstable part along discontinuities and shearing through intact rock. It is most likely to be seen when *in situ* stresses are higher than the compressive strength of rock, making buckling failure impossible. Some severe field examples are reported by Sperry and Heuer (1979), who observed in Navajo irrigation tunnels in shale and sandstone and by Hill and Bauer (1984), who observed in mine openings in shale. In the model tests of circular openings in jointed coal carried out by Kaiser (1979), this type of failure was observed dominantly, even though the samples were loaded hydrostatically.

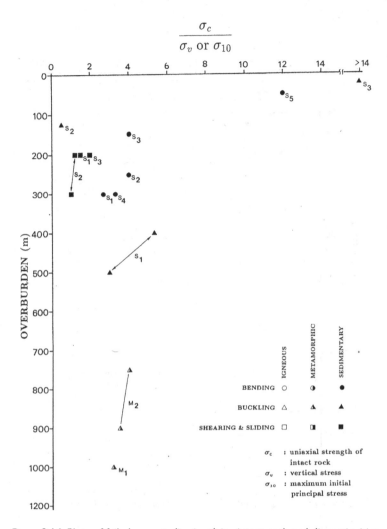

Figure 2.14 Plots of failed case studies involving intact rock and discontinuities.

iii) Failure modes involving only discontinuities (blocky medium only)

These types of failure can occur at any depth, as long as the rock mass has discontinuity sets of two or more (Fig. 2.15):

Block falls: This type of failure is observed in the roofs of openings due to gravitational forces. Some examples were observed in the field and model tests were done in the laboratory, shown in Figure 2.10 (Isaac and Bubb, 1981; Dezhen and Sijing, 1982; Weiss-Malik and Kuhn, 1979; Pistone and del Rio, 1982; Detzlhofer, 1968, etc.).

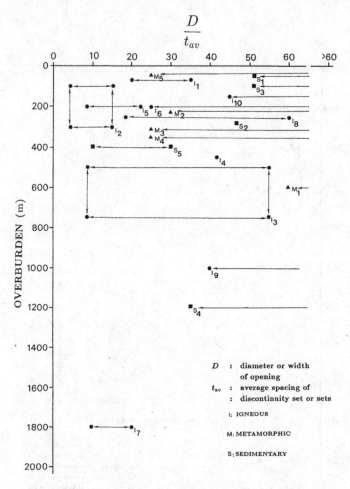

Figure 2.15 Plots of failed case studies involving discontinuities only.

Sliding: This type of failure is observed when one of the discontinuity sets daylights near the toe of sidewalls and the disturbing forces are greater than its shear resistance. Some examples of such failures in field and model tests are shown in Figure 2.11 (Pistone and del Rio, 1982; Kamemura *et al*., 1986; Reik and Soetomo, 1986, etc.).

Toppling: The inclination of the critical discontinuity set, on which toppling will occur, should be such that no sliding failure is possible. Some examples of such failures in field and model tests are shown in Figure 2.11 (Pistone and del Rio, 1982; Isaac and Bubb, 1981, etc.).

Sliding and toppling: This type of failure is observed when the conditions for the two types of failures are satisfied. Some examples for such failures are shown in Figure 2.11.

2.3   MODES OF INSTABILITY OF SLOPES

As in the case of underground openings, a similar type of classification can be made for rock slopes (Fig. 2.16) (Aydan *et al.*, 1988b). Pictures of some slope failures observed *in situ* and in laboratory tests are shown in Figures 2.17 and 2.18.

i)   **Failure modes involving only intact rock**

  Shear failure: This type failure is observed in cases such that the slope angle and height are sufficient to cause shearing of the intact medium in continuous, tabular, or blocky medium. In tabular or blocky medium, the internal structure and slope geometry should be such that no other forms of instabilities are possible. Some examples observed in field and laboratory model tests are shown in Figures 2.17 and 2.18 (Hutchinson, 1971; Hoek and Bray, 1977; Tokashiki and Aydan, 2010). Depending upon the slope angle, tensile cracks at the top of slopes may appear, and the failure of slopes, therefore, can be due to a combination of shearing and tensile stresses.

  Bending failure: This type of failure is likely to be seen in the case of slopes with a toe eroded. The mode of failure is similar to that of cantilevers. Some examples for such failure observed in model tests are shown in Figure 2.18. The failure is often observed in cliffs near sea sides or river embankments (Skudrzyk *et al.*, 1986; Tharp, 1983; Okagbue and Abam, 1986, etc.). For this type of failure, the ratio of the erosion depth to the slope height should be sufficient to cause bending failure rather than shear failure.

ii)   **Failure modes involving discontinuities and intact rock**

  Combined shear and sliding failure: This type of failure can occur when one of the discontinuity sets has an inclination equal to the slope angle and no other forms of failure is possible. This failure manifests itself as sliding along a critical plane and the shearing of intact rock near the toe of the slope (Fig. 2.16) (Brawner *et al.*, 1971; Aydan *et al.*, 1992).

  Buckling: This type of failure occurs when the slope angle is equal to that of the discontinuity set and the ratio of discontinuity spacing to the slope height is relatively small. It is a recently recognized form of instability and reported case studies are rare (Walton and Coates, 1980; Cavers, 1981, etc.). A field example for such a failure at the Elbistan open-pit mine is shown in Figure 2.17 (Aydan *et al.*, 1996).

  Flexural toppling: This type of failure occurs in the case of slopes excavated in sedimentary or metamorphic rocks. Although this type of failure is a local one in the case of underground openings, it is a global form of failure in the case of slopes. Flexural toppling was first recognized by Erguvanlı and Goodman (1972) and Hoffmann (1974), and some fundamental studies on this failure form were undertaken by Aydan and Kawamoto (1987, 1992) and Aydan *et al.* (1988c). Some *in situ* and laboratory examples for such a failure are shown in Figures 2.17 and 2.18.

ii)   **Failure modes involving only discontinuities**

  Sliding failure: There are two types of sliding failure (Fig. 2.16). These are:

   Planar sliding: This involves only one set, the strike of which is parallel or nearly parallel to the slope axis, and occurs along a critical plane, daylighting near the toe

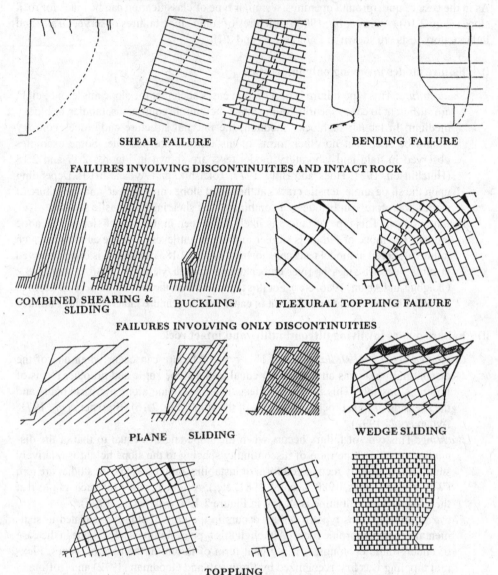

FAILURES INVOLVING ONLY INTACT ROCK

SHEAR FAILURE   BENDING FAILURE

FAILURES INVOLVING DISCONTINUITIES AND INTACT ROCK

COMBINED SHEARING &   BUCKLING   FLEXURAL TOPPLING FAILURE
SLIDING

FAILURES INVOLVING ONLY DISCONTINUITIES

PLANE   SLIDING   WEDGE SLIDING

TOPPLING

Figure 2.16 Classifications of modes of instability in slopes.

of the slope (Hoek and Bray, 1977; Aydan *et al.*, 1989). Some examples of failed slopes in field and model tests are shown in Figures 2.17 and 2.18.

Wedge sliding: This involves two throughgoing discontinuity sets and occurs when the intersections of two sets daylight near the toe of the slope (Wittke, 1964; Shimizu *et al.*, 1988; Kumsar *et al.*, 2000; Aydan and Kumsar, 2010). An example of failed slopes in the field is shown in Figure 2.17.

Shear sliding   Buckling (Elbistan)

Shearing & Sliding (Elbistan)   Sliding (Selçuk)

Toppling (Susuzdede tepe - İzmir)   Flexural Toppling (Bayındır)

Figure 2.17 Pictures of failures observed in slopes in field.

Figure 2.18 Pictures of failures observed in slopes in laboratory tests.

Toppling failure: This occurs when one of the discontinuity sets, the strike of which is parallel or nearly parallel to the axis of slope, has an inclination such that no sliding is possible (Goodman and Bray, 1976; Aydan and Kawamoto, 1987; Aydan *et al.*, 1989). Some field and laboratory examples are shown in Figures 2.17 and 2.18.

Combined toppling and sliding failure: This type of failure is observed when both conditions for toppling and sliding are satisfied (Aydan *et al.*, 1989; Aydan *et al.*, 1992). An example of failed slopes in model tests in the laboratory is shown in Figure 2.17.

Failure modes of sliding and toppling are also global forms of failure, as compared to the local character in the case of underground openings.

2.4 MODES OF INSTABILITY OF FOUNDATIONS

The modes of failure and the classification for foundations would be similar to those of slopes. Therefore, the repetition is avoided, but some pictures and illustrations of foundation failure together with their classifications are shown in Figures 2.19, 2.20, and 2.21 under

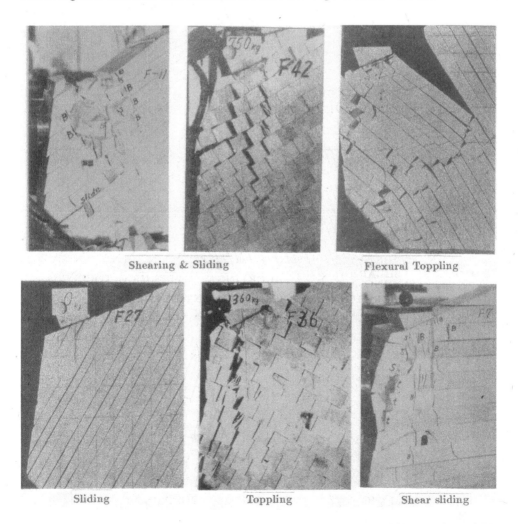

Shearing & Sliding   Flexural Toppling

Sliding   Toppling   Shear sliding

Figure 2.19 Pictures of modes of instability in foundations under compressive and tensile stress fields.

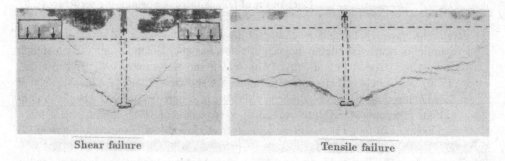

Shear failure   Tensile failure

Figure 2.19 (Continued)

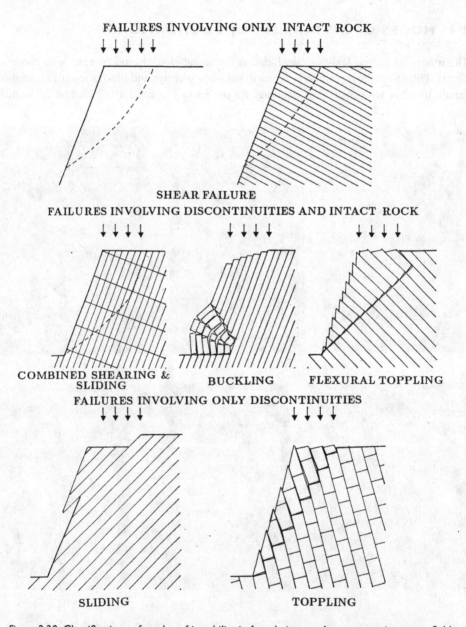

FAILURES INVOLVING ONLY INTACT ROCK

SHEAR FAILURE

FAILURES INVOLVING DISCONTINUITIES AND INTACT ROCK

COMBINED SHEARING &   BUCKLING   FLEXURAL TOPPLING
SLIDING

FAILURES INVOLVING ONLY DISCONTINUITIES

SLIDING   TOPPLING

Figure 2.20 Classifications of modes of instability in foundations under compressive stress field.

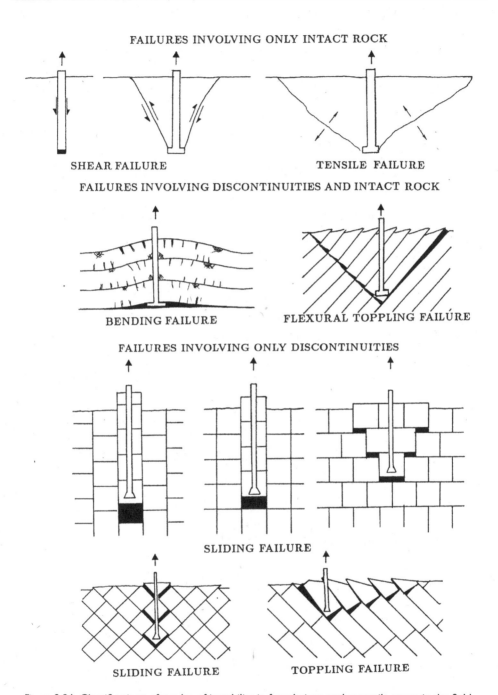

Figure 2.21 Classifications of modes of instability in foundations under tensile stress in the field.

compressive and tensile stress fields. The reported examples of failures of foundations in the field and the laboratory are presently few, and most of the tests are associated with model tests (Bernaix, 1966; Krsmanovic *et al.*, 1965; Hayashi and Fujiwara, 1963; Good-man, 1976; Ebisu *et al.*, 1994a, 1994b).

Chapter 3

Design philosophy of rock support and rock reinforcement

3.1 INTRODUCTION

The primary concerns regarding the support and reinforcement design of rock engineering structures, as apparent from the term, is whether the structure under consideration is self-supporting, and if it is not, what kind of strategy must be followed for the overall stability of the structure during and after excavation. The selection of support/reinforcement members is not only closely associated with their mechanical functions but also with their advantages and disadvantages related to environmental, constructional, and economic conditions. Nevertheless, as this book is more concerned with the mechanical functions of the support/reinforcement members, the discussions are herein restricted mainly to the mechanics of support/reinforcement members and supporting procedures, with occasional references made to the environmental, constructional, and economic aspects. The discussions are mainly concerned with the supporting/reinforcement philosophy used in the design of underground openings, as they are more generalizable than surface structures. Nevertheless, considerations are given to other structures from time to time.

The present support/reinforcement philosophy mainly consists of two fundamental steps:

Step 1: Determination of the magnitude of unbalanced loads to be resisted by the chosen single or combination of support/reinforcement members
Step 2: Selection of the support/reinforcement members suitable not only from the mechanical point of view but also from the constructional, economic, and environmental points of view

In rock engineering, the design approaches can be categorized into three groups:

- Empirical
- Analytical
- Numerical

In the empirical methods, rock mass classification systems are extensively used for feasibility and pre-design studies, and often also for the final design.

In the design of rock engineering structures, the anticipated form of instability is of great importance. The instabilities around underground openings in rock may be categorized as global instability and local instability, defined as (Aydan, 1989, 2016):

Global instability: This is defined as when the excavated space cannot be kept open and the failure of the surrounding mass continues to take place indefinitely unless any supportive and/or reinforcement measure is undertaken. The global instability would be as a result of exceeding the strength of surrounding rocks due to the redistribution of initial ground stresses.

Local instability: After clearance of the failed zone and without taking any supportive measures, if the remaining space can be kept open, the form of instability is termed local instability. The main cause of failure is the dead weight of rock in a particular zone about the cavity, defined by the geometry of underground openings and the spatial distribution of discontinuities.

The design of support/reinforcement systems of large underground openings and tunnels in rock engineering is of great importance, as these structures are required to be stable during their service lifetime (Aydan, 1989). Provided that the elements of support/reinforcement systems are resistant against chemical actions due to environmental conditions and their long-term behavior is satisfactory, the support systems must be designed against anticipated load conditions. As rock masses have many geological discontinuities and weakness zones, the load acting on support systems may be due to the dead weight of potential unstable blocks formed by rock discontinuities, which may be designated as structurally controlled or local instability modes and independent of *in situ* stress state or inward displacement of rock mass due to elasto-plastic or elasto-visco-plastic behavior induced by *in situ* stresses (Fig. 3.1). Therefore, the main purpose of the design of support/reinforcement systems must be well established with due considerations of these situations.

Rock mass classifications are commonly used for various engineering design and stability assessments and they are initially proposed for the design of a given rock structure. However, this trend has been changing, and the main objectives of rock mass classifications have become to identify the most significant parameters influencing the behavior of rock masses, to divide a particular rock mass formulation into groups of similar behavior, to provide the characterizations of each rock mass class, to derive quantitative data and guidelines for engineering design, and also to provide a common

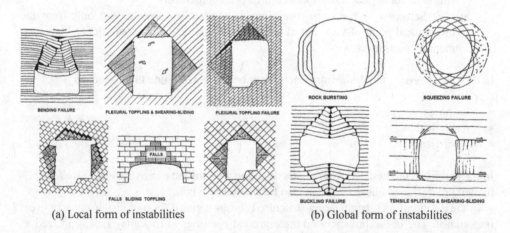

(a) Local form of instabilities   (b) Global form of instabilities

Figure 3.1 Instability modes of underground openings (re-arranged from Aydan, 1989).

basis for engineers and engineering geologists. These are based on empirical relations between rock mass parameters and engineering applications, such as tunnels and other underground caverns.

Although the history of rock classifications for a given specific structure is old, the rock mass classification system proposed by Terzaghi in 1946 for tunnels with steel set support has become the basis for the follow-up quantitative rock mass classifications. Currently, there are many rock classification systems in rock engineering, particularly in the tunneling area, such as Rock Mass Rating (RMR) (Bieniawski (1973, 1989), Q-system (Barton *et al.*, 1974), RSR (Wickham *et al.*, 1972), and Rock Mass Quality Rating (RMQR) by Aydan *et al.*, 2014. In addition, rock mass classifications of NEXCO (known as DORO-KODAN) and JR (KYU-KOKUTETSU) are commonly used to design tunnels in Japan. Nevertheless, utilizing these systems to characterize complex rock mass conditions is a challenge for engineers.

In this chapter, several classification systems have been briefly explained, and quantitative assessments have been done based on RMQR. Because it is required for an engineer to select the most appropriate method for determining design parameters in rock engineering, a brief overview of the kinds of rock loads and the procedures to determine their magnitude, empirical, analytical, and numerical techniques has been provided and discussed, with the objective of unifying the present methods of design.

3.2 EMPIRICAL DESIGN METHODS

As mentioned in the introduction, Terzaghi (1946) considered steel ribs as the main support member and visualized a loosened region of rock mass in the roof and sidewalls, as illustrated in Figure 3.2. His main idea originates from his trapdoor experiments with soils, and he visualized that the support load (pressure) on steel ribs as a fraction of the weight of the potentially unstable ground, which is given as:

$$p_i^r = \gamma B \tag{3.1}$$

where B is tunnel width and γ is the unit weight of potentially unstable ground.

Since then, this concept has been utilized in visualizing and calculating the pressure on support members in geotechnical engineering, including both soil and rock tunnels. As noted from Figure 3.2, the load on tunnel support results from the surrounding ground (which may be soil or rock mass), which is a function of assumed ground properties and loosening zone around the tunnel. Although this concept is quite simple to use, major issues arise how to relate this pressure in rock mass to the true *in situ* stresses in rock mass. Although their main formulations differ, Protodyakonov and Terzaghi proposed independently the following relationship:

$$\frac{p_i^r}{\gamma B} = \frac{1}{\tan \phi} \tag{3.2}$$

where ϕ is the friction angle of potentially unstable ground.

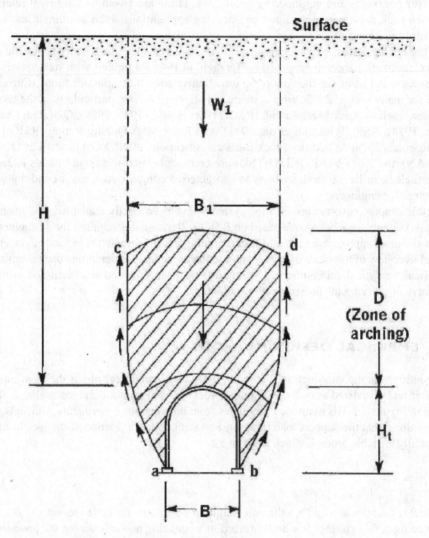

Figure 3.2 Terzaghi's rock load concept (from Terzaghi, 1946).

As rock mass always has discontinuities, the rock layers and/or blocks of rocks may become detached or loosened due to gravity, blasting, or groundwater seepage and act on support members as rock load. Such failures in rock mass may be classified as local failures (Aydan, 1989; Kawamoto *et al.*, 1991). They may loosen more if the ground is shaken further, such as by earthquakes (Fig. 3.3). The original concept of Terzaghi is utilized in many rock classification systems, which may be categorized as an empirical approach.

In the following subsection, the empirical approaches are briefly explained.

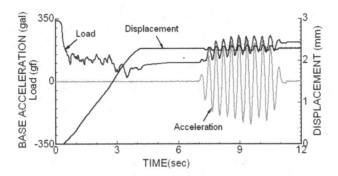

Figure 3.3 Load and displacement response of a trapdoor experiment subjected to shaking (note that the pressure on the trapdoor is increased after shaking despite no further downward displacement of the trapdoor).

3.2.1   Rock Quality Designation (RQD) method

Deere *et al.* (1969) suggested the following relationship between the roof pressure and the Rock Quality Designation (RQD), which is a percentage of rock cores whose length is greater than 10 cm for a given 1 m length of cores.

$$\frac{p_i^r}{\gamma B} = 0.2 + 0.025RQD \tag{3.3}$$

The length and number of rockbolts and rockanchors and the thickness of shotcrete are computed using their load-bearing capacity, loosened load height, and required anchorage length. This concept is followed in other rock classification systems, such as RMR and Q-system.

3.2.2   Rock Mass Rating (RMR)

Bieniawski (1973, 1976) published the details of a rock mass classification called the Geomechanics rock classification or the Rock Mass Rating (RMR) system. Over the years, this system has been refined as more case records have been examined, and the reader should be aware that Bieniawski (1989) has made significant changes in the ratings assigned to different parameters and he suggests that the 1989 version be used. In this section, support design according to RMR has been briefly described.

Bieniawski (1989) published a set of guidelines for the selection of support in tunnels in rock using the value of RMR for rock mass. These guidelines are reproduced in Table 3.1. Note that these guidelines have been published for a 10-m-span horseshoe-shaped tunnel, constructed using drill and blast methods, in a rock mass subjected to a vertical stress < 25 MPa (equivalent to a depth below surface of < 900 m). It should be noted that Table 3.1 has not had a major revision since 1973. In many mining and civil engineering applications, steel-fiber-reinforced shotcrete may be considered in place of wire mesh and shotcrete.

Table 3.1 Guidelines for excavation and support of 10-m-span rock tunnels in accordance with the RMR system (after Bieniawski, 1989).

Rock mass class	Excavation	Rockbolts (20 mm diameter, fully grouted)	Shotcrete	Steel sets
I – Very good rock RMR 81–100	Full face, 3 m advance.	Generally no support required except spot bolting.		
II – Good rock RMR 61–80	Full face, 1–1.5 m advance. Complete support 20 m from face.	Locally, bolts in crown 3 m long, spaced 2.5 m with occasional wire mesh.	50 mm in crown where required.	None.
III – Fair rock RMR 41–60	Top heading and bench 1.5–3 m advance in top heading. Commence support after each blast. Complete support 10 m from face.	Systematic bolts 4 m long, spaced 1.5–2 m in crown and walls with wire mesh in crown.	50–100 mm in crown and 30 mm in sides.	None.
IV – Poor rock RMR 21–40	Top heading and bench 1.0–1.5 m advance in top heading. Install support concurrently with excavation. 10 m from face.	Systematic bolts 4–5 m long, spaced 1–1.5 m in crown and walls with wire mesh.	100–150 mm in crown and 100 mm in sides.	Light to medium ribs spaced 1.5 m where required.
V – Very poor rock RMR < 20	Multiple drifts 0.5–1.5 m advance in top heading. Install support concurrently with excavation. Shotcrete as soon as possible after blasting.	Systematic bolts 5–6 m long, spaced 1–1.5 m in crown and walls with wire mesh. Bolt invert.	150–200 mm in crown. 150 mm in sides, and 50 mm on face.	Medium to heavy ribs spaced 0.75 m with steel lagging and forepoling if required. Close invert.

Ünal (1983, 1992) suggested the relationship below between the normalized support pressure and RMR, and this relationship is used by Bieniawski (1989) and his followers:

$$\frac{p_i^r}{\gamma B} = \left(1 - \frac{RMR}{100}\right)$$ (3.4)

3.2.3   Q-system (rock tunneling quality index)

Barton *et al.* (1974) analyzed various case history data in Norway and other major projects around the world. Rockbolts, rockanchors, and shotcrete with or without wire mesh are the fundamental elements of the support system. They prepared several charts to determine the geometrical parameters of the support members in Figure 3.4. Barton *et al.* (1974) suggested the following relationship between the normalized support pressure and Q-value with the consideration of joint roughness parameter (J_r), which is a measure of the friction angle of discontinuities:

$$p_i^r = \frac{200}{J_r} \frac{1}{\sqrt[3]{Q}}$$ (3.5)

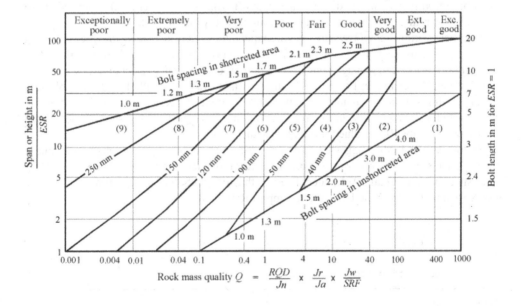

REINFORCEMENT CATEGORIES
1) Unsupported
2) Spot bolting
3) Systematic bolting
4) Systematic bolting with 40-100 mm
   unreinforced shotcrete

5) Fibre reinforced shotcrete, 50 - 90 mm, and bolting
6) Fibre reinforced shotcrete, 90 - 120 mm, and bolting
7) Fibre reinforced shotcrete, 120 - 150 mm, and bolting
8) Fibre reinforced shotcrete, > 150 mm, with reinforced
   ribs of shotcrete and bolting
9) Cast concrete lining

Figure 3.4 Estimated support categories based on the tunneling quality index Q. (After Grimstad and Barton, 1993).

Based upon analyses of case records, Grimstad and Barton (1993) recently modified Equation (3.5) and suggested the following relationships:

$$p_i^r = \frac{2\sqrt{J_n}}{J_r}\frac{1}{\sqrt[3]{Q}}$$

(3.6)

Barton *et al.* (1974) defined an additional parameter, which they called the Equivalent Dimension, D_e, of the excavation to relate the value of the index Q to the stability and support requirements of underground excavations to consider their importance. This dimension is obtained by dividing the span, diameter, or wall height of the excavation by a quantity called the Excavation Support Ratio (ESR) as:

$$D_e = \frac{\text{Excavation span diameter or height (m)}}{\text{Excavation Support Ratio ESR}}$$

(3.7)

The value of ESR is related to the intended use of the excavation and to the degree of safety. Barton *et al.* (1974) suggest the values in Table 3.2. It should be noted that ESR might be interpreted as the safety factor.

The equivalent dimension, D_e, plotted against the value of Q, is used to define a number of support categories in a chart published in the original paper by Barton *et al.* (1974). This chart has recently been updated by Grimstad and Barton (1993) to reflect the increasing use of steel-fiber-reinforced shotcrete in underground excavation support.

The stress factor in the Q-system also attempts to consider the true *in situ* stress-induced stability problems, such as rockburst, squeezing, and rock spalling. Furthermore, the effect of the blasting-induced damage zone around underground openings is counted by reducing the Q value. For example, Løset (1997) suggests that, for rocks with $4 < Q < 30$, blasting damage will result in the creation of new "joints" with a consequent local reduction in the value of Q for the rock surrounding the excavation. He suggests that this can be accounted for by reducing the RQD value for the blast-damaged zone.

Barton *et al.* (1980) suggested additional information on rockbolt length, maximum unsupported spans, and roof-support pressures to supplement the support recommendations published in the original 1974 paper. The length L of rockbolts can be estimated from the excavation width B and the Excavation Support Ratio ESR:

$$L = 2 + \frac{0.15B}{ESR}$$

(3.8)

Table 3.2 Values of ESR for various structures.

Excavation category	ESR
A Temporary mine openings	3–5
B Permanent mine openings, water tunnels for hydro power (excluding high pressure penstocks), pilot tunnels, drifts, and headings for large excavations	1.6
C Storage, access tunnels	1.3
D Power stations, major road and railway tunnels, civil defiance chambers, portal intersections	1.0
E Underground nuclear power stations, railway stations, sports and public facilities, factories	0.8

Barton *et al.* (1980) suggest that the maximum unsupported span, which can be interpreted as the distance between tunnel face and the nearest support ring, can be estimated from:

$$\text{Maximum span (unsupported)} = 2ESRQ^{0.4} \qquad (3.9)$$

3.2.4 Rock Mass Quality Rating (RMQR)

Rock Mass Quality Rating (RMQR) is a new rock classification, developed by Aydan *et al.*, 2014. This new rock classification quantifies the state of rock mass and assists in estimating the geomechanical properties (UCS, cohesion, friction angle, deformation modulus, Poisson's ratio, and tensile strength) of rock masses using a unified formula that considers RMQR together with intrinsic geomechanical properties of intact rock.

Tunnels, which are also becoming larger in recent years (with widths up to 14 m), are relatively smaller in size (10–11 m wide, 7–9 m high) and are long linear structures. There is rich worldwide experience in tunneling under diverse rock conditions. Tunnels may be excavated in various rock masses, which may be subjected to squeezing and rockbursting, as well as structurally induced failure. Except for new large tunnels, the support/reinforcement system of tunnels generally consists of rockbolts, shotcrete, and steel ribs as primary support members, and concrete lining to smoothen the airflow, to prevent direct seepage of groundwater into the tunnel, and to provide an auxiliary safety measure against rock loads after the introduction of New Austrian Tunneling Method (NATM). When rock mass is not competent against stress-induced yielding, tunnels may be lined with an invert concrete liner. When tunnels are excavated by tunnel-boring machines (TBMs), rockbolts and shotcrete may totally disappear. Using the databases mentioned above and adopting the approach of Aydan and Kawamoto (1999), several interrelationships have been established for the dimensions of support members and the size parameters of the underground openings with the consideration of structurally controlled and stress-induced instability modes as given in Table 3.3 (Aydan, 2016). It may also be used for preliminary support/reinforcement design when the surrounding rock mass is subjected to even stress-induced yielding.

Table 3.3 Empirical relationships between rock mass quality rate (RMQR) and the dimensions of support members normalized by arch span (L_a) or sidewall height (H_s) (* for large underground caverns).

Support Member	Size	Roof-Arch	Sidewall
Rock Anchors	Length	$\dfrac{L}{L_a} = 0.8 - \dfrac{RMQR}{200}$	$\dfrac{L}{H_s} = 0.7 - \dfrac{RMQR}{200}$
	Spacing (m)	$e_{av} = 2 + 0.02RMQR$	$e_{av} = 2 + 0.03RMQR$
Rockbolts	Length	$\dfrac{L_b}{L_a} = 0.35 - \dfrac{RMQR}{500}$	$\dfrac{L_b}{H_s} = 0.30 - \dfrac{RMQR}{500}$
	Spacing (m)	$e_{av} = 1 + 0.015RMQR$	$e_{av} = 1 + 0.015RMQR$
Shotcrete	Thickness	$\dfrac{t^{sr}}{L_a} = 0.0125 - \dfrac{RMQR}{10000}$	$\dfrac{t^{sw}}{H_s} = 0.0075 - \dfrac{RMQR}{18000}$
Concrete Liner*	Thickness (mm)	1000	none

Table 3.4 Support systems for tunnels (D or B, 10-m span) (Aydan and Ulusay, 2013).

RMQR range	Rockbolts		Shotcrete	Steel ribs		Wire mesh	Concrete Lining	Invert	
	L_b (m)	e_b (m)	t_s (mm)	Type	e_r (m)		(mm)	Lining (mm)	Bolt L (m)
100 ≥ RMQR > 95	–	–	–	–	–	–	–	–	–
95 ≥ RMQR > 80	2–3	2.5	50	–	–	–	–	–	–
80 ≥ RMQR > 60	3–4	2.0	100	Light	1.5	Yes	200	–	–
60 ≥ RMQR > 40	4–5	1.5	150	Medium	1.2	Yes	300	300	–
40 ≥ RMQR > 20	5–6	1.0	200	Heavy	1.0	Yes	500	500	5–6
20 > RMQR	6–7	0.5	250	Very heavy	0.8	Yes	800	800	6–7

On the basis of the relations between RMQR and the support members derived from the databases, past experiences, and empirical, analytical, and numerical methods, Aydan and Ulusay (2013) proposed Table 3.4 for the empirical design of support systems for tunnels, which may be subjected to even stress-induced failure modes, such as squeezing and rock-bursting, respectively. In the case of tunnels, when the RMQR < 20, the UCS of intact rock is less than 20 MPa, and the overburden is greater than 100 m, squeezing problems may be encountered. Under such circumstances, forepoles, face bolting, and shotcreting may be required.

3.3   ANALYTICAL APPROACH

The analytical approaches in tunneling are based on the closed-form solutions of the equation of motion without inertia for static case. In some cases, time dependency of surrounding rock may also be taken into account. The simplest condition for deriving analytical solutions is a circular opening subjected to hydrostatic *in situ* stress. However, there are several solutions for non-hydrostatic conditions (Kastner, 1961; Gerçek, 1988, 1993, 1996, 1997; Gerçek and Geniş, 1999).

3.3.1   Hydrostatic in situ stress state

A huge number of studies were performed for the determination of stress and strain fields about cylindrical (circular) and spherical openings excavated in a hydrostatic far-field stress field, as it is much easier to obtain analytical solutions for this particular situation. An analytical solution for elasto-plastic behavior of rocks is important for providing fundamental information for assessing the stability of openings, as well as for support design and excavation stage of underground openings. The first analytical solution was developed by Fenner (1938) by assuming that rock mass exhibits an elastic-perfectly plastic behavior. Talobre (1957) also developed his own solution. Since these earlier studies, a large number of solutions was proposed and used for the design. The major differences among these methods are associated with the assumed elasto-plastic behavior, the yield function, and how support members are considered. One can find a summary of solutions in the article by Brown *et al.* (1983), including their own solutions. Egger (1973a,b) also discussed the utilization of analytical solutions for spherical openings to infer the stress distributions and displacement of rock mass near tunnel faces.

Aydan (1989) developed analytical solutions for tunnels supported by rockbolts, shotcrete, steel ribs, and concrete linings, and his solutions also consider the interaction of rockbolts and the surrounding ground.

Aydan *et al.* (1993, 1996) developed solutions for determining stress and strain fields around cylindrical tunnels in squeezing ground, and they were extended to spherical openings to obtain a unified solution for the radially symmetric problem. Rock mass around an opening is assumed to obey the Mohr-Coulomb yield criterion, and the solution presented in study is developed for the elastic-perfect-residual plastic material behavior. Although it is possible to develop solutions for the Hoek-Brown criterion, it is not done on purpose, as the generalized Hoek-Brown criterion violates the Euler theorem used in the classical theory of plasticity for constitutive modeling of rocks. In this proposed unified solution, a shape coefficient (n) is defined –it is 1 for cylindrical openings and 2 for spherical openings. Aydan and Geniş (2010) have unified analytical methods and presented some procedures for the consideration of effects of support systems and long-term properties based on the original proposal by Aydan (1989). The details of this method can be found in Chapter 7.

3.3.2   Non-hydrostatic in situ stress state

The analytical solution developed by Kirsch (1898) for stress and deformation distribution in elastic medium about a circular opening in a biaxial far-field stress state was widely used by engineers to understand stress concentration and estimation of possible yield zones. Kastner (1961) was a pioneer in applying it to tunneling, and his method has been also now used to infer the stress state from borehole breakouts (Zoback *et al.*, 1980). The analytical solutions for other geometries of underground openings are also developed by several pioneers, such as Inglis (1913); Mindlin (1940, 1949) and Muskhelishvili (1953). These solutions and their applications to deep underground openings in rock mass are summarized in various textbooks and articles (e.g. Obert and Duvall, 1967; Jaeger and Cook, 1979; Terzaghi and Richart, 1952; Mindlin, 1940; Sokolnikoff, 1956; Timoshenko and Goodier, 1951; Muskhelishvili, 1953; Verruijt, 1997). Gerçek (1988, 1996, 1997) developed a semi-analytical method to evaluate the stress distribution around underground openings with various shapes based on the complex variable method. Gerçek and Geniş (1999) applied this technique to obtain possible yield zones around underground openings. Aydan and Geniş (2010) extended this method to obtain a potential yield zone due to slippage of discontinuities in surrounding rock mass. This method can be used to determine the length of rockbolts/rockanchors to prevent local instabilities.

There were also some attempts to consider the nonlinear behavior of surrounding medium. The solution obtained by Galin (see Savin, 1961, for an English description) for a medium behaving in an elastic-perfectly plastic manner with the use of the Tresca yield criterion was the first of its kind. Detournay (1983) attempted to extend his solution to the Mohr-Coulomb material. However, the solutions are not always unique and theoretical derivations become extremely cumbersome.

3.4 NUMERICAL METHODS

Numerical methods are used to solve the governing equations written in the form of ordinary and partial differential equations, approximately. These are basically the finite difference method (FDM), the finite element method (FEM), and the boundary element method (BEM). The momentum conservation law is given in the following form (Eringen, 1980):

$$\rho \frac{\partial \mathbf{v}}{\partial t} = -\nabla \cdot \boldsymbol{\sigma} + \mathbf{b} \tag{3.10}$$

where ρ, $\mathbf{v}$, $\boldsymbol{\sigma}$, and $\mathbf{b}$ are density, velocity, stress tensor, and body force, respectively. The discretized form of Equation (3.10) takes the following form irrespective of solution technique:

$$[M]\{\ddot{U}\} + [C]\{\dot{U}\} + [K]\{\phi\} = \{F\} \tag{3.11}$$

The specific forms of matrices $[M]$, $[C]$, and $[K]$, and vector $\{F\}$ in Equation (3.11) will only differ depending upon the method of solution chosen and the dimensions of physical space. Viscosity matrix $[C]$ is associated with the rate dependency of the geomaterials. For static analysis, Equation (3.11) reduces to:

$$[C]\{\dot{U}\} + [K]\{U\} = \{F\} \tag{3.12}$$

The equation above, which is of the parabolic type, is utilized to evaluate the time-dependent behavior of surrounding rock mass in rock excavations. If time dependency is negligible, Equation (3.12) is reduced to the following form:

$$[K]\{U\} = \{F\} \tag{3.13}$$

Equation (3.13) is of the elliptical type, which is commonly used in rock excavations for checking the preliminary design or problematic sections.

When this method is applied to excavations in rock mass, special elements are introduced for evaluating the response of support and reinforcement members, such as rockbolts, shotcrete, steel ribs, and concrete lining (e.g. Aydan, 1989; Aydan and Kawamoto, 1991; Aydan et al., 1992), and joint, interface, or contact elements to simulate discontinuities (Goodman et al., 1968; Ghaboussi et al., 1973; Aydan et al., 1996b), which are explained in detail in the following chapters. Constitutive laws together with various yield criteria are used to assess the response and stability of tunnels in rock masses.

If material behavior involves nonlinearity, the equation systems of Equations (3.11)–(3.13) must be solved iteratively with the implementation of required conditions associated with the constitutive law chosen. The iteration techniques may be broadly classified as initial, secant, or tangential stiffness method (e.g. Owen and Hinton, 1980).

The existence of discontinuities in rock mass has special importance on the stability of rock engineering structures, directional seepage, diffusion, or heat transport, and its treatment in any analysis requires special attention. Various types of finite element methods with joint or interface elements – the discrete element method (DEM), discontinuous deformation analysis (DDA), the discrete finite element method (DFEM), and the displacement discontinuity method (DDM) –have been developed so far. Although these methods are mostly concerned with the solution of the equation of motion, they can be used for seepage, heat transport, or diffusion problems. The fundamental features of the available methods are described in the following by quoting a recent review on these methods by Kawamoto and Aydan (1999).

i) **No-tension finite element method**

The no-tension finite element method was proposed by Valliappan in 1969 (Zienkiewicz *et al.*, 1969). The essence of this method lies with the assumption of no tensile strength for rock mass, as it contains discontinuities. In the finite element implementation, the tensile strength of media is assumed to be nil. It behaves elastically when all principal stresses are compressive. The excess stress is redistributed to the elastically behaving media using a similar procedure adopted in the finite element method with the consideration of elastic-perfectly plastic behavior.

ii) **Pseudo discontinuum finite element method**

This method was first proposed by Baudendistel *et al.* in 1970. In this method, the effect of discontinuities in the finite element method is considered through the introduction of directional yield criterion for the elasto-plastic behavior. Its effect on the deformation characteristics of the rock mass is not taken into account. If there is any yielding in a given element, the excess stress is computed and the iteration scheme for elastic-perfectly plastic behavior is implemented. If there is more than one discontinuity set, the excess stress is computed for the discontinuity set that yields the largest value (Aydan and Kawamoto, 2001).

iii) **Smeared crack element**

The smeared crack element method within the finite element method was initially proposed by Rashid (1968) and adopted by Pietruszczak and Mroz (1981) in media having weakness planes or developing fracture planes. This method evaluates the equivalent stiffness matrix of the element and allows the directional plastic yielding within the element. This approach is adopted by Tang (1997) in the solutions scheme called rock progressive failure analysis (RFPA) with the use of a fine finite element mesh.

iv) **Discrete finite element method (DFEM)**

Finite element techniques using contact, joint, or interface elements have been developed for representing discontinuities between blocks in rock masses. The simplest approach for representing joints is the contact element, which was originally developed for bond problems between steel bars and concrete. The contact element is a two-noded element having normal and shear stiffnesses. This model is recently used to model block systems by Aydan and Mamaghani (e.g. Mamaghani *et al.*, 1994, Aydan *et al.*, 1996b) by assigning a finite thickness to the contact element and employing an updated Lagrangian scheme to deal with large block movements. The contact element can easily deal with sliding and separation movements.

v) **Finite element method with joint or interface element (FEM-J)**

Goodman *et al.* (1968) proposed a four-noded joint element for joints. This model is a four-noded version of the contact element of Ngo and Scordelis (1967) and it has the following characteristics. In a two-dimensional domain, joints are assumed to be tabular with zero thickness. They have no resistance to the net tensile forces in the normal direction, but they have high resistance to compression. Joint elements may deform under normal pressure, especially if there are crushable asperities. The shear strength is presented by a bilinear Mohr-Coulomb envelope. The joint elements are designed to be compatible with solid elements. Ghaboussi *et al.* (1973) proposed a four-noded interface element for joints. This model is a further improvement of the joint element by assigning a finite thickness to joints. Zienkiewicz and Pande (1977) modified the formulation of rectangular elements to model joints and introduced an elastic-visco-plastic-type constitutive law for joints. The thin layer element proposed by Desai *et al.* (1984) is also similar to that of Ghaboussi (1988). These models are widely used for rock engineering structures in fractured and jointed media.

vi) Displacement discontinuity method (DDM)
This technique is generally used together with the boundary element method (BEM). The discontinuities are modeled as a finite length segment in an elastic medium with a relative displacement. In other words, the discontinuities are treated as internal boundaries with prescribed displacements. As an alternative approach to the technique of Crouch and Starfield (1983) and Crotty and Wardle (1985) use interface elements to model discontinuities, and the domain is discretized into several subdomains.

vii) Discontinuous deformation analysis (DDA)
Shi (1988) proposed a method called discontinuous deformation analysis (DDA). Intact blocks were assumed to be deformable and are subjected to constant strain and stress due to the order of the interpolation functions used for the displacement field of the blocks. In the original model, the inertia term was neglected so that the damping becomes unnecessary. For dynamic problems, although damping is not introduced into the system, the large time steps used in the numerical integration in time-domain results in artificial damping. It should be noted that this type of damping is due to the integration technique for time domain and has nothing to do with the mechanical characteristics of rock masses (i.e. frictional properties). Although the fundamental concept is not very different from Cundall's model, the main difference results from the solution procedure adopted in both methods. In other words, the equation system of blocks and its contacts are assembled into a global equation system in Shi's approach. Recently, Ohnishi *et al.* (1995) introduced an elasto-plastic constitutive law for intact blocks and gave an application of this method to rock engineering structures.

viii) Discrete element method (DEM)
The distinct element method (rigid block models) for jointed rocks was developed by Cundall in 1971. In Cundall's model, problems are treated as dynamic from the very beginning of formulation. It is assumed that the contact force is produced by the action of springs, which are applied whenever a corner penetrates an edge. Normal and shear stiffness were introduced between the respective forces and displacements in his original model. Furthermore, to account for slippage and separation of block contacts, he also introduced the law of plasticity. For the simplicity of calculation of contact forces due to the overlapping of the block, he assumed that the blocks do not change their original configurations. To solve the equations of the whole domain, he never assembled the equilibrium equations of blocks into a large equation system but solved them through a step-by-step procedure, which he called a marching scheme. His solution technique has two main merits:

1) Storage memory of computers can be small (note that computer technology was not so advanced during the late 1960s); therefore, it could run on a microcomputer.

2) The separation and slippage of contacts can be easily taken into account since the global matrix representing block connectivity is never assembled. If a large

assembled matrix is used, such a matrix will result in zero or very nearly zero diago-
nals, which subsequently cause singularity or ill conditioning of the matrix system.

As the governing equation is of hyperbolic type, the system could not become stabilized
even for static cases unless a damping is introduced into the equation system. In recent years,
he improved the original model by considering the deformability of intact blocks and their
elasto-plastic behavior. Cundall's model has been actively used lately in rock engineering
structures designed by the NGI group (e.g. Barton *et al.*, 1986, 1987).

How to select the appropriate dimension in numerical analyses of geoengineering structures is always an important issue. Every structure is three-dimensional in physical space.
If time is considered, the problem becomes four-dimensional. During the time of doing a
numerical analysis of an advancing tunnel in 1986, the data preparation and visualization
were extremely difficult (Aydan *et al.*, 1988). Furthermore, the memory size and computation speed were also severe problems.

The present tiny notebook computers now have a storage capacity of several TBs.
Although the memory size and computation speed of computers have increased and pre- and
post-processing of computational results have become much more convenient and less labor
intensive, it is still very difficult to select the appropriate dimension in numerical analyses
of structures. Also, decisions in engineering design are still based on rules of thumb and/or
one-dimensional analytical or numerical analysis of structures.

Another important aspect in numerical analyses is how to simulate the excavation procedure. Aydan (2011, 2017) performed a numerical analysis of a circular underground opening
excavation using a dynamic procedure. The effect of the impulsive application of excavation force is evaluated for an axisymmetric cylindrical opening under initial hydrostatic
stress using a dynamic visco-elastic finite element method developed by Aydan (2011). The
responses of displacement, velocity, and acceleration of the tunnel surface with a radius
of 5 m are plotted in Figure 3.5a. As noted from the figure, the sudden application of the
excavation force, in other words, the sudden release of ground pressure results in 1.6 times
the static ground displacement at the opening perimeter, and shaking disappears almost at
2 seconds. As time progresses, it becomes asymptotic to the static value and velocity and
acceleration disappear.

The resulting tangential and radial stress components near the opening perimeter (25 cm
from the opening surface) are plotted in Figure 3.5b as a function of time. It is of great

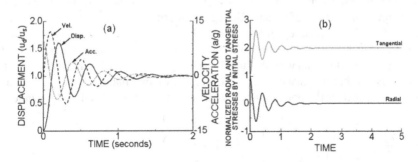

Figure 3.5 (a) Responses of displacement, velocity, and acceleration of the circular underground opening surface; (b) Responses of radial and tangential stress components near the tunnel surface.

interest that the tangential stress is greater than that under static condition. Furthermore, a very high radial stress of tensile character occurs near the tunnel perimeter. This implies that the opening may be subjected to a transient stress state, which is quite different from that under static conditions. However, if the surrounding rock behaves visco-elastically, they become asymptotic to their static equivalents. In other words, the surrounding rock may become plastic even though the static condition may imply otherwise.

3.5   METHODS FOR STABILIZATION AGAINST LOCAL INSTABILITIES

Local instabilities about underground openings in a layered or blocky rock mass generally involve block falls, block slides, flexural or block toppling, and combined block toppling and slides. For the sake of simplicity, only falls in the roof, sliding in sidewalls (Fig. 3.1), and flexural toppling in the roof and sidewalls are considered, and procedures to determine rock loads to be carried by support/reinforcement members in relation to idealized discontinuity patterns are described (Aydan, 2016). More complex cases can be found in Aydan (1989) and Kawamoto et al. (1991, 1994). Though discontinuity patterns in rock mass are closely associated with rock type, the patterns in a blocky rock mass can be generally classified (Aydan et al., 1989) as cross-continuous pattern and intermittent pattern (see Chapter 2). The intermittent pattern actually represents the most likely pattern in all kinds of rock. If an intermittency parameter introduced by Aydan et al. (1989) is used, it is seen that the cross-continuous pattern is a special case of the intermittent pattern.

First, how to determine the boundaries of the potentially unstable region in relation to the spatial distribution of discontinuity sets and the opening geometry is described. Though the geometry differs from a circular shape to a horseshoe shape, the shape of the opening is assumed to have vertical sidewalls with a circular-arched roof, as shown in Figure 3.6a. Next, it is assumed that the problem is two-dimensional and the maximum size of potentially unstable regions are calculated, although block theories may take into account the three-dimensional nature of the problem. This assumption is justified as far as the safety and the speed of calculations are concerned. Nevertheless, it may sometimes be necessary to carry out more detailed calculations when the cost of support elements is found to be quite high.

3.5.1   Estimation of suspension loads

Suspension loads arise when any frictional resistance on the critical bounding planes cannot be mobilized during movements of an unstable region towards the opening. In other words, when the apex angle of the unstable region is greater than 90° and there is no possibility of this angle becoming less than 90° during movements due to asperities that may exist on the critical bounding planes, the load due to the dead weight of the unstable region is referred to as a suspension load.

Let us consider that the unstable region is defined by two planes α_1 and α_2 as shown in Figure 3.6a. The area of the potentially unstable region can be obtained using geometry as:

$$A^r = \frac{L_a}{2}\left[L_a \frac{\tan\alpha_1 \tan\alpha_2^*}{\tan\alpha_1 + \tan\alpha_2^*} - R\left\{ 2\theta \frac{R}{L_a} - \cos\theta \right\} \right] \tag{3.14}$$

where L_a is width, R, θ are radius and angle of arch $\alpha_2^* = \alpha_2 - \xi$

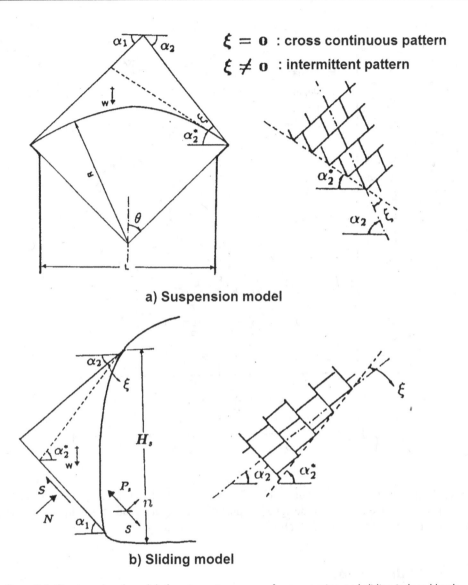

$\xi = 0$: cross continuous pattern

$\xi \neq 0$: intermittent pattern

a) Suspension model

b) Sliding model

Figure 3.6 Computational models for support pressures for suspension and sliding-induced loads.

For a given thickness t, the suspension load can be written in the following form:

$$F_{sus} = \gamma A^r \cdot t \qquad (3.15)$$

Figure 3.7 shows plotted results for the required support pressure against rock falls for various intersection angles of discontinuity sets. As the natural discontinuity intersection angle lies between 60° and 120°, the presently utilized arch support pressures provided by

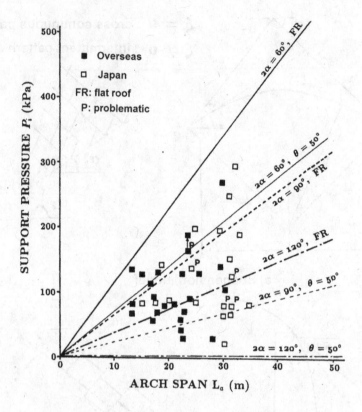

Figure 3.7 Support pressures for roofs and comparison with actual examples.

rockbolts and anchors for the roof caverns fall into the calculated range. In other words, if the supports of cavern roofs are designed against the suspension load, it should be sufficient and safe, provided that intact rock about the cavern does not yield.

3.5.2   Sliding loads

Sliding failure is possible when the frictional resistance of the critically orientated discontinuity set satisfies the following condition:

$$\tan \alpha_i < \tan \phi_i \qquad (3.16)$$

where subscript i represents the discontinuity set on which sliding is likely. The sliding load can be determined from the limiting equilibrium approach in the following form:

$$F_{slid} = \gamma S_{slid} \cdot t \frac{\sin(\alpha_i - \phi_i)}{\cos \phi_i} \qquad (3.17)$$

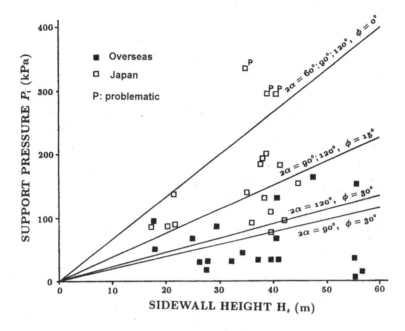

Figure 3.8 Support pressures for sidewalls and comparison with actual examples.

The area S_{slid} of the sliding region can be easily determined from the geometry of the region prone to sliding in relation to the geometry of opening. For example, the area of sliding body shown in Figure 3.6b can be specifically obtained as follows:

$$S_{slid} = \frac{H_s}{2}\left[H_s \frac{\tan\alpha_1 \tan\alpha_2^*}{\tan\alpha_1 + \tan\alpha_2^*}\right]$$

(3.18)

Figure 3.8 shows plotted results for the required support pressure against sliding for various intersection angles of discontinuity sets. As the natural discontinuity intersection angle lies between 60° and 120°, the presently used support pressures provided by rockbolts and/or anchors for the sidewall of caverns fall into the calculated range. In other words, if the support design of sidewalls of caverns is designed against the sliding load, it should be sufficient and safe, provided that intact rock about the cavern does not become plastic.

The above formulations are based on the static loads. If dynamic loads resulting from earthquakes or turbines are present, for example, the method suggested by Aydan *et al.* (2012) can be utilized.

3.5.3   Loads due to flexural toppling

When the rock mass is thinly layered, the layers may not be strong enough to resist the tensile stresses due to bending under the gravitational forces. Seismic forces may also induce additional loads, easing flexural toppling failure. In such cases, tensile stresses in layers

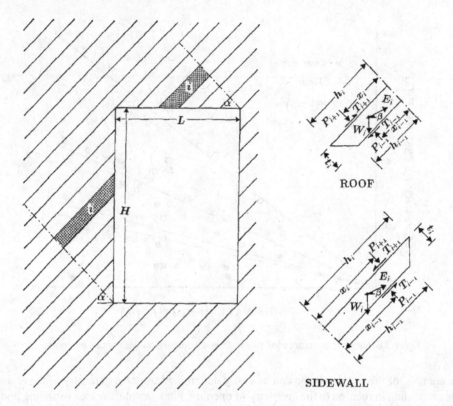

Figure 3.9 Model for limiting equilibrium analysis of flexural toppling of an underground opening (from Aydan and Kawamoto, 1992).

should be reduced below their tensile strength, if stability is required. If the roof consists of n layers (Fig. 3.9), the outer fiber stresses for layers in the sidewall and in the roof take the following forms (Aydan, 1989; Aydan and Kawamoto, 1992):

$$\sigma_t^i = \pm \frac{N_i}{A_i} + \frac{6t_i}{I_i}\left[P_{i+1}\eta h_i - T_{i+1}\frac{t_i}{2} - P_{i-1}\eta h_{i-1} - T_{i-1}\frac{t_i}{2} + S_i\right] \tag{3.19}$$

where $N_i = W_i \cos \alpha - E_i \sin (\alpha + \beta)$; $S_i = W_i \sin \alpha + E_i \cos(\alpha + \beta)$; $W_i = \gamma_i t_i b(h_{i-1} + h_i)/2$; $A_i = t_i b$; γ_i: unit weight of the layer; h_i and h_{i+1} are side lengths of the interfaces between layers $i - 1$, i and layers i, $i + 1$, respectively; P_{i-1}, P_{i+1} and T_{i-1}, T_{i+1} are normal and shear forces acting on of the interfaces between layers $i - 1$, i and layers i, $i + 1$, respectively; t_i: thickness of layer I; b: width; a: layer inclination; and η: coefficient of load action location. Sign (+) stands for layers in roof and (−) for layers in sidewalls.

Introducing the yield condition such that the outer fiber stress of the layer is equal to the tensile strength σ_T of the rock with a factor of safety SF as:

$$\sigma_t^i \leq \frac{\sigma_T}{SF} \tag{3.20}$$

and assuming the normal and shear forces acting interfaces of layers through frictional yielding condition (ϕ: is friction angle)

$$T_{i+1} = P_{i+1}\mu; \quad T_{i-1} = P_{i-1}\mu; \quad \mu = \tan\phi \tag{3.21}$$

the normal forces acting on layer $i-1$ can be easily obtained as

$$P_{i-1} = \frac{P_{i+1}\left(\eta h_i - \mu\frac{t_i}{2}\right) + S_i \frac{h_i}{2} - \frac{2I_i}{t_i}\left(\frac{\sigma_T}{SF} \pm \frac{N_i}{A_i}\right)}{\left(\eta h_{i-1} + \mu\frac{t_i}{2}\right)} \tag{3.22}$$

The equation above is solved by a step-by-step method and rock load P_s is obtained from the following criterion:

$$P_0 > 0 \text{ and } P_0 = P_S \tag{3.23}$$

If $P_0 > 0$, it is interpreted that some support measures are necessary.

The stabilization of underground openings against flexural toppling failure requires reduction in the magnitude of the moment and an increase in the compressive normal forces acting on interfaces. There are a number of ways to provide such an effect through the use of artificial support. Pre-stressed cables and/or fully grouted rockbolts are effective solutions. When the pre-stressed cables are used, they should be anchored beyond the basal plane, otherwise pre-stress forces may cause much higher bending stresses in layers. The alternative is to use fully grouted rockbolts or "dowels." The fully grouted rockbolts would be more economical than rockanchors, as they are shorter and do not need to be pre-stressed. In Chapter 7, a procedure proposed by Aydan (Aydan, 1989; Aydan and Kawamoto, 1992) introduces how to consider the reinforcement effect of the fully grouted rockbolts against flexural toppling failure.

3.6   INTEGRATED AND UNIFIED METHOD OF DESIGN

The design of underground structures was generally done using empirical methods, such as rock classifications and/or with the help of some theoretical methods and model tests, before computers became available to geoengineers. The present tendency is also to use numerical techniques such as finite element, finite difference, and boundary element methods for design purposes. As the rock mass always contains numerous discontinuities, there is no

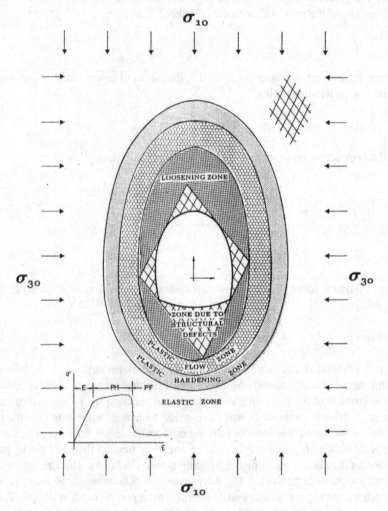

Figure 3.10 An illustration of the plastic zones, the loosening zone, and the zone due to structural defects.

unified method of design of underground structures. Furthermore, how to assess the global and local stabilities around underground openings is always an important issue. Figure 3.10 shows a proposal suggested by Aydan (1989) for assessing the global and local stabilities around underground openings.

Aydan (1989) suggested that the structural defect approach and the rock-support interaction approach could be unified for a more generalized approach. This unification assumes that there are three zones about the opening, namely, the plastic zone, the loosening zone, and the zone due to structural defects. The conditions for the appearance of such zones are

- The *structural defect zone* can occur when rock has one discontinuity set or more, and at least one of which daylights on the surface of the opening.

- The *loosening zone* can occur when the initial *in situ* stress ratio and the geometry of the effective opening shape defined by the geometrical orientation of structural defects (discontinuity sets) are such that tensile stress regions about the opening are to occur. Note that the term of loosening zone is herein associated with the zone caused by tensile stresses, which the rock mass may not sustain and becomes free from the true ground stress field about the opening. It is distinguished from the zone created by the plastification of rock due to compressive stress field.
- The *plastic zone* can occur if the redistributed stress state is such that it is sufficient to cause the yielding of rock in the region outside the possible above two other zones.

All three of these zones may not be observed in every excavation, and their occurrence would depend upon the conditions, such as the geologic structure of rock mass, the geometry of the opening, the initial *in situ* stress field, and the mechanical properties of rock. Therefore, there may be a number of varieties in real rock engineering practices.

The present philosophy of underground structures design may be outlined as follows (Kawamoto *et al.*, 1991; Aydan and Kawamoto, 2001; Kawamoto and Aydan, 1999; Aydan, 2016) (Fig. 3.11):

Stage I: *Geological and geophysical investigations and testing*: This stage involves the investigations of the geological structure of the site and geological structural defects, such as faults, shear zones, and joints. Geophysical explorations are done to characterize the rock mass. The mechanical properties of intact rocks, rock mass, faults, joints, and seismic wave velocities are measured by means of laboratory and *in situ* tests. *In situ* stress state is evaluated through measurement and/or inference techniques.

Stage II: *General evaluation*: A general evaluation of the properties and structure of the rock mass is carried out based on geological and geophysical investigations, experiments, and *in situ* testing, considering past case histories and experiences.

Stage III: *Global and local stability assessment*: Possibility of global and local instability is checked on the basis of rock mass classifications, kinematic stability models, and simplified numerical analysis (i.e. Cording, 1973; Barton *et al.*, 1974; Bieniawski, 1989; Aydan, 1989; Kawamoto *et al.*, 1991; Aydan *et al.*, 2013; Aydan and Ulusay, 2014). Preliminary support/reinforcement system is designed. The design of support/reinforcement systems is not well established, and various approaches include the following:

 i) The first approach is based on restraining the development of plastic zone or assuming that it acts as a dead weight on the support members and they should be carried by the members or transferred to the elastic zone. The plastic zone is calculated by the closed-form solutions or numerical analysis.

 ii) The second approach is based on the defining the maximum size of potentially unstable zone on the basis of discontinuity surveying and making the support members either to suspend that zone to the stable zone in the roof or prevent its sliding or toppling into the opening. The block theory also belongs this group.

 iii) Third approach is based on the consideration and modeling of support/reinforcement members in numerical or theoretical analysis. Nevertheless, this type analysis is rarely applied.

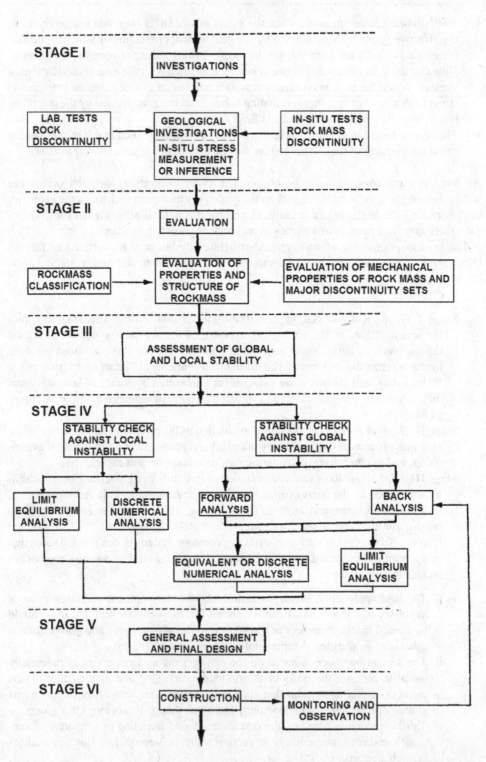

Figure 3.11 Flow chart of design and construction of an underground structure (from Aydan, 2016).

Stage IV: *Detailed stability analysis against global and local instability modes*: First, rock masses are generally modeled by three models for both global and local instability modes:

i) *Equivalent mass approach:* The rock mass is assumed to have mechanical properties that are fractions of intact rocks (Singh, 1973; Aydan and Kawamoto, 2000, 2001; Aydan *et al.*, 2013). This reduction in the properties is done by measuring the mechanical properties directly, rock mass classifications, or the squared ratio of the elastic wave velocity of the rock mass to that of intact rock.

ii) *Semi-explicit continuum models:* The effect of discontinuities of finite length is considered a damage in the body and modeled by some tensorial methods (i.e. fabric, damage tensors) to consider the effect of discontinuities within the framework of continuum mechanics (Oda *et al.*, 1993; Kawamoto *et al.*, 1988).

iii) *Explicit approaches:* Intact rocks and discontinuities are modeled individually (Cundall, 1971; Goodman *et al.*, 1968; Shi, 1988; Aydan *et al.*, 1996b).

Then, the methods of analysis of the stability of tunnels are chosen:

i) *Closed-form methods:* Although closed-form solutions are appropriate for some simple geometry and material behaviors, they are often used in the preliminary design of structures, since the error in modeling the rock mass is much larger than the error caused by the difference between the theoretical model and actual structure (Aydan, 1989; Aydan and Geniş, 2010).

ii) *Numerical methods:* Once the mechanical model for rock mass is chosen and the constitutive relations are established, it is a simple matter of calculations. The general tendency is to use the elasto-plastic type of constitutive relations in analyses (Kawamoto and Aydan, 1999). Nevertheless, the general tendency is to restrict the calculations to elastic case only as the nonlinear models cost more. The numerical models based on the explicit models of rock mass are used less because of the difficulties of presenting the discontinuous nature of rock mass and the huge effort required to input the data.

iii) *Limiting equilibrium approaches:* The limiting equilibrium approaches are generally restricted to investigating the local instability of underground openings. Block theory and other methods are available for this purpose (Shi, 1988; Kawamoto *et al.*, 1991; Aydan and Kawamoto, 2001; Aydan and Tokashiki, 2011).

With the advance of computers in recent decades, the design procedures for underground openings also utilize computational techniques, such as FEM or BEM. Nevertheless, the material properties specification for the surrounding mass is the most difficult aspect of such designs, as the existence of discontinuities in rock mass present considerable restrictions on the mechanical models used for the rock mass. Most of the techniques model the rock mass on an equivalent continuum with reduced mechanical properties in order to check the global stability of underground structures. Because of the limitation of the equivalent continuum modeling, additional stability analyses are carried out to check the stability of tunnels against some modes of instabilities of the local kind, which are due to discontinuities.

Stage V: *General assessment and final design*: Based on past experiences, rock classifications, and preliminary and detailed analyses of the stability of the tunnel and support system, a general assessment is performed and the final design is decided on.

Furthermore, a detailed plan of excavation, construction, and monitoring procedures is laid out. In addition, some criteria for a monitored response of the surrounding rock mass in conjunction with excavation stages are established.

Stage VI: *Construction and back analyses:* The construction of the underground structures is carried out according to the final design. However, the safety and efficiency of the construction are checked from time to time to monitor results and the overall advance rate of the excavation. If monitored results are somewhat different from the anticipated results, back analyses are carried out and the final design is checked or modified if necessary.

It is almost a common procedure to monitor the behavior and response of the surrounding rock mass during excavation. The monitoring schemes generally involve the continuous measurement of displacement of the surrounding rock mass and acoustic emissions. Recent procedures include viewing boreholes next to boreholes, in which extensometers are installed (Uchida *et al.*, 1993). Additionally, seismic tomography, borehole deformability tests, and permeability tests are performed on the surrounding rock mass to assess its material property changes. The back-analysis technique proposed by Sakurai and his co-workers (Sakurai, 1993) has also become a common procedure for assessing the response of the surrounding rock mass to excavation steps.

3.7 CONSIDERATIONS ON THE PHILOSOPHY OF SUPPORT AND REINFORCEMENT DESIGN OF ROCK SLOPES

Rock slopes are generally associated with the construction of railways, highways, and dams in civil engineering, open-pit mines in mining engineering, and natural slopes in the form of mountains and cliffs. As a rock mass near the ground surface is more fractured, weathered, and prone to degradation due to atmospheric agents, the instabilities associated with structural defects (e.g. discontinuities) are much more common than those involving the yielding of rock mass and intact rock. Figure 3.12 illustrates the concept of zones of yielding of discontinuities, discontinuities and intact rock and yielding within rock slopes. Compared to underground openings, the failures associated with discontinuities are much more likely and may be more problematic. Therefore, any design scheme for rock slopes must consider various possible failure modes and evaluate the stability of designed slopes in relation to the above elements.

3.7.1 Empirical design systems

One can find some guidelines or standards for designing rock slopes. The most important parameter in rock slope design is the determination of the stable slope angle. However, the evaluation of rock masses in many available standards are too rough and they do not count the essential parameters of rock masses. Within the knowledge of the author, there is only rock mass classification proposed for rock slopes (by Romana, 1985) and it is called Slope Mass Rating (SMR). This system is fundamentally based on Rock Mass Rating (RMR) proposed by Bieniawski (1989), and it adjusts RMR value for a given rock mass. The fundamental table for adjustment is shown in Figure 13.13. Although this system has been utilized and applied in Spain and other countries, it is not widely used.

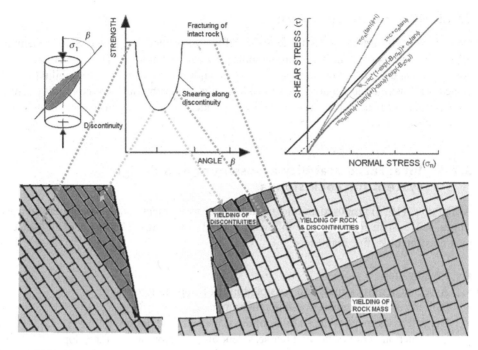

Figure 3.12 Yield zones associated with discontinuities, rock and discontinuities, and rock mass in rock slopes.

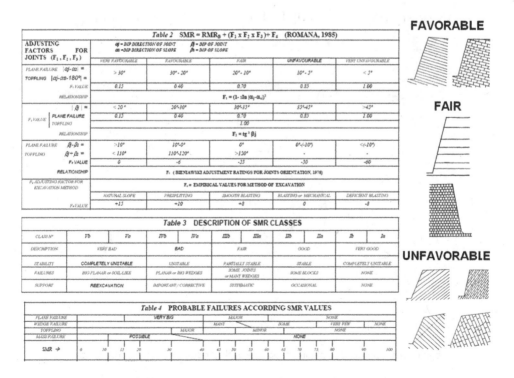

Figure 3.13 Fundamental features of Slope Mass Rating approach.

3.7.2   Kinematic approach

One of the earliest approaches for slope stability assessment and design is the kinematic approach. This approach utilizes stereographic projection techniques. The orientation of discontinuities and their frictional properties play a major role in assessing the stability of rock slopes, as illustrated in Figure 3.14. The kinematic approach could not cover all possible failure modes of rock slopes. Nevertheless, it provides a quick, easy evaluation of potential modes of rock slopes.

3.7.3   Integrated stability assessment and design system for rock slopes

Aydan *et al.* (1991) proposed an integrated stability assessment and design system for rock slopes. This system consists of the following several subsystems: The design of any slope in rock mass involves five main steps:

- Investigation
- Assessment of rock mass and possible forms of slope failure
- Stability analyses against possible failure modes
- General assessment of the stability of the slope and final design
- Construction, monitoring, and reassessment of the performance of the slope

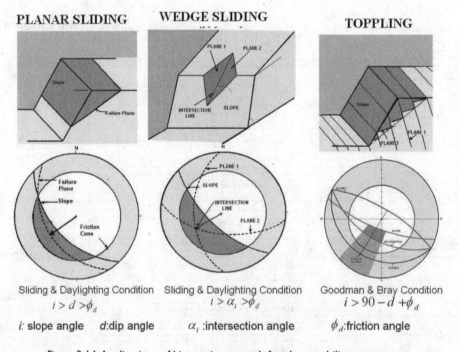

PLANAR SLIDING   WEDGE SLIDING   TOPPLING

Sliding & Daylighting Condition   Sliding & Daylighting Condition   Goodman & Bray Condition
$$i > d > \phi_d$$ $$i > \alpha_i > \phi_d$$ $$i > 90 - d + \phi_d$$

i: slope angle   *d*:dip angle α_i :intersection angle ϕ_d:friction angle

Figure 3.14 Applications of kinematic approach for slope stability assessment.

The system is designed to cover the first four main steps of the overall design scheme of rock slopes. These following four steps, designed as subsystems, are described.

1) Subsystem for investigation

This step involves the geological investigation of the site, laboratory, and *in situ* mechanical testing on intact material and discontinuities and documentation surveying. The geological investigation of the site generally involves the structural features, such as faults, main discontinuity sets, their orientation, spacing, continuity, etc. For the processing of field measurements on the orientation, spacing, and continuity of discontinuity sets, this subsystem uses subprograms written in FORTRAN. When field measurements on these structural features could not be directly made, the data on the structural features could be obtained through the processing of images on three different planes with known unit vectors. The initial data for the processing are read from the images through a digitizer on each plane. Then, the statistical distribution of the digitized data is evaluated and the spectra of processed datum on orientations and spacings on each image are determined (Fig. 3.4). From this processing, the orientation of each discontinuity set and its mean spacings are determined. If no field measurements by direct methods or photographic means could be done, the subsystem has another means of determination, which is based on the geologic information on the type of rock, angular relations on its blocky nature and its block size, and the orientation of a characteristic geological structural feature, i.e. flow plane in igneous rocks, schistosity or foliation plane in metamorphic rocks, or bedding plane in sedimentary rocks.

The mechanical properties of intact rocks and discontinuity sets are assumed to be given as input data, from direct tests either in the laboratory or *in situ*. If no measurements are available, the necessary data for similar types of rocks could be obtained from a database system available in the development environment.

2) Subsystem for the assessment of rock mass and possible forms of slope failure

In this step, the structure of the rock mass is assessed on the basis of data obtained in the previous subsystem. The rock mass is classified into three main groups (Fig. 2.6):

- Continuous mass (no discontinuity sets)
- Layered mass (one discontinuity set: bedded rock mass, schistose rock mass)
- Blocky mass (number of discontinuity sets are two or more)

The possible forms of rock slope instability depending upon the structure of rock mass are evaluated and possible modes of failure are determined according to the inclination of throughgoing discontinuity (e.g. bedding plane, schistosity plane, flow plane), as illustrated in Figures 3.15 and 3.16. The possible forms of failure of a given slope are assessed by the system based on data on the slope's geometry, orientations of discontinuity sets, the frictional angle of respective sets, and the structural nature of the rock mass.

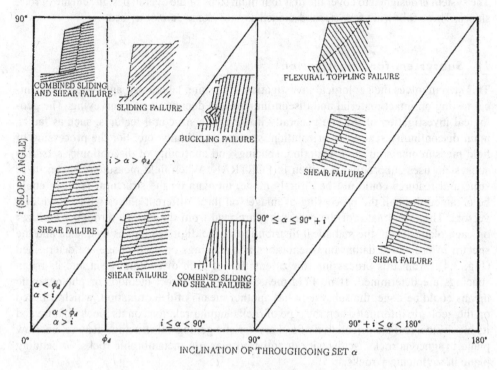

Figure 3.15 Selection of possible failure modes for layered rock mass.

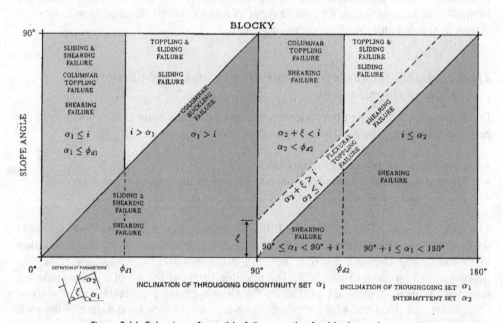

Figure 3.16 Selection of possible failure modes for blocky rock mass.

3) Subsystem for the stability analyses against possible failure modes

This subsystem provides stability analysis methods for various kinds of rock slope failure, based on the limiting equilibrium concept. The following methods of stability analysis for rock slopes are installed in the system:

- Slip circle methods: Fellenius (1936), Bishop (1965), Spencer (1967), Aydan *et al.* (1992)
- Planar failure
- Wedge failure: Wittke (1964)
- Combined sliding and shearing failure: Aydan *et al.* (1992)
- Flexural toppling failure: Aydan and Kawamoto (1992)
- Columnar toppling: Aydan *et al.* (1989)
- Combined columnar toppling and sliding: Aydan *et al.* (1989)

Once the possible forms of stability are determined based on the given criteria, the system automatically selects the stability analysis methods and displays the output in terms of safety factor for each respective failure mode. The subsystem is also designed to have a function for the parametric studies for a given form of failure of the slope on demand.

4) Subsystem for the general assessment and design

This subsystem compares the outputs of the stability analysis for each specific form of slope failure and evaluates the most likely form. If further analysis is required to determine the optimum slope angle and height, this subsystem activates the function for parametric studies for the most likely form of failure and for a chosen factor of safety, and then displays the output in graphical form. If it is required, the effect of reinforcement and support systems is evaluated.

Figure 3.17 shows an application of this integrated stability and design system to rock slopes and compares with actual observations on stable and unstable rock slopes. As noted from the figure, it is possible to consider various possible modes of instability in this approach.

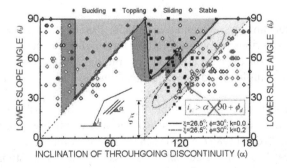

Figure 3.17 An application of the integrated stability and design system to rock slopes and comparisons with actual observations on stable and unstable rock slopes (from Aydan, 2008).

3.8   CONSIDERATIONS ON PHILOSOPHY OF SUPPORT DESIGN OF PYLONS

There is an increasing demand for the use of rockanchors as foundations in many geotechnical engineering structures, such as transmission towers. Particularly, designing rockanchors as foundations for transmission towers is likely to be one of the intensive fields of research, since the present design concept is utilizes piles and treat rock mass as if soil, and it is over-conservative and costly. The typical design of transmission towers utilizing rockanchors as foundations should resist against uplift loads due mainly to wind loading. The use of rockanchors as foundations is cost effective, since the volume of excavation is small compared to the commonly used caisson-type pile or inverted T-shaped foundations. This is an important aspect of foundation design, because many transmission towers are constructed in mountainous locations where construction difficulties include material transport, excavation, and the disposal of the excavated materials. Nevertheless, experience with using rockanchors as foundations is limited and the data on their long-term performance is scarce.

As anchors are human-made structures, some surfaces of weakness are inevitably introduced, and these surfaces have lower adhesive strength than that of the medium (Aydan, 1989, Aydan *et al.*, 1990). As applied loads are transferred into the medium through these interfaces, failure by shearing occurs at the interface with the lowest strength. Interfaces in anchors are:

- Grout-bar interface (GB)
- Grout-rock interface (GR)
- Grout-sheath-grout interface (GSG) (if a corrosion protection sheath is used)

The mechanism and fracturing states along these interfaces were previously investigated and discussed in detail by Aydan (Aydan, 1989; Aydan *et al.*, 1990).

There are, in general, four fundamental approaches for the design of anchors:

- Standards and empirical approaches
- Limiting equilibrium approaches
- Closed-form solutions
- Numerical methods

For the design of rockanchors, Aydan *et al.* (1985a, 1985b) and Aydan (1989) developed several theoretical solutions for various failure forms of interfaces and yielding of steel bar. A simplified version of these solutions for the design purposes is briefly given in Table 3.5. For more complex conditions, use the numerical techniques also developed by Aydan (1989).

The closed-form solutions given in Table 3.5 consider the possible shear failure along one of the interfaces, which are the most critical surfaces within the system of rockanchors/rockbolts. Although Aydan (1989) extended these solutions for the elasto-plastic behavior of not only interfaces but also of steel bar, and as the problem is to estimate the anchorage capacity of interfaces, not the resistance of the steel bar which is very well-known, the solutions for elastic and elasto-plastic behaviors are only given herein for brevity. Details of the derivations can be found in Chapter 4.

Table 3.5 Simplified version of formulas to estimate pull-out capacity of rockanchors.

Behavior of interfaces	Anchor-head displacement (u_0)	Anchor-head axial stress (σ_0)
Elastic limit	$\dfrac{\tau_p}{K_g}$	$\dfrac{\tau_p E_b \alpha}{K_g}\tanh(\alpha L)$
Elastic-perfectly plastic	$\dfrac{\tau_p}{K_g}\left[L_p\alpha\tanh\!\big(\alpha(L-L_p)\big)+1\right]+\dfrac{\tau_p L_p^2}{E_b r_b}$	$\dfrac{\tau_p}{K_g}\left[L_p\alpha\tanh\!\big(\alpha(L-L_p)\big)+\dfrac{2K_g L_p}{r}\right]$
Elastic-perfect-brittle-plastic	$(\xi-1)\dfrac{\tau_p}{K_g}\dfrac{L_r}{L_p-L_r}+\dfrac{\tau_p}{K_g}(L_p-L_r)L_r+\eta\dfrac{\tau_p L_r^2}{E_b r_a}+\xi\dfrac{\tau_p}{K_g}$	$(\xi-1)\dfrac{\tau_p}{K_g}\dfrac{E_b}{L_p-L_r}+\dfrac{\tau_p}{r_b}(L_p-L_r)+2\eta\dfrac{\tau_p L_r}{r_b}$

$$\alpha=\sqrt{\frac{2K_g}{E_b r_b}}\,;\ K_g=\frac{G_g}{r_a\ln(r_h/r_b)}\cdot\left|\frac{\chi-1}{\chi}\right|;\ K_m=\frac{G_m}{r_h\ln(r_0/r_h)};\ \chi=\frac{G_m}{G_g}\frac{\ln(r_h/r_b)}{\ln(r_0/r_h)}+1;\ \eta=\frac{\tau_r}{\tau_p};\ \xi=\frac{\gamma_r}{\gamma_p}$$

3.8.1   Geological, geophysical, and mechanical investigations

Several exploration techniques for the geology of a site can be used. Investigations concentrate on the type of rocks and the geologic structure of rock mass, such as the density and orientation of discontinuities, RQD, Q value, RMR, RMQR, CRIEP's classification, etc., using borehole drilling and outcrop surveying. Outcrop surveying is generally suggested, but detailed information is usually difficult to obtain as the rock mass is covered with top-soil. Therefore, borehole drilling is necessary to get more information. The investigations show that most of the available assessment techniques are not satisfactory for assessing rock masses as anchor foundations. It seems that the density and orientation of discontinuities are probably the most important parameters for the assessment of the rock mass. Geophysical exploration techniques seem to be appropriate for assessing the rock mass (Ebisu *et al.*, 1992). Among various geophysical techniques, the exploration technique based on the Rayleigh waves is the most sensitive to the structure of rock masses, and it is suitable for the site investigation of near-surface masses because it yields the most detailed information on the physical state of rock mass. Besides these reasons, the technique is also the most cost-effective and easy to operate on the site, since it does not require a borehole. We have obtained a good correlation between the deformability of rock masses by the borehole jack test and the Rayleigh wave velocity V_r.

The design procedure briefly involves the following steps and is explained in the following subsections.

3.8.2   Specification of material properties

The following parameters are important for the design of rockanchors:

- Deformability of rock mass
- Mechanical properties of steel and grouting material
- Shear strength of interfaces

i)   Specification of deformability of rock mass

For measuring the deformability of rock mass, it is presently suggested to use the borehole jack test or the pressuremeter test in principle at each transmission tower site. However, in the future, we have been considering using the correlation between the Rayleigh wave velocity V_r and deformability of rock masses, as an indirect assessment of the deformability of rock mass, which is expected to reduce the cost of site investigation.

ii)   Specification of mechanical properties of steel and grouting material

Mechanical properties of steel and grouting material can be evaluated from standard tests. The data on mechanical properties of steel and grouting material will be generally the same unless the type of steel and grouting material are varied.

iii)   Specification of the shear strength of interfaces

The shear strength of interfaces is not the same as that of the grouting material and is variable depending upon the surface morphology of the interfaces. For the specification of the shear strength of interfaces in boreholes, the direct shear test results and the radial response of the surrounding medium about the borehole should be used as proposed by Aydan *et al.* (1995).

iv)   Estimation of uplift capacity of anchor systems

For estimating the uplift capacity of anchors, one of the formulas presented in Table 3.5 can be used depending upon the designer's choice. The design of rockanchors involves not only mechanical but also economic, constructional, and environmental considerations, as well as each country's safety concerns regarding failure. From the mechanical point of view, the design of anchors involves the following:

* Determination of diameters of bars and boreholes
* Determination of spacing of anchors
* Determination of anchorage length

Borehole diameter is governed by the constructional conditions and the existence of the corrosion protection sheath. Diameters of the bars and spacing are governed by the produced sizes of the bars and the allowable intensity of pull-out load per anchor. From the experience of the author, it is found that if the anchor spacing is 30 times the anchor diameter, the interaction between anchors almost disappears. Once the diameters of the borehole and bar and the spacing of anchors are specified, then the problem is to determine the anchorage length. As stated at the beginning of this chapter, the long-term performance of rockanchors under cyclic loading has to be investigated, as the experience is very scarce. This is important, as the displacement of the super structure is limited. Cyclic pull-out tests on rockanchors, which are 3 m long in highly jointed sandstone and granite, indicated that displacement responses increase as a function of cycle number N at various loading levels. It is therefore important to keep the load level of anchors below the elastic limit of interfaces and steel bar.

3.9   CONSIDERATIONS ON THE PHILOSOPHY OF FOUNDATION DESIGN OF DAMS AND BRIDGES

As pointed out previously, rock mass near the ground surface is quite fractured and may have undergone heavy weathering and degradation. The foundations of dams and large bridges are carefully investigated and evaluated. In any case, the assessment of rock masses and their mechanical resistance with the consideration of rock mass structure is an essential part of their design. Regarding dams, the seepage through rock mass and/or stability of rock slopes near reservoirs are another important parameter for their design. Particularly, the slope failures at the Vaiont (Semenza and Ghirotti, 2000) and Aratozawa dams (Aydan, 2015) and the foundation failure of the Mallpaset dam are always taken into account as the worst fatal incidents in dam engineering. A typical dam design requires elastic response of foundation rock and the resistance of the dam body against translation failure and/or overturning. Rockanchors are particularly used to reduce overturning moment and increase the shearing capacity of gravity or arch dams. The load-bearing capacity of the foundations of dams and large bridges requires a careful assessment of the response of the foundation rock mass and a careful evaluation of the effect of discontinuities.

Tunnel-type anchorages may be adopted for suspension bridges. Despite the utilization of numerical methods for assessing their response to possible loading conditions, the current practice always requires a check of the uplift capacity of anchorage, based its own weight and the friction between anchorage and surrounding rock mass. Similarly, the pull-out capacity of gravity-type anchorages of suspension bridges are based on the dead weight of the anchorage and the frictional resistance of the foundation ground.

Chapter 4

Rockbolts (rockanchors)

4.1 INTRODUCTION

Rockbolts are nowadays one of most popular support members in rock engineering works. They are generally grouped into two types: (i) mechanically anchored rockbolts and (ii) grout-anchored rockbolts (Fig. 4.1) (Aydan, 1989). The mechanically anchored bolts are usually regarded as a special form of grout-anchored rockbolts from the mechanical point of view (Aydan *et al.*, 1986d, 1987d). While the mechanically anchored bolts mobilize only frictional forces, grout-anchored bolts mobilize both frictional and adhesive forces.

A rockbolt system consists of a cylindrical bar with high tensile, shear, and compressive strength and a connecting medium between the bar and the surrounding medium. The connecting medium is either a mechanical fixture utilizing a frictional type of anchorage by applying a compressive radial pressure towards the borehole or a cement- or resin-based grouting-type fixture utilizing an adhesion and frictional type of anchorage. The bars are generally made of steel. However, the bars made of fiberglass or carbon fibers are also

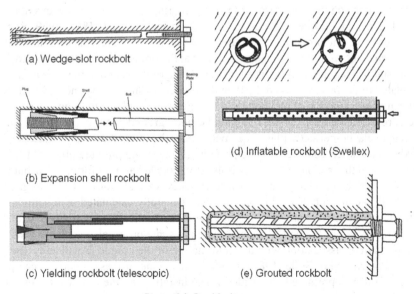

(a) Wedge-slot rockbolt

(b) Expansion shell rockbolt

(c) Yielding rockbolt (telescopic)

(d) Inflatable rockbolt (Swellex)

(e) Grouted rockbolt

Figure 4.1 Rockbolt types.

starting to be used in a corrosive embedment environment. Grouting material is either resin or cement. Though resin-type grouting materials are preferable because of their quickly developing bonding strength, cement type of grouting material is more widely used because of its much lower cost.

The earliest form of rockbolts was made of wood and was used to prevent rock falls between the face and the supports in longwall mining by Beyl in 1912; he tried later to introduce the bolting in British mines. Nevertheless, the fundamental studies of rockbolts and rockbolting were not undertaken until 1948 (Rabcewicz, 1955). The first such studies were initiated by the U.S. Bureau of Mines. The research studies undertaken by Panek (1956a, 1956b, 1962a, 1962b) were mainly concerned with the suspension and beam building effects of rockbolts. Noting the arching observed in Panek's model tests, Evans (1960) had pointed out the arch formation effect of rockbolts following the fracturing of layered roofs and suggested a simple theoretical formula based on the failure of arch by compression at the limit state. This suggestion was later elaborated by Cox (1974) by including the possibility of failure at abutments. While both experimental and theoretical studies were taking place in the field of mining engineering, Rabcewicz and his co-workers were also trying to introduce rockbolts into tunneling, together with the tunneling method, which they named the New Austrian Tunneling Method – NATM. Rabcewicz (1957a) did also some model tests, and using his field experiences, he established the principles of his method. In his model tests, he observed that by compressing a gravel-like material by rockbolts with surface wire meshing, it was possible to create a carrying rock arch that can resist loads corresponding to several times the arch height. This experimental result made him realize that if the intrinsic strength of rock could be mobilized, it was possible to deal with rock loads more easily. Then, he tried to find the theoretical basis for the reinforcement effects of rockbolts (Rabcewicz, 1957b). His first proposal was also based on the conventional arching theory, but he tried to consider a more complex case. In his arching theory, a layered rock mass was considered and its stability investigated by considering the sliding along a critical plane within the created arch. Following this initial trial, Rabcewicz and his co-workers did more elaborate model tests and proposed a dimensioning procedure for the carrying ring, based on conclusions drawn from the model tests and Talobre's proposal (1957) (Rabcewicz, 1964, 1965, 1969; Rabcewicz and Golser, 1973). The proposed procedure is for calculating the resistance of a circular ring under a given external uniformly distributed load at the limiting state. The proposed solution does not establish any relation between the resistance of the ring and the *in situ* stress state. If it is related to the *in situ* stress field, then the procedure simply becomes equivalent to the case, in which the function of rockbolts and shotcrete is considered to be an equivalent internal pressure exerted on the surface of a circular underground opening situated in a medium exhibiting an elastic-perfectly plastic behavior under the hydrostatic state of stress.

Rockbolts are often considered to be providing a confining pressure effect at the excavation surface, and this action of the bolts is usually modeled as a uniformly distributed internal pressure acting on underground surface (Rabcewicz and Golser, 1973; Egger, 1973a, 1973b, etc.). The magnitude of this pressure was generally calculated by assuming it to be equal to that of the applied pre-stress in the case of mechanically or partially anchored bolts or the yield strength of the bolt. Recently, there are some proposals to calculate the offered internal pressure in association with the deformation of the ground (Adali and Rösel, 1980; Hoek and Brown, 1980; Stille, 1983). Nevertheless, none of these proposals can properly evaluate the internal pressure by considering the installation time of bolts, the effect of the bearing plate, etc.

Besides the internal pressure (surface pressure) action of bolts, it was suggested that the bolts can also be used to increase the apparent shear resistance and the apparent deformation modulus of rock (Horino *et al.*, 1971; Feder, 1976; Egger 1973a,b,c; Tsuchiya, 1981; Stille, 1983; Wullschläger and Natau, 1983; Aydan *et al.*, 1987b) and reinforce the discontinuity planes. The earliest analytical studies on the shear reinforcement of discontinuity planes were done by Rabcewicz (1957b); Lang (1961); Panek (1962b) and Elfman (1969), and experimental studies were undertaken by Horino *et al.* (1971); Bjüström (1974); Haas (1976); Hibino and Motojima (1981); Ludvig (1983); Egger and Fernandes (1983) and Yoshinaka *et al.* (1986).

The reinforcement effects of rockbolts are generally classified as follows (Panek, 1962b; Egger 1973a,b,c, etc.):

- Internal pressure effect
- Improving the physical properties of ground
- Suspension effect
- Shear reinforcement
- Beam building effect
- Arch formation effect

This type of classification is merely associated with the particular reinforcement function required from rockbolts in a given problem of reinforcement and has nothing to do with the reinforcement mechanism of the bolts. The reinforcement effects of rockbolts are, in a real sense, associated with their axial and shear responses and the interaction taking place between themselves and the surrounding medium under given loading and boundary conditions (Aydan *et al.*, 1986d, 1987a, 1987c, 1987d). These responses manifest themselves in a number of ways, which were the primary cause of confusion on the reinforcement effect of rockbolts. In the following sections, the rockbolt system will be carefully investigated from the mechanical point of view, and it will be shown that it is possible to evaluate a number of the reinforcement effect of rockbolts just by a proper modeling of their axial and shear responses (Aydan *et al.*, 1986c).

Rockanchors are a special form of rockbolts. The rockanchors are generally made of steel cables consisting of several wires spirally bundled together. They have a certain anchorage length and are pre-tensioned to provide active support load on the excavation surface. They are generally preferred over rockbolts in view of limited installation space and commonly used in large underground caverns, rock slopes, and dam and bridge foundations.

4.2   ROCKBOLT/ROCKANCHOR MATERIALS AND THEIR MECHANICAL BEHAVIORS

The main component of rockbolt (rockanchor) systems is made of steel in the form of a cylindrical bar/cable. However, the bolts made of glass or carbon fibers is also starting to be used. As the bolt material is the principal load-bearing element in the system, it is usually required to satisfy the following conditions:

- To have high tensile and shear resistances
- Not to shed its load suddenly and to have a considerable large strain capacity following the yielding

- Little volumetric change due to stress and thermal changes
- To be anti-corrosive, fire-proof, and highly durable
- Cheap and easy to handle, transport, store, and install

The steel seems to satisfy most of the conditions stated above.

4.2.1   Yield/failure criteria of rockbolts

The mechanical behavior of the bolting material for steel and glass-fiber bars in tension and shear are shown in Figures 4.2 and 4.3. The yield/failure criteria of the main load-bearing element of the rockbolts are a function of axial and shear stresses. The yield/failure criteria

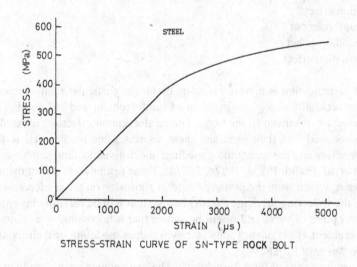

STRESS–STRAIN CURVE OF SN-TYPE ROCK BOLT

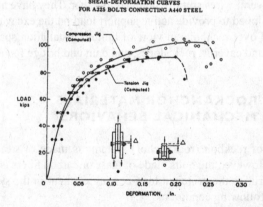

Figure 4.2 Behavior of steel bars under tensile and shear loadings.

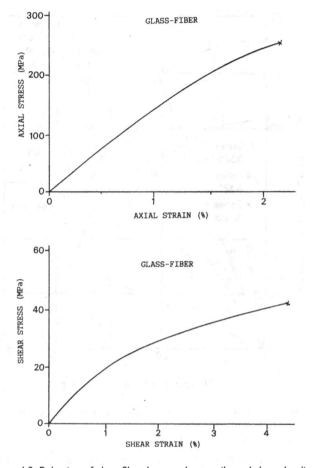

Figure 4.3 Behavior of glass-fiber bars under tensile and shear loadings.

of the steel bar under a combined stress state is also shown in Figure 4.4, and it may be mathematically expressed as:

$$\sigma_e = \sqrt{\sigma_a \cos^2 \theta + \tau_s \sin^2 \theta}$$
(4.1)

where σ_a is the absolute resistance under pure tension/compression and τ_s is shearing resistance. θ is the angle between direction of loading and the longitudinal axis of the rockbolt.

The shear resistance (τ_s) of the rockbolt materials is generally given by the following empirical relationship:

$$\tau_s = \beta \sigma_a$$
(4.2)

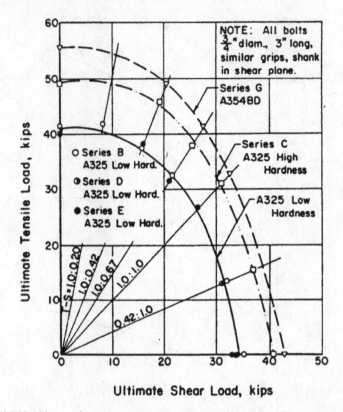

Figure 4.4 Yield locus of steel bars under combined loading (after Chesson *et al.*, 1965).

where β is an empirical relationship and its value is about 0.58 or greater for steel. However, it may be much less than 0.58 for fiberglass or carbon fibers. When the above equations are used for steel cables, it is strongly required that the axial stress should be always in tension.

Steel generally exhibits yield at about 0.7 times its ultimate tensile/compression resistance and its behavior is strain hardening.

4.2.2 Constitutive modeling of rockbolt material

The mechanical behavior of steel, which is commonly used as a bolt material, has been well investigated, and its constitutive law is also well established. Steel is a typical representative material for the non-dilatant plastic behavior, and the constitutive laws, based on the classical plasticity theory, are appropriate. A brief outline of the constitutive law for the steel bar is given below.

The derivation of an incremental elasto-plastic constitutive law, based on the conventional elasto-plastic theory, starts with the following:

- Yield function

$$F(\sigma, \kappa) = f(\sigma) - K(\kappa) = 0 \tag{4.3}$$

- Flow rule

$$d\varepsilon^p = \lambda \frac{\partial G}{\partial \sigma} \tag{4.4}$$

- Prager's consistency condition

$$dF = \frac{\partial F}{\partial \sigma} \cdot d\sigma + \frac{\partial F}{\partial \kappa} \frac{\partial \kappa}{\partial \varepsilon^p} \cdot d\varepsilon^p = 0 \tag{4.5}$$

- Linear decomposition of the strain increment $d\varepsilon$ into its elastic and plastic components $d\varepsilon^e$ and $d\varepsilon^p$ (Fig. 4.5).

$$d\varepsilon = d\varepsilon^e + d\varepsilon^p \tag{4.6}$$

- Hoek's law

$$d\sigma = \mathbf{D}^e d\varepsilon^e \tag{4.7}$$

where
σ = stress tensor
$K(\kappa)$ = hardening function
G = plastic potential
λ = proportionality coefficient
κ = hardening parameter
ε = strain tensor
ε^e = elastic strain tensor
ε^p = plastic strain tensor
$\mathbf{D}^e$ = elasticity tensor
$(\cdot)$ = inner product

Substituting Equation (4.4) into Equation (4.5), and rearranging the resulting expression, together with denoting its denominator by h (hardening modulus):

$$h = -\frac{\partial F}{\partial \kappa} \frac{\partial \kappa}{\partial \varepsilon^p} \cdot \frac{\partial G}{\partial \sigma} \tag{4.8}$$

we have λ as:

$$\lambda = \frac{1}{h}\frac{\partial F}{\partial \sigma} \cdot d\sigma \tag{4.9}$$

Now, when we insert the above relationship into Equation (4.4), we have the constitutive relationship between the plastic strain increment $d\varepsilon^p$ and the stress increment $d\sigma$, which is also known as the Melan's formula, as:

$$d\varepsilon^p = \frac{1}{h}\frac{\partial G}{\partial \sigma}\left(\frac{\partial F}{\partial \sigma} \cdot d\sigma\right) = \frac{1}{h}\left(\frac{\partial G}{\partial \sigma} \otimes \frac{\partial F}{\partial \sigma}\right)d\sigma = \mathbf{C}^p d\sigma \tag{4.10}$$

where $(\otimes)$ denotes the tensor product. The inverse of the above relationship could not be done, as the determinant of the plasticity matrix $|\mathbf{C}^p| = 0$ irrespective of whether the plastic potential G is of the associated or nonassociated type. Therefore, the following technique is used to establish the relationship between $d\sigma$ and $d\varepsilon$. Using the relationships (4.6), (4.7), and (4.10), one can write the following:

$$d\sigma = \mathbf{D}^e d\varepsilon - \mathbf{D}^e \frac{1}{h}\frac{\partial G}{\partial \sigma}\left(\frac{\partial F}{\partial \sigma} \cdot d\sigma\right) \tag{4.11}$$

Taking the dot products of the both sides of the above expression by $\partial F/\partial \sigma$ yields:

$$\frac{\partial F}{\partial \sigma} \cdot d\sigma = \frac{\dfrac{\partial F}{\partial \sigma} \cdot (\mathbf{D}^e d\varepsilon)}{1 + \dfrac{1}{h}\dfrac{\partial F}{\partial \sigma} \cdot \left(\mathbf{D}^e \dfrac{\partial G}{\partial \sigma}\right)} \tag{4.12}$$

Substituting the above relationship in (4.11) gives the incremental elasto-plastic constitutive law as:

$$d\sigma = \left(\mathbf{D}^e - \frac{d\mathbf{D}^e \dfrac{\partial G}{\partial \sigma} \otimes \dfrac{\partial F}{\partial \sigma}\mathbf{D}^e}{h + \dfrac{\partial F}{\partial \sigma} \cdot \left(\mathbf{D}^e \dfrac{\partial G}{\partial \sigma}\right)}\right)d\varepsilon \tag{4.13}$$

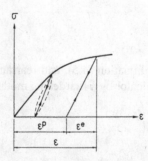

Figure 4.5 Definition of elastic and plastic strain (one-dimensional case).

The hardening modulus h is generally determined as a function of a hardening parameter κ by employing either a work-hardening model or strain-hardening model. The hardening parameter κ is defined for both cases as follows:

$$\kappa = W^p = \int \boldsymbol{\sigma} \cdot d\boldsymbol{\varepsilon}^p \qquad \text{work hardening} \tag{4.14}$$
$$\kappa = \int \|d\boldsymbol{\varepsilon}^p\| \qquad \text{strain hardening} \tag{4.15}$$

where W^p is plastic work.

The materials (i.e. steel, fiberglass) used as a bar material exhibit a non-dilatant plastic behavior and harden isotropically. Therefore, a work-hardening model is generally used together with the effective stress-strain concept, which are defined as:

$$\sigma_e = \sqrt{\frac{3}{2}(\mathbf{s} \cdot \mathbf{s})} \quad d\varepsilon_e^p = \sqrt{\frac{2}{3}(d\boldsymbol{\varepsilon}^p \cdot d\boldsymbol{\varepsilon}^p)} \tag{4.16}$$

where $\mathbf{s}$ is deviatoric stress tensor.

Since the volumetric plastic strain increment $d\bar{\varepsilon}_v^p = 0$ together with the co-axiality of the stress and plastic strain, the hardening parameter κ of work-hardening type can be rewritten in the following form:

$$d\kappa = dW^p = \boldsymbol{\sigma} \cdot d\boldsymbol{\varepsilon}^p = \mathbf{s} \cdot d\mathbf{e}^p = \sigma_e d\varepsilon_e^p \tag{4.17}$$

where $d\mathbf{e}^p$ is the deviatoric strain increment. The hardening modulus h for this case takes the following form with the use of Euler's theorem[1] by taking a homogeneous plastic potential G of order m:

$$h = -\frac{\partial F}{\partial \kappa}\frac{\partial \kappa}{\partial \boldsymbol{\varepsilon}^p} \cdot \frac{\partial G}{\partial \boldsymbol{\sigma}} = \frac{\partial K}{\partial W^p}\boldsymbol{\sigma} \cdot \frac{\partial G}{\partial \boldsymbol{\sigma}} = m\frac{\partial K}{\partial W^p}G \tag{4.18}$$

If $F = G$ and $f(\boldsymbol{\sigma}) = \sigma_e$, then the hardening modulus h becomes:

$$h = m\frac{\partial K}{\partial W^p}f(\boldsymbol{\sigma}) = m\frac{\partial \sigma_e}{\partial \varepsilon_e} \tag{4.19}$$

The hardening modulus can then be easily obtained from a gradient of the plot of a uniaxial test in σ_1 and ε_1^p space, since the effective stress and strain in the uniaxial state become:

$$\sigma_e = \sigma_1 \quad \varepsilon_e^p = \varepsilon_1^p \tag{4.20}$$

Figure 4.6 shows the plots of the predicted and experimental hardening functions of steel bar.

---

1 $\mathbf{x} \cdot \partial f / \partial \mathbf{x} = mf$.

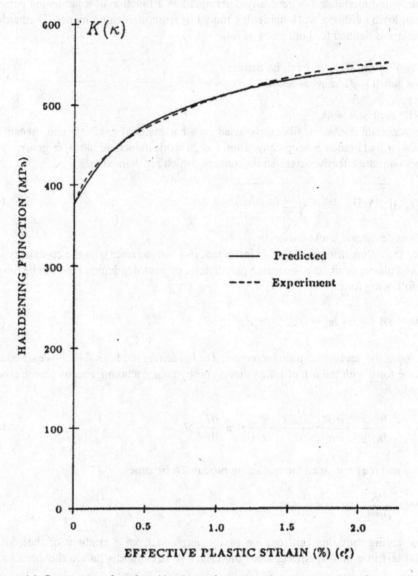

Figure 4.6 Comparison of predicted hardening function with experimental one for steel bar.

4.3   CHARACTERISTICS AND MATERIAL BEHAVIOR OF BONDING ANNULUS

4.3.1   Push-out/pull-out tests

In situ and laboratory push-out or pull-out tests are carried out with the objective of determining the bonding strength/bearing capacity of rockbolts and investigating the effect of the geometrical parameters of the bolt and borehole on the bearing capacity.

4.3.1.1   Experimental setup

An experimental program was undertaken by Aydan (1989) to investigate the anchorage mechanism of the grouted rockbolts and the effect of various parameters, such as bolt-borehole diameter ratio and surface configurations of steel bars under triaxial stress state. For this purpose, a special triaxial cell, which is capable of accommodating a sample of 12 cm in diameter and 20 cm in height, was constructed. Steel bars of 13 mm and 19 mm in diameter with smooth or ribbed surfaces (Fig. 4.7) were used. The yielding strength and ultimate strength of the bars were 300 MPa and 470–540 MPa with an elastic modulus of 200–210 GPa. The steel bars were instrumented with axially directional electrical resistant gauges spaced on a machined surface (Fig. 4.8).

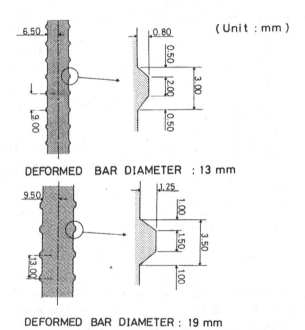

Figure 4.7 Surface configurations of ribbed steel bars.

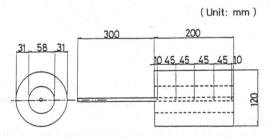

Figure 4.8 Steel bars with instrumented strain gauges.

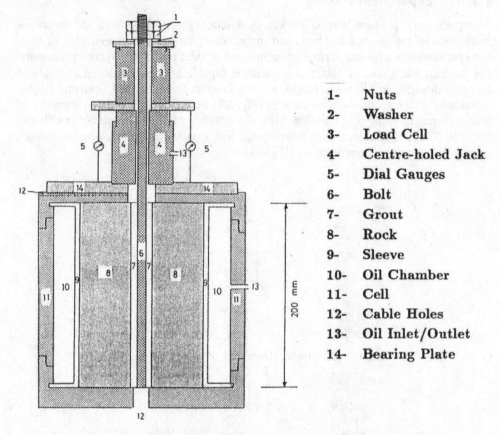

Figure 4.9 Experimental setup (pull-out tests).

1-   Nuts
2-   Washer
3-   Load Cell
4-   Centre-holed Jack
5-   Dial Gauges
6-   Bolt
7-   Grout
8-   Rock
9-   Sleeve
10-   Oil Chamber
11-   Cell
12-   Cable Holes
13-   Oil Inlet/Outlet
14-   Bearing Plate

The samples of rocks had the bolts cast into centrally drilled holes of various diameters with a cement-based grouting material. Figure 4.9 shows the experimental setup in the case of pull-out tests. The loads on bars were applied by the hydraulic ram of a servo-control compression-testing machine in the case of push-out tests and by a central-holed jack in the case of pull-out tests.

In push-out and pull-out tests, the loads on the bars were measured by the attached load cells, while the displacement of the bolt head was measured by two linear variable displacement transducers (LVDT). The measured loads, displacements, and strains were monitored at certain intervals of time using a microcomputer through a GP-IB interface and recorded onto a floppy disk.

4.3.1.2 Test results

Tests results are summarized in terms of bearing capacity, failure modes, and distributional characteristics of the axial strain and stress. Figure 4.10 shows the ultimate load-bearing capacity of bolts tested in push-out and pull-out tests. The load-bearing capacity of rockbolts is 25% higher in the case of push-out tests than that of pull-out tests. This fact is attributable to the so-called Poisson effect (the radial stress is of compressive character in the push-out case, whereas it tends to become tensile in the case of pull-out tests, particularly if the bolt

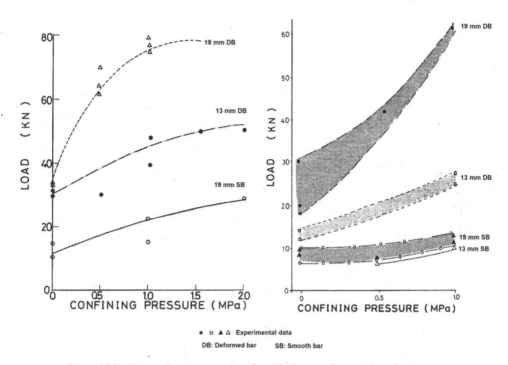

Figure 4.10 Ultimate bearing capacity of rockbolts in pull-out and push-out tests.

surface is smooth). The bearing capacity of ribbed bars is much higher than that of smooth bars. The increase of the bearing capacity is attributable to the normal stress of compressive character resulting from the geometric dilatancy of the bolt surface. The geometrical dilatancy of the surface is probably one of the most important parameters in determining the overall load-bearing capacity (the geometrical dilatancy is defined in Figure 4.19).

Stress and strain distributions in rockbolts seem to be closely influenced by the geometrical configurations of the tested sample and the mechanical properties of the bolt, grouting material, and rock, as well as boundary conditions (restraint and confining stress conditions). Figure 4.11 shows typical axial strain distributions during loading in the case of ribbed bar and smooth bar of 19 mm in diameter under the confining pressure of 1.0 MPa. As noted in both figures, there is a stress concentration at the beginning of loading near the loading end, then the concentration travels towards the unloaded end of the bar. This is associated with the shearing along one of the surfaces of weakness in the rockbolt system (i.e. grout-rock interface and bolt-grout interface). In these particular cases, the failures both took place along the bolt-grout interface.

Failure modes observed in tests can generally be classified as:

i) Failure along the bolt-grout interface: This type of failure was observed in all tests on steel bars with a smooth surface and in deformed bars installed in large boreholes (Fig. 4.12a).
ii) Failure along the grout-rock interface: This type of failure was observed in the case of deformed bars only installed in smaller diameter boreholes.
iii) Failure by splitting of grout and rock annulus: Although shearing failure along one of the interfaces is the main cause of failure, some samples were failed by splitting at a confining pressure of 0.0 MPa in the case of deformed bars (Fig. 4.12b,c). This is

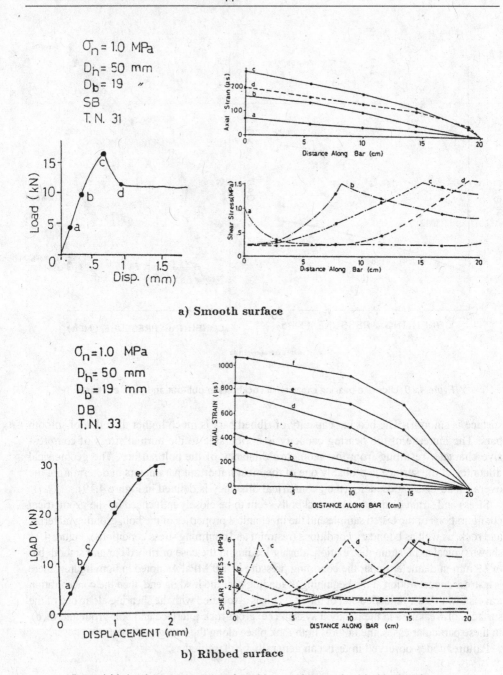

a) Smooth surface

b) Ribbed surface

Figure 4.11 Axial strain response of steel bars with smooth and ribbed surfaces.

thought to be due to the geometrical dilatancy of the bolt-grout interface during shearing, which caused an internal pressure on the borehole. This, in turn, lead to tensile splitting of cylindrical samples of rock. The strain history of the outer perimeter of samples during a push-out test is shown in Figure 4.13. As is apparent from the strain history, the

(a) Pull-out Test Set-up

(b) Shearing along bolt-grout interface (SB)

(c) Shearing along bolt-grout interface (DB)

(d) Splitting failure

Figure 4.12 Photographs of observed failure modes.

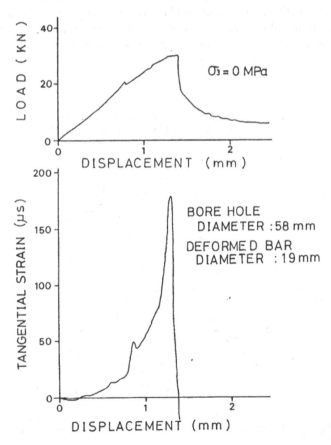

$\sigma_3 = 0$ MPa

BORE HOLE
   DIAMETER : 58 mm

DEFORMED BAR
   DIAMETER : 19 mm

Figure 4.13 Tangential strain history of the outer perimeter during a push-out test.

geometrical dilatancy of the interface creates very large tangential tensile stresses about the perimeter of the borehole. If the rock is too weak to resist to these stresses, a tensile rupturing takes place (Fig. 4.12c). When a confining pressure of more than 0.5 MPa was applied, this type of failure disappeared.

The pull-out and push-out tests were instructive in understanding the effect of various parameters on the mechanical behavior of the bolt system. Nevertheless, it is extremely difficult to determine the material properties for the evaluation of the mechanical performance of rockbolts under various states of stress. Therefore, it was concluded that another type of test, described in the following subsection, is necessary for this purpose and that the pull-out tests, though important for checking the effect of several parameters, are not suitable for determining such properties.

4.3.1.3   Cracking in the close vicinity of the bolt-grout interface

Observations on the cracking state in the close vicinity of the bolt-grout interface has revealed that there are four distinct cracks (Fig. 4.14):

i)   High-angle tension crack (denoted as HATC in Figure 4.14): This crack initiates at the tips of ribs or asperities in the grout annulus and is inclined at an angle ranging between 60° and 80° to the global shearing direction. This is likely to be seen when the surrounding medium has a low deformation modulus compared with that of the bolt and is a result of tensile stresses.

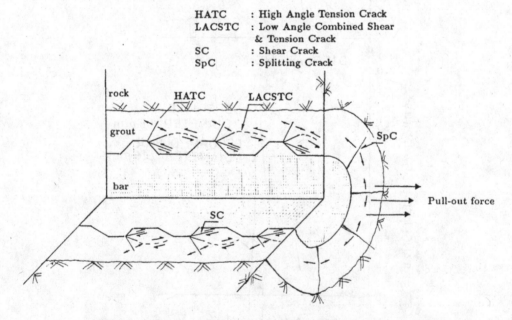

Figure 4.14 Cracking in the close vicinity of the bolt-grout interface.

ii) Low-angle combined tension-shear crack (denoted as LACTSC in Figure 4.14): This crack also initiates near the tips of ribs or asperities and is inclined at an angle ranging between 10° and 35° to the global shearing direction. The crack is first initiated by tensile stresses and propagates up to the middle of the distance between two rib tips, mainly due to the same kind of stress state. Then, the orientation of the crack starts to change and is directed towards the rear tip of the grout asperity. Close examination of the second part of the crack indicates an intense shearing state. When the failure takes place along the bolt-grout interface, it is thought that this type of cracking mainly governs the failure.

iii) Shear crack (denoted as SC in Figure 4.14): This crack occurs just in front of the steel rib and is caused by shearing stresses. This crack is inclined at an angle ranging between 10° and 35° to the rib surface in contact with the grout annulus.

iv) Splitting crack (denoted as SpC in Figure 4.14): This crack is parallel to the global shearing direction and occurs due to the geometrical dilatancy of the bolt-grout interface, causing tensile tangential stresses in the annulus.

The occurrence of all of these four distinct cracks in a pull-out test may or may not be observed, depending upon the material properties of materials in the system and their geometry.

4.3.2   Shear tests

As discussed in the previous section, the shear resistance of interfaces rather than the grouting material is of great importance in the overall resistance of the rockbolt system. The pull-out tests in this respect are not suitable for determining the resistance of interfaces, as stress distributions in the system are highly influenced by the geometry of the bar, borehole, and the embedment sample and their material properties. Therefore, it was concluded that shear tests on each interface, as well as on the grouting material, are necessary. For this purpose, a testing program has been undertaken on

- Bolt-grout interface
- Grout-rock interface
- Grouting material

4.3.2.1   Experimental setup, measurements, and testing procedure

Test samples of the bolt-grout interface consisted of a steel part and a grout part. The surface of the steel part had the same surface configuration as that of the reinforced bars used in pull-out and push-out tests, as shown in Figure 4.15. As for the grout-rock interface, the tests samples also consisted of two parts – a grout part and a rock part. The dimensions of the samples were 100 mm high, 150 mm wide, and 150 mm long. Vertical and horizontal displacements were measured using LVDTs, and shear and normal loads were measured by load cells. During tests, the normal stress on the samples was kept constant. All data were recorded onto floppies using a microcomputer through a GP-IB interface board. First, the normal load is applied on the samples to a preset value, then the shear load was applied in increments.

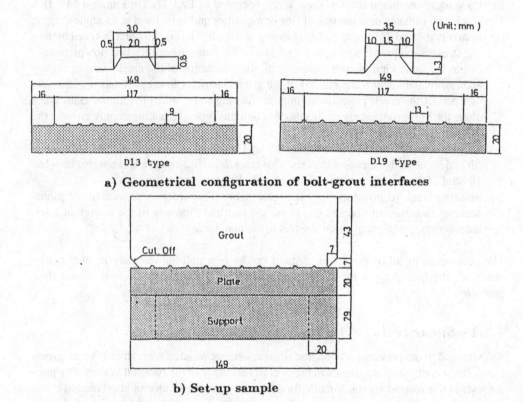

a) Geometrical configuration of bolt-grout interfaces

b) Set-up sample

Figure 4.15 Sketch of bolt-grout interface samples for shear tests.

4.3.3.2   Results and discussions

Shear stress versus relative shear displacement curves and normal displacement versus shear displacement curves obtained under various normal loads are shown in Figure 4.16. As apparent from each test, the shear resistances of the interfaces and the grouting material differ from each other. The least shear resistance was offered by the grout-smooth steel interface, followed by the grout-rock interface. The grout-steel interface of ribbed type corresponding to 19-mm ribbed bolt surface had offered the largest shear resistance (Fig. 4.17). Figure 4.18 shows the resistance of various interfaces and grouting material at normal stress levels of 0.5 MPa and 1.0 MPa. As the shear resistances of the bolt-grout interface and the grout-rock interface are less than those of grouting material and rock, they are called the surfaces or the bands of weakness in the rockbolt system.

The tests showed that the dilatational behavior of interfaces is of great importance. This dilatational behavior consists of two components; one resulting from the geometrical configuration of the interfaces and the second resulting from the plastification of the material near the interfaces.

The geometrical dilatancy is directly related to the configuration of the surface, along which shearing is imposed and distinguished from the material dilatancy. The geometrical dilatancy can be illustrated by a simple model, as shown in Figure 4.19. Let us consider two

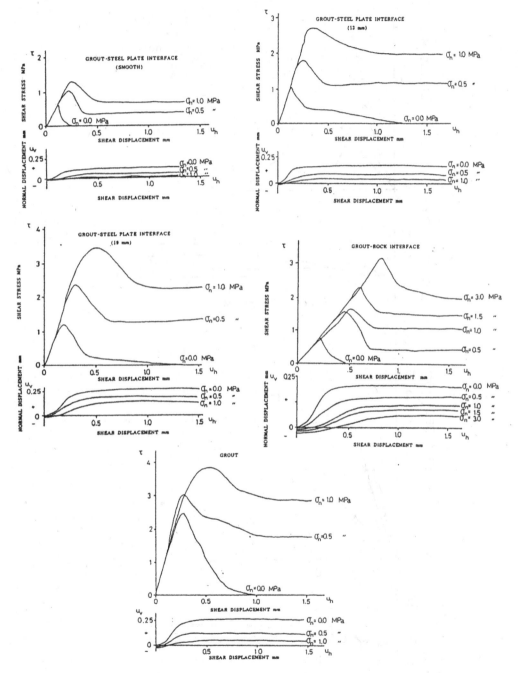

Figure 4.16 Shear responses of interfaces and grouting material.

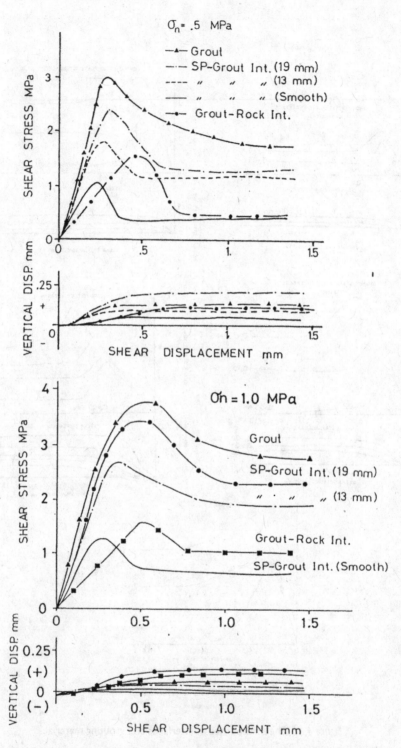

Figure 4.17 Comparison of shear responses of interfaces and grouting material.

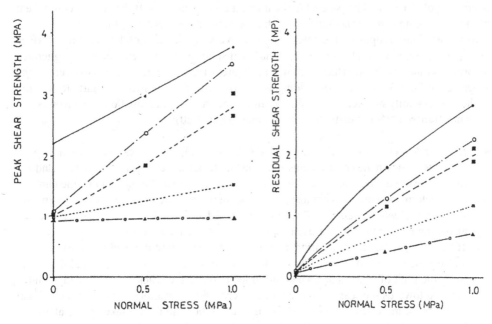

—•— Grout —..—o— Steel plate-Grout Interface (19mm) —•— Rock-Grout Interface
---■--- Steel plate-Grout Interface (13mm) —o—▲ Steel plate-Grout Interface(Smooth)

Figure 4.18 Comparison of shear resistances of interfaces and grouting material.

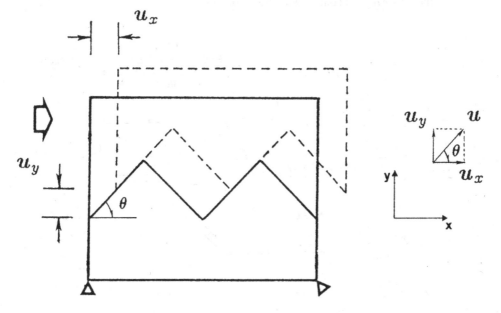

Figure 4.19 Illustration of the geometrical dilatancy concept.

mating blocks with surfaces having regularly shaped asperities. Under a constant normal loading, if the applied shear load is sufficiently high to cause the sliding of the upper block, provided that the lower half of the block is constrained, the displacement vector due to the overriding of blocks will be parallel to the inclination of asperities. If the normal component of the displacement vector is dilative, it is herein termed the geometric dilatancy.

The adhesion strength of the interfaces is of very small magnitude, and the interface becomes separated even at a very small relative displacement. A sketch of the exaggerated deformation state of the interface is shown in Figure 4.19. The material tends to override the asperities of interfaces. If this overriding is constrained, then the material starts to fracture in the close vicinity of interfaces within a finite zone. A close examination of interfaces has shown that there are three distinct types of cracks (Fig. 4.20):

- *High-angle tension crack* (denoted as HATC in Fig. 4.20): This crack initiates at the tips of ribs or asperities in the grout annulus and is inclined at an angle ranging between 60° and 80° to the global shearing direction. This is likely to be seen when the surrounding medium has a low deformation modulus compared with that of the bolt and is a result of tensile stresses.
- *Low-angle combined tension-shear crack* (denoted as LACTSC in Fig. 4.20): This crack also initiates near the tips of ribs or asperities and is inclined at an angle ranging between 10° and 35° to the global shearing direction. The crack is first initiated by tensile stresses and propagates up to the middle of the distance between two rib tips, mainly due to the same kind of stress state. Then, the orientation of the crack starts to change and is directed towards the rear tip of the grout asperity. Close examination of the second part of the crack indicates an intense shearing state. When the failure takes place along the bolt-grout interface, it is thought that this type of cracking mainly governs the failure.
- *Shear crack* (denoted as SC in Fig. 4.20): This crack occurs just in front of the rib and is caused by shearing stresses. The crack is inclined at an angle ranging between 10° and 35° to the rib surface in contact with the grout annulus.

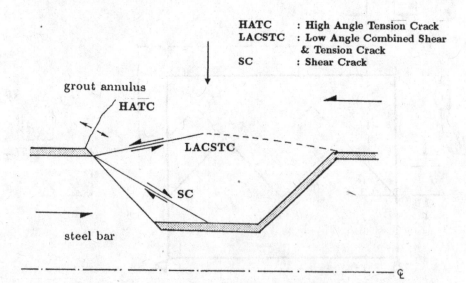

HATC	: High Angle Tension Crack
LACSTC	: Low Angle Combined Shear & Tension Crack
SC	: Shear Crack

Figure 4.20 Deformation state and fracturing in the close vicinity of bolt-grout interface.

The above three distinct cracks observed in the direct shear tests were in good agreement with those observed in pull-out tests, except the splitting crack. This similarity also confirms that the direct shear tests are the most suitable means for the determination of properties for the analysis.

4.3.3.3　Constitutive modeling of grout annulus and interfaces

The grout annulus or interfaces are herein considered bands with a finite thickness. The thickness of the bands is assumed to be associated with the thickness of shear bands observed in tests or in nature, and if asperities exist, their height along the plane. Assigning a finite thickness to such bands also makes the physical meaning of parameters clear and their determination from tests easy.

The original multi-response theory proposed by Ichikawa (1985) basically separates the deviatoric and hydrostatic responses of materials and are represented by the response functions expressed in terms of elastic and plastic deviatoric and hydrostatic components of strains. In the following presentations, the deviatoric and hydrostatic terms are replaced by the normal and shear terms. The response functions of the grout annulus or interfaces for shear and normal stresses are:

$$
\begin{aligned}
\tau_{ga} &= \Phi_{ga}^e(\gamma_{ga}^e), \quad \tau_{ga} = \Phi_{ga}^p(\gamma_{ga}^p, \varepsilon_{ga}^p), \\
\sigma_{ga} &= \Psi_{ga}^e(\varepsilon_{ga}^e), \quad \sigma_{ga} = \Psi_{ga}^p(\gamma_{ga}^p, \varepsilon_{ga}^p).
\end{aligned} \tag{4.21}
$$

where γ_{ga}, τ_{ga} are the shear strain and stress, and ε_{ga}, σ_{ga} are the normal strain and stress. Superscripts e and p stand for adjectives elastic and plastic, respectively.

The incremental constitutive equation for the plastic behavior is given in the matrix form as:

$$
\begin{Bmatrix} d\tau_{ga} \\ d\sigma_{ga} \end{Bmatrix} = \begin{bmatrix} \partial\Phi_{ga}^p / \partial\gamma_{ga}^p & \partial\Phi_{ga}^p / \partial\varepsilon_{ga}^p \\ \partial\Psi_{ga}^p / \partial\gamma_{ga}^p & \partial\Psi_{ga}^p / \partial\varepsilon_{ga}^p \end{bmatrix} \begin{Bmatrix} d\gamma_{ga}^p \\ d\varepsilon_{ga}^p \end{Bmatrix} \tag{4.22}
$$

Denoting

$$
G_1^p = \frac{\partial\Phi_{ga}^p}{\partial\gamma_{ga}^p}, \quad G_2^p = \frac{\partial\Phi_{ga}^p}{\partial\varepsilon_{ga}^p}, \quad K_1^p = \frac{\partial\Psi_{ga}^p}{\partial\varepsilon_{ga}^p}, \quad K_2^p = \frac{\partial\Psi_{ga}^p}{\partial\gamma_{ga}^p}
$$

and taking the inverse of the above expression and after some rearrangements we have:

$$
\begin{Bmatrix} d\gamma_{ga}^p \\ d\varepsilon_{ga}^p \end{Bmatrix} = \begin{bmatrix} 1/(G_1^p - G_2^p \dfrac{K_2^p}{K_1^p}) & -\dfrac{G_2^p}{K_1^p}/(G_1^p - G_2^p \dfrac{K_2^p}{K_1^p}) \\ -\dfrac{K_2^p}{G_1^p}/(K_1^p - K_2^p \dfrac{G_2^p}{G_1^p}) & 1/(K_1^p - K_2^p \dfrac{G_2^p}{G_1^p}) \end{bmatrix} \begin{Bmatrix} d\tau_{ga} \\ d\sigma_{ga} \end{Bmatrix} \tag{4.23}
$$

Denoting,

$$
h_s = G_1^p - G_2^p \frac{K_2^p}{K_1^p}, \quad h_n = K_1^p - K_2^p \frac{G_2^p}{G_1^p}, \quad \mu = -\frac{G_2^p}{K_1^p}, \quad \beta = -\frac{K_2^p}{G_1^p}
$$

the above relationship can be rewritten as:

$$\left\{\begin{array}{c} d\gamma^p_{ga} \\ d\varepsilon^p_{ga} \end{array}\right\} = \left[\begin{array}{cc} 1/h_s & \mu/h_s \\ \beta/h_n & 1/h_n \end{array}\right] \left\{\begin{array}{c} d\tau_{ga} \\ d\sigma_{ga} \end{array}\right\} \tag{4.24}$$

In the above expression, parameters h_s and h_n are physically interpreted as the hardening moduli for the respective responses. The hardening modulus for each respective response is distinct. It should be noted that the hardening modulus for each respective response cannot be distinctly evaluated in the case of the conventional plasticity theory, which is based on the scalar potential theory. The terms denoted by μ and β are the friction coefficient and the dilatancy factor, respectively. These terms are interrelated to each other in conventional plasticity. On the other hand, the present theory evaluates these terms independently from each other. As a result, while the direct inverse of the plasticity matrix using the conventional plasticity theory is not possible as the matrix is singular, the direct inverse of the plasticity matrix obtained using the multi-response theory is possible, since its determinant exists. An explanation for this can be found when one compares the Eigen values of the matrices. In the case of the matrix by the conventional plasticity theory, there exists one non-zero Eigen value at most, and the other one is zero. On the other hand, the matrix by the multi-response theory has two distinct non-zero Eigen values.

The yield function is vectorial and is written in the following form:

$$\mathbf{f} = \left\{\begin{array}{c} f_1 \\ f_2 \end{array}\right\} = \mathbf{0} \qquad \begin{array}{l} f_1 = \tau_{ga} - \Phi^p_{ga}(\gamma^p_{ga}, \varepsilon^p_{ga}) \\ \\ f_2 = \sigma_{ga} - \Psi^p_{ga}(\gamma^p_{ga}, \varepsilon^p_{ga}) \end{array} \tag{4.25}$$

The determination of the hardening function and the response functions from the experimental data is simply a fitting problem from the mathematical point of view. The functions may involve a single variable or several variables and can be classified as (Hamming, 1973; Lancaster and Salkauskas, 1986):

- Polynomial functions

  - Orthogonal functions
  - Chebyshev polynomials
  - Rational functions

- Fourier series

  - Periodic functions
  - Non-periodic functions

- Exponential functions

  - Sums of exponentials
  - The Laplace transform

In data-fitting problems, the choice of one of the above functions, which is suitable for the proper representation of the experimental data, is a subjective matter and a difficult task. Once the function is chosen, the problem is then how to determine its coefficients (and powers). The chosen functions for the plastic responses of the bar and grout annulus and interfaces are written in the following forms:

- Shear stress response

$$\Phi(\gamma^p, \varepsilon_n^p) = \frac{A_1\gamma_{ga}^p + A_2\gamma_{ga}^p\varepsilon_{ga}^p}{1 + B_1\gamma_{ga}^p + B_2\varepsilon_{ga}^p + B_3\gamma_{ga}^p\varepsilon_{ga}^p + B_4\gamma_{ga}^{p\,2} + B_5\gamma_{ga}^{p\,2}\varepsilon_{ga}^p} = \frac{N_\Phi(\gamma^p, \varepsilon_n^p)}{D_\Phi(\gamma^p, \varepsilon_n^p)} \qquad (4.26a)$$

- Normal stress response

$$\Psi(\gamma^p, \varepsilon_n^p) = \frac{A_1\gamma_{ga}^p + A_2\gamma_{ga}^p\varepsilon^p + A_3\varepsilon_{ga}^p}{1 + B_1\gamma_{ga}^p} = \frac{N_\Psi(\gamma^p, \varepsilon_n^p)}{D_\Psi(\gamma^p, \varepsilon_n^p)} \qquad (4.26b)$$

Together with the condition that one coefficient must be non-zero in the denominator function, the coefficients involved in the chosen function can be evaluated by either the direct approach or the least square method. The least square method has been employed to determine the coefficients of the fractional functions. The least square method basically is the minimization of an error function given as:

$$MinE = \sum_i^N \left[f(x_i, y_i) - \frac{N(x_i, y_i)}{D(x_i, y_i)} \right]^2 \qquad (4.27)$$

where N is the number of data points. The above expression can also be written in the following form:

$$MinE = \sum_i^N \frac{1}{D^2(x_i, y_i)} \left[D(x_i, y_i) f(x_i, y_i) - N(x_i, y_i) \right]^2 \qquad (4.28)$$

When the above expression is minimized, the unknown coefficients occur nonlinearly in the resulting expressions. Therefore, the use of an iterative procedure becomes necessary to determine the coefficients. The iterative procedure used in this study is the one suggested by Hamming (1973), as follows:

$$MinE_k = \sum_i^N \frac{1}{D_{k-1}^2(x_i, y_i)} \left[D_k(x_i, y_i) f(x_i, y_i) - N(x_i, y_i) \right]^2 \quad k = 1, 2, 3, \cdots \qquad (4.29)$$

When nothing is known, for a start, D_0 can be chosen as 1 so that the unknowns occur linearly in the above expression. The iteration continues by increasing k until a minimum error E_k is obtained.

The coefficients for the functions are determined from the described procedure for bolt-grout interfaces. The specific values of the respective functions for the bolt-grout interface (13 mm) are given below.

i)　Shear stress response:

$$A_1 = 4.23,\ A_2 = -16.68,\ B_1 = -.24,\ B_2 = -4.97,\ B_3 = 20.48,\ B_4 = 3.155,\ B_5 = -15.51$$

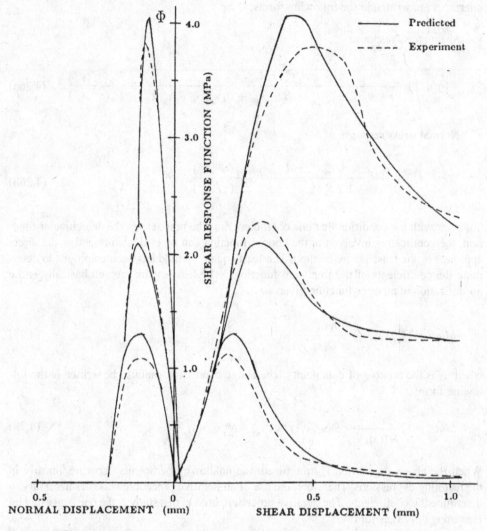

Figure 4.21 Comparison of predicted shear stress response function with experimental ones for bolt-grout interface.

Figure 4.21 shows the plots of the predicted and experimental response functions.

ii)   Normal stress response:

$$A_1 = -2.5, \quad A_2 = -25.77, \quad A_3 = 7.24, \quad B_1 = 2.1$$

Figure 4.22 shows the plot of the normal response function in the space of plastic shear and normal displacements.

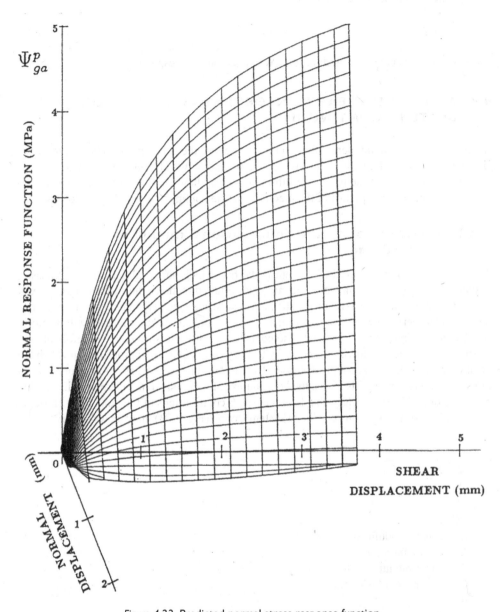

Figure 4.22 Predicted normal stress response function.

The response functions for the elastic behavior of grout annulus and interfaces are assumed to be linear functions, given as:

- Elastic shear stress response function:

$$\Phi^e = G_{ga}\gamma^e_{ga} \tag{4.30a}$$

- Elastic normal stress response function:

$$\Phi^e = E_{ga}\varepsilon_{ga}^e \tag{4.30b}$$

where G_{ga} and E_{ga} are shear modulus and Young's modulus, respectively.

4.4   AXIAL AND SHEAR REINFORCEMENT EFFECTS OF BOLTS IN CONTINUUM

In continuum, it is often reported that rockbolts improve the apparent mechanical properties (i.e. deformation modulus, strength, etc.) of the medium in which they are installed and offer a confinement effect. In the following subsections, it is shown how to evaluate the contributions to the apparent properties of the medium and the confinement effect of rockbolts.

4.4.1   Contribution to the deformational moduli of the medium

The main element in the rockbolt system is the bar, which has axial and shear stiffnesses. Although these stiffnesses may be disregarded in certain directions, the stiffness against axial loading parallel to the bolt axis and the stiffness against shearing perpendicular to the bolt axis have to be properly taken into account. In any given problem, these stiffnesses will be mobilized, to a certain extent. To evaluate the contribution of these stiffnesses to deformational properties of a medium, it will be necessary to carry out an averaging procedure in which the geometrical dimensions of a representative element must be specified, as in the theory of mixtures or composite materials. Thus, the contribution of rockbolts to the medium will become a relative quantity. Specifically, the axial and shear contributions of the bolts to an isotropic medium, in a local Cartesian coordinate system (oxyz) (Fig. 4.23), may be written as:

$$\Delta E_x^* = (E_b - E_r)n$$
$$\Delta G_y^* = (G_b - G_r)n \tag{4.31}$$
$$\Delta G_z^* = (G_b - G_r)n$$

where
E_r = elastic modulus of medium
E_b = elastic modulus of bar
G_r = shear modulus of medium
G_b = shear modulus of bar
$n = A_b/A_t$
A_b = cross-section of bar
A_t = cross-section of representative volume

4.4.2   Contribution to the strength of the medium

Although the contribution of deformational properties of the bar to a continuum can be deterministically evaluated, the contribution of bolts to the strength of the continuum could not be done straightforwardly. This is because of the difference in the yielding strain of the geomaterials compared with that of the bar. In other words, the axial and shear contribution of the bar could not be fully mobilized at the time of yielding of geomaterials, as seen from Figure 4.24, in which the axial response of various rocks and steel are plotted. Since the response of bolts is closely associated with the response of the minimum principal strain ε_3 in real engineering situations, such as in underground excavations and slopes, the resistance and confinement will be of very small magnitude at the time of yielding of the rock. Therefore, the strength of the bar will become mobilized after an appreciable amount of straining of the geomaterials has taken place, provided that the medium behaves as a continuum. As a result, this fact will make the apparent behavior of bolted geomaterial bodies more ductile, compared with that of unbolted geomaterials.

To visualize the contribution of rockbolts to the strength of rock, let us consider a bolted rock pillar subjected to a triaxial stress state, as shown in Figure 4.25. The normal stress σ_α^* and shear stress τ_α^* on a given plane α can be shown to be (stresses with asterisk refer to the stresses in bolted state):

$$\sigma_\alpha^* = \frac{\sigma_1^* + \sigma_3^*}{2} + \frac{\sigma_1^* - \sigma_3^*}{2}\cos 2\alpha \qquad (4.32)$$

$$\tau_\alpha^* = \frac{\sigma_1^* - \sigma_3^*}{2}\sin 2\alpha \qquad (4.33)$$

where

$$\sigma_3^* = \sigma_3 + \Delta\sigma_3$$
$$\sigma_1^* = \sigma_1 + \Delta\sigma_1$$

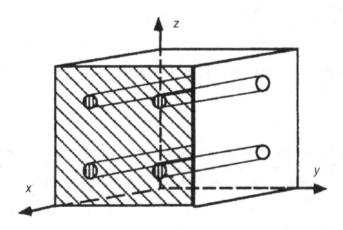

Figure 4.23 Coordinate system for bolted rock mass.

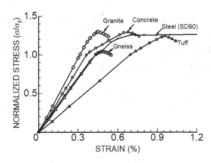

Figure 4.24 Comparison of the uniaxial behavior of rocks with that of steel.

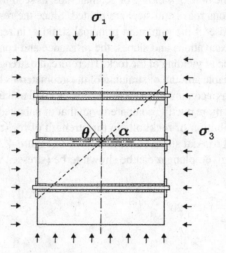

Figure 4.25 Notations for a bolted rock pillar.

The bolts crossing the plane α, the axial component σ_a and shear σ_s component are related to the stress σ_b in bolt in the direction of plane α by:

$$\sigma_a = \sigma_b \cos \alpha \quad \sigma_s = \sigma_b \sin \alpha \tag{4.34}$$

The contribution of the axial response of the bolts to the lateral confining stress σ_3 may be written in the following form:

$$\Delta \sigma_3 = \sigma_a n \cdot m \tag{4.35}$$

where n is the bolting density parameter and m is the number of bolts. The action of bolts will tend to reduce the magnitude of the shear stress on the plane α. This reduction may be evaluated by considering the projection of the stress components in the bolt on the plane α in the following form:

$$\Delta \tau_a = -(\sigma_s \sin \alpha + \sigma_a \cos \alpha) n \cdot m \tag{4.36}$$

where $\Delta \tau_a$ stands for the shear stress decrement due to bolt resistance.

The geomaterials are generally cohesive and frictional and obey the Mohr-Coulomb yield criterion. The failure may take place along a single plane or two conjugate planes. Introducing the Mohr-Coulomb criterion, which is given by:

$$\tau = c + \sigma_n \tan\phi \tag{4.37}$$

where c and ϕ are cohesion and friction angle, respectively, and rearranging the resulting expressions yields:

$$\sigma_1^* = q\sigma_3 + \sigma_c + q\Delta\sigma_3 + \Delta\sigma_c \qquad (4.38)$$

where σ_c and $\Delta\sigma_c$ is the uniaxial strength of rock and the increment in uniaxial strength due to the strength of the bolt, respectively. q is the triaxial coefficient, given as:

$$q = \frac{1+\sin\phi}{1-\sin\phi}$$

Thus, the contributions of rockbolts to the strength of rock pillar are:

$$\Delta\sigma_1 = q\Delta\sigma_3 + \Delta\sigma_c, \quad \Delta\sigma_c = n \cdot m\frac{2\sigma_b\cos\phi}{1-\sin\phi} \qquad (4.39)$$

As noted from the above expression, the first and the second terms on the right-hand side of Equation (4.39) correspond to a confining pressure effect due to axial straining of the bar and the contribution of the strength of the bar under the combined loading, respectively. The strength σ_b of the bar at yielding can be calculated from the von Mises criterion. For this particular case, the von Mises criterion is written in the following form:

$$\sigma_b = \frac{\sigma_t}{\sqrt{\sin^2\theta + 3\cos^2\theta}} \qquad (4.40)$$

where
σ_t = tensile yield strength of bar

θ should be taken as the angle between the bolt axis and the normal of plane α for translational-type movements and as the angle between the bolt axis and the plane α for separation-type movements.

4.4.3   Improvement of apparent mechanical properties of rock and confining pressure effect

As stated in Chapter 3, the contributions of the bolts to the apparent mechanical properties of rock from the equivalent material point of view is closely associated with the intrinsic properties of bolt, rock, and its density. To illustrate these contributions more specifically for different type of rocks, parametric studies are done. Figure 4.26 shows the contribution of the elastic modulus of rockbolt to that of rock of various type (for rock classes, refer to ISRM suggested methods; Brown, 1981). The equivalent elastic modulus normalized by the elastic modulus of rock (plotted in Fig. 4.26) is defined as:

$$\frac{E_{eq}}{E_r} = 1 + \left(\frac{E_b}{E_r} - 1\right)\frac{A_b}{A_t} \qquad (4.41)$$

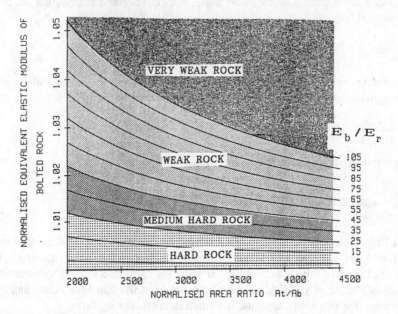

Figure 4.26 Contribution of the elastic modulus of rockbolts to the elastic modulus of intact rock.

where
E_r = elastic modulus of intact rock
E_b = elastic modulus of rockbolt
A_b = cross-section of rockbolt
$A_t = e_t \cdot e_l$
e_t = transverse spacing of rockbolt
e_l = longitudinal spacing of rockbolt

As noted from the figure, the contribution will be more pronounced in the weak rocks as compared with those in hard rocks. Nevertheless, the maximum contribution of a rockbolt of 25 mm in diameter together with an elastic modulus of 180 GPa, installed in a pattern 1×1 m^2 and $A_t/A_b = 2037$ cannot be greater than 5% in weak rocks.

Next, a similar type of parametric study on the confining pressure effect of rockbolts was investigated. Figure 4.27 shows the normalized contribution of the bolts by the strength of rock. The normalized internal pressure (confining pressure) to the compressive strength of intact rock is defined as:

$$\frac{\Delta P_i}{\sigma_r} = \frac{\sigma_b}{\sigma_r} \frac{A_b}{A_t} \tag{4.42}$$

where
σ_b = tensile strength of rockbolt
σ_r = compressive strength of intact rock
ΔP_i = increment of internal pressure P_i

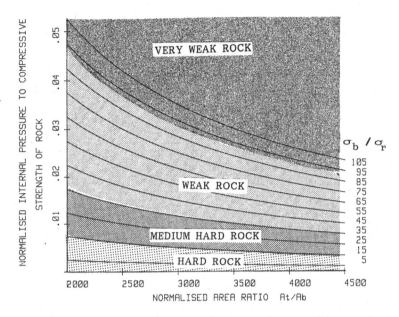

Figure 4.27 Contribution of rockbolts to normalized internal pressure by its tensile strength.

A_b = cross-section of rockbolt
$A_t = e_t \cdot e_l$
e_t = transverse spacing of rockbolts
e_l = longitudinal spacing of rockbolt

Once again, we note that the contributions are more pronounced in the case of weaker rocks than in the case of stronger rocks. It should be noted that the strength herein represents the strength of intact rocks and that the magnitude of contributions is also low in this case.

4.5   AXIAL AND SHEAR REINFORCEMENT EFFECTS OF BOLTS IN MEDIUM WITH DISCONTINUITIES

In a medium with discontinuities, the discontinuity planes may exhibit various displacement patterns. The fundamental behaviors are:

*   Separation (opening-up)
*   Closing
*   Translation

These displacement patterns will be observed either individually or combined. Some of these are illustrated in Figure 4.28. Rockbolts will be mainly required to prevent the opening

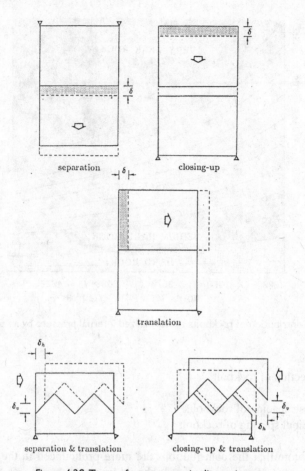

separation   closing-up

translation

separation & translation   closing-up & translation

Figure 4.28 Types of movements in discontinuum.

and/or translational movements of discontinuity planes. Depending upon the direction and the nature of displacements of discontinuities and the installation pattern of rockbolts, the reinforcement effects will result from the axial and/or shear responses of the bolts to these displacement fields.

Discontinuities are usually planar in large scale but wavy in small scale. While the discontinuities cannot transfer tensile loads, they are usually capable of transferring very high compressive normal loads. As for the shear loads, their resistance will be largely influenced by their surface configurations, the properties of wall rock, and its frictional properties. The required reinforcement effects of the bolts will be

• to provide a tensile response to transfer the load from one side to another in the case of the separation type movements, and/or
• to contribute to the shear resistance of the discontinuity by providing an additional shear strength through its own strength and resisting shear and frictional loads.

4.5.1   Increment of the tensile resistance of a discontinuity plane by a rockbolt

Cases that necessitate a tensile resistance arise when the discontinuity walls tend to separate from each other. The tensile resistance T_{sr} offered by a rockbolt can be given in the following form (Fig. 4.29):

$$T_{sr} = \sigma_b A_b \tag{4.43}$$

where A_b is the cross-section of the bar. The resistance σ_b of the bolt can also be obtained from Equation (4.40). The angle θ in Equation (4.40) must be taken as the angle between the direction of separation movement and the bolt axis. The maximum resistance will be obtained when the angle becomes 90°, since the tensile strength of the bar is greater than its shear strength.

4.5.2   Increment of the shear resistance of a discontinuity plane by a rockbolt

Rockbolts contribute to the shear resistance of discontinuities depending upon the magnitude and the character of stress components in the bolts and their orientation with respect to the sense of movement along the discontinuity plane. These contributions are:

*   Reinforcement due to axial response of bolts

    *   Frictional component
    *   Shear component

*   Reinforcement due to shear response of bolts (dowel effect)

To investigate all these effects one by one in detail, consider two reinforced blocks of rock put together so as to create a thoroughgoing discontinuity plane that has a constant friction angle ϕ, as shown in Figure 4.30. Let the angle between the bolt axis and the normal of the

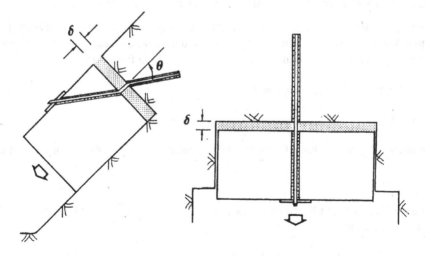

Figure 4.29 Typical examples where the tensile resistance of bolts is required.

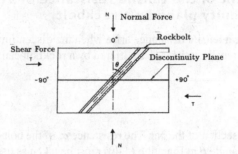

Figure 4.30 Notations for a bolted discontinuity.

discontinuity plane be θ. The axial and shear stresses in the bolt resulting from the translational movement along the discontinuity can be given in terms of the stress σ_b in the bolt, acting in the direction of movement as:

Axial stress σ_a

$$\sigma_a = \sigma_b \sin \theta \tag{4.44}$$

Shear stress σ_s

$$\sigma_s = \sigma_b \cos \theta \tag{4.45}$$

In the following, the reinforcement effects due to the axial response and shear responses of a bolt are separately discussed; first, so that the similarity and the difference between the mechanically anchored bolts, partially grouted bolts, and fully grouted bolts can be clearly shown. Then, the total reinforcement effects of bolts are presented.

4.5.2.1   Reinforcement due to the axial response of bolts

The axial stress will create normal and shear force components on the discontinuity plane. Assuming that these forces are uniformly distributed over the shearing area A_s, one gets the following expressions for normal and shear stresses resulting from bolts as:

$$\sigma_{nb} = n\sigma_a \cos \theta, \qquad \tau_{nb} = n\sigma_a \sin \theta \tag{4.36}$$

where $n = A_b/A_s$.

Introducing the friction law of Amonton (Bowden and Tabor, 1964), which is given as:

$$\tau/\sigma_n = \tan \phi \tag{4.37}$$

the total shear resistance offered by the axial stress component of a rockbolt can be written in the following form:

$$\tau_{bt} = \tau_{nb} + \sigma_{nb} \tan \phi = \pm n\sigma_a (\cos \theta \tan \phi + \sin \theta) \tag{4.38}$$

where signs (+) and (−) stand for adjectives tensile and compressive, respectively.

To see how these effects can contribute to the shear resistance of discontinuity, it is neces-sary to know the character of stresses in the bolt and its installation angle. A sample calcula-tion is carried out by varying the installation angle of the bolt for various friction angles of discontinuity plane for two cases, as shown in Figure 4.31.

When the axial stress in bolts is tensile for the range of $-90° \leq \theta \leq 0°$, the frictional contri-bution is positive while the shear contribution is negative. On the other hand, when the stress is compressive, a reverse situation appears. For the range $0° \leq \theta \leq +90°$, both effects will

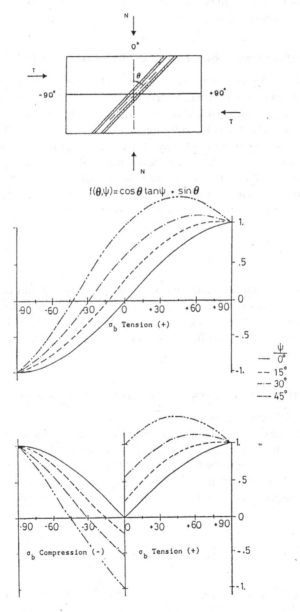

Figure 4.31 Effect of installation and the character of axial stress in bolts on their reinforcement effect.

be contributing positively to the shear resistance of the discontinuity. Thus, the installation angle of rockbolts must be between 0° and +90° for effective use. Figure 4.31 implies that bolts will be most effective when they are installed at an angle of 90° − ϕ, as long as their axial stress contribution is considered.

When the bolts are installed at an angle between −90° ≤ θ ≤ 0°, there will be a tendency of developing compressive stresses in the bolts due to the direction of loading. If the bolts are of the mechanical type and are pre-stressed, after a certain amount of displacement the bolts loosen in the hole. However, if the bolts are of the grouted type, the axial stress in the bolts will be compressive. On the other hand, when they are installed at an angle between 0° ≤ θ ≤ +90°, the tendency will be of the tensile type. Mechanical and grouted rockbolts will all experience tensile stresses within this range.

4.5.2.2   *Reinforcement due to shear response of rockbolts*

The effect of shear response of the rockbolts may be considered in a similar manner to its axial contribution. However, such a consideration will make the resistance offered by rockbolts to become independent of the character of the axial load caused in rock. As experimentally confirmed that the reinforcement effect is mostly affected by the character of the net axial stress, that is to say, if the stress is of tensile character, the largest resistance is offered; otherwise, the least resistance is offered. The implication is that the frictional component due to the shear stress in the bolt should be omitted, as it cannot be distributed over the large area of the discontinuity plane, compared with that of the axial response. Therefore, the shear response contribution should always be taken into account as:

$$\sigma_{rs} = \sigma_s \cos \theta \tag{4.49}$$

The shear response of mechanically anchored rockbolts comes into effect after a certain amount of displacement occurs, since a space exists between the bar and the borehole. On the other hand, the situation is different in the case of grouted rockbolts, in which the response of the bolt comes into effect as soon as the relative shear displacement along the plane occurs.

4.5.2.3   *Total shear reinforcement offered by rockbolts*

The total shear reinforcement offered by rockbolts to a discontinuity plane will depend upon the character of the axial stress in bolts, the installation angle, and rockbolt type, as well as its dimensions. For mechanically anchored rockbolts, the contribution of the shear resistance will be due to the axial load in the bolt, and its mathematical form will be the same as Equation (4.48). On the other hand, the grouted rockbolts will offer a shear resistance from not only their axial response but also their shear response, and it will take the following form for a rockbolt:

$$T_t = nA_b\sigma_b\left(1+\frac{1}{2}\tan\phi\sin 2\theta\right) \tag{4.50}$$

where n is bolting density parameter and σ_b is the strength of the bolt given by Equation (4.40).

4.5.2.4 Shear reinforcement of a discontinuity by a rockbolt

Figure 4.32 shows the sample of tests reported by Egger and Fernandes (1983). They tested a concrete pillar, reinforced by two bolts of 6 mm in diameter with a total tensile resistance of 14 kN, and had a thoroughgoing discontinuity plane inclined at an angle of 45° with respect to the pillar axis. The inclination of the bolts with respect to the normal of the discontinuity was varied between 0° and +65° during tests. Aydan (1989) compared the ultimate resistances predicted by available other proposals listed below:

Dowel effect only (Aydan, 1989):

$$F_b = \eta \frac{nA_b\sigma_t}{\sqrt{\sin^2\theta + \Omega\cos^2\theta}} \tag{4.51}$$

Combined friction and dowel effects (Aydan, 1989):

$$F_b = \eta \frac{nA_b\sigma_t}{\sqrt{\sin^2\theta + \Omega\cos^2\theta}} \left[1 + \frac{1}{2}\tan\phi\sin 2\theta \right] \tag{4.52}$$

Formula by Aydan et al. (1987):

$$F_b = \eta n A_b \sigma_t [(1 - \xi)|\sin\theta + \cos\theta\tan\phi| + \xi] \tag{4.53}$$

Formula by Egger and Fernandes (1983):

$$F_b = \eta \frac{nA_b\sigma_t}{\sqrt{\sin^2\theta + \Omega\cos^2\theta}} [\sin\theta + \cos\theta\tan\phi] \tag{4.54}$$

where

$$\eta = \frac{A_p}{\sin 2\alpha - \tan\phi(1 - \cos 2\alpha)}, \quad \xi = \cos\left(45 - \frac{\phi_i}{2}\right), \quad \Omega = \left[\frac{\sigma_t}{\sigma_s}\right]^2$$

σ_t = tensile strength of steel bar
σ_s = shear strength of steel bar
θ = inclination of bolt from the normal of discontinuity
α = inclination of discontinuity
ϕ = friction angle of discontinuity
ϕ_i = friction angle of host rock
A_b = cross-section of bolt
A_p = cross-section of pillar
n = Number of rockbolts

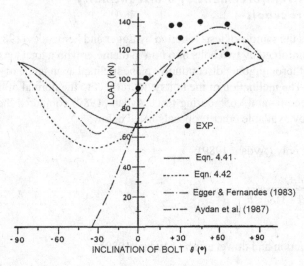

Figure 4.32 Comparison of various formula with experimental results reported by Egger and Fernandes (1983).

4.5.3   Response of rockbolts to movements at/along discontinuities

As discussed, the movements of discontinuities are either normal to the discontinuity plane (in the form of separation or closing up) or translation (sliding) or both. The responses of the bolts to these types of movement of discontinuities will be given in more detail in the following:

1)   Response of bolts against separation movement of a discontinuity

Let us consider a case that a rockbolt crosses a discontinuity plane perpendicularly and the displacement of rock in the direction of bolt axis is given in the form as shown in Figure 4.33a. For various end conditions, the axial and shear stress distributions in/along the bolt are calculated for the given material properties as shown in Figure 4.33b and 4.33c (see Aydan, 1989 for details). As noted from the figure, the axial stress has a peak at the location where the bolt crosses the discontinuity, and large shear stress concentrations are observed in the close vicinity of the discontinuity. It should be noted that the character of the axial stress is tensile.

2)   Response of bolts against closing movement of a discontinuity

Similarly, let us consider a case that a rockbolt crosses a discontinuity plane perpendicularly and assume that the displacement of rock in the direction of bolt axis is given in the form as shown in Figure 4.34a. For various end conditions, the axial and shear stress distributions in/along the bolt can be calculated for the given material properties as shown in Figure 4.34b and 4.34c. As noted from the figure, the axial stress has a peak at the location where the bolt crosses the discontinuity, and large shear stress concentrations are observed in the close vicinity of the discontinuity. However, in

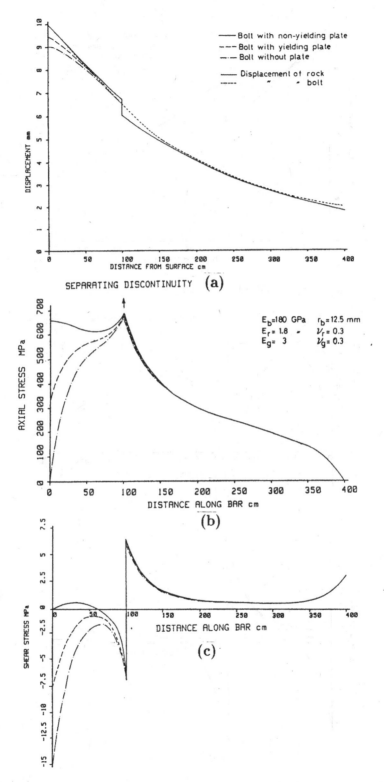

Figure 4.33 Response of a bolt against the separation movement of a discontinuity.

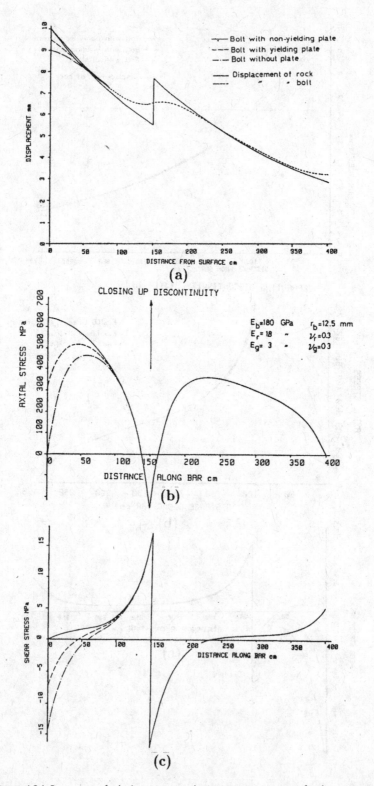

Figure 4.34 Response of a bolt against to closing-up movements of a discontinuity.

this case, the axial stress in the bolt tends to become compressive at the location of discontinuity undergoing a closing-up type of movement.

3) Response of bolts against translational movement of a discontinuity

When the movement along the discontinuity is translational, the installation angle of rockbolts with respect to the sense of translation is of paramount importance, as the stress state created in rockbolts is a function of the sense of movement with respect to the installation orientation of the bolt. For this reason, three installation angles were considered, namely, −45°, 0°, and +45° with respect to the normal of discontinuity (Fig. 4.35a). The calculated axial stress distribution along the bolts for the installation angle of +45° is shown in Figure 4.35b. Also in this case, a spike in the axial stress distribution is observed at the location where the bolt crosses the discontinuity. The axial stress responses of the bolt at the location where the peak is observed are shown in Figure 4.35c for three different installation angles of the bolt in relation to the relative displacement along the discontinuity. It is clear that the development of the axial stress is closely associated with the installation angle of the bolt with respect to the sense of translational movement. The shear stress development along the bar is just proportional to the axial stress distributions.

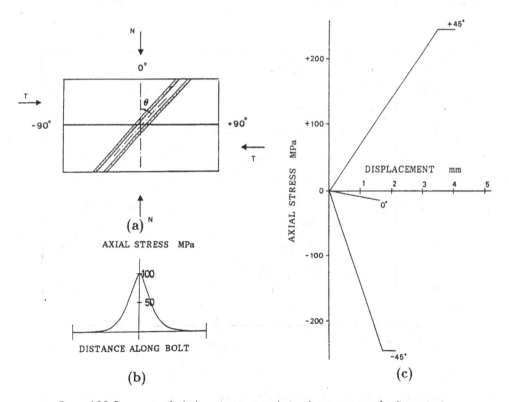

Figure 4.35 Response of a bolt against to translational movements of a discontinuity.

4.6   ESTIMATION OF THE CYCLIC YIELD STRENGTH OF INTERFACES FOR PULL-OUT CAPACITY

The yield strength of interfaces under cyclic loading conditions can be taken into account as a convolution of the short-term strength yield criterion τ_s and a function $S(N)$ for cyclic loading of interfaces:

$$\tau(N) = \tau_s * S(N) \tag{4.55}$$

The function τ_s can be one of the yield criteria proposed in the preceding section. The function $S(N)$ is suggested to be of the following form:

$$S(N) = A + Be^{-\frac{N-1}{N^*}} \tag{4.56}$$

The specific values of the constants in the above formula determined from tests for the grout-tendon interface are:

$$A = 0.75 \qquad B = 2.05 \qquad N^* = 22,500 \tag{4.57}$$

With the use of above approach, Figure 4.36 shows the variation of the grout-tendon interface as a function of the number of cycles.

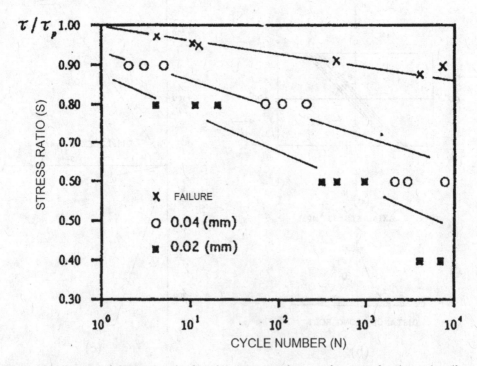

Figure 4.36 Variation of shear strength of tendon-grout interface as a function of cycle number (from Aydan et al., 1994).

4.7 ESTIMATION OF THE YIELD STRENGTH OF INTERFACES IN BOREHOLES

The normal (radial) stress on interfaces during pull-out tests of rockanchors or rockbolts is an important element for estimating the shear strength of interfaces in boreholes. The normal stress is likely to be a function of the stiffness of the surrounding medium, since experimental results indicate that the effect of the stiffness on *in situ* tests plays a big role on the interface strength. However, the assessment of the normal stress is still an unsolved problem. For the specification of the shear strength of interfaces in boreholes, we suggest using the direct shear tests results and the radial response of the surrounding medium around the borehole. The fundamental concept of the procedure is illustrated in Figure 4.37 (Aydan *et al.*, 1995). When both curves shown in the figure intersect, that intersection point is assumed to correspond to the effective normal stress to be observed in rockanchor systems. The normal stress (radial)–displacement response is a function of the deformability of surrounding medium. If the interface yield criterion is a function of normal stress (i.e. Mohr-Coulomb criterion), the interface strength will be a function of the deformability of the surrounding medium. As noted from the figure, the shear strength of interface will increase as the rock deformation modulus of rock increases. In the following, we demonstrate this conceptual model in terms of mathematical expressions.

As it is well known from the theory of elasticity, the following relationship holds between the internally applied radial pressure and radial deformation of the hole in an infinite medium:

$$u = A\sigma_n, \quad A = \frac{1+v}{E}a \tag{4.58}$$

where v: Poisson's ratio; E: elastic modulus; a: radius of hole. Let us assume also that the functional form of peak dilatancy of interface is of the following forms:

$$u_d = \frac{B}{1+C\sigma_n^{av}} \tag{4.59}$$

where B and C are experimental constants and σ_n^{av} is averaged normal stress over a typical wave length of the interface. Using following conditions:

$$u_d = u_o \quad at \quad \sigma_n^{av} = 0 \quad \frac{du_d}{d\sigma_n^{av}} = -\tan i_a \alpha \quad at \quad \sigma_n^{av} = 0$$

where i_a is asperity inclination; u_o is height of asperity; α is a parameter to adjust the units, the coefficients B and C are obtained as:

$$B = u_o \qquad C = \frac{\tan i_a \alpha}{u_o}$$

If u_o is given in meters and average normal stress in MPa and $\alpha = 0.001$, then the final expression is:

$$u_d = \frac{u_o^2}{u_o + \alpha \tan i_a \sigma_n^{av}} \tag{4.60}$$

The contact state of an interface with a periodic asperity pattern can be illustrated as shown in Figure 4.37. The contact area decreases as the interface dilates and the contact area can be written as:

$$L_i^c = L_i - \frac{u_d}{\tan i_a} \tag{4.61}$$

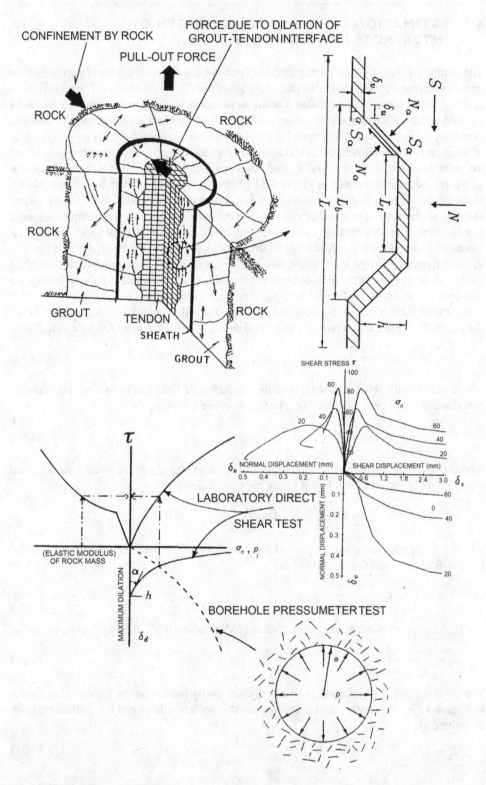

Figure 4.37 Modeling of deformation state in borehole and estimation model of interface shear strength in boreholes.

Average normal stress on the interface in terms of actual normal stress is expressed as:

$$\sigma_n^{av} = \frac{L_i^c}{L_b}\sigma_n \tag{4.62}$$

Inserting relationships given by Equation (4.62) in Equation (4.59), we finally obtain the following identity:

$$\alpha^2 \tan^2 i_a \sigma_n^{av3} + 2u_o \alpha \tan i_a \sigma_n^{av2} + (1 - A^* L_i \alpha)u_o^2 \tan i_a \sigma_n^{av} - A^*(L_i \tan i_a u_o^3 - u_o^4) = 0 \tag{4.63}$$

where $A^* = 1/(A \cdot L_b)$. The solution of the above identity yields the desired values of σ_n^{av}.

As a specific application, the formulation above to the tendon-grout interface is given herein. The asperity inclination of the tendon is 45°. Consequently, L_i is equal to asperity height u_o from the geometry. Inserting these values in expression yields:

$$\left(\alpha^2 \sigma_n^{av2} + 2u_o \alpha \sigma_n^{av} + (1 - A^* L_i \alpha)u_o^2\right)\sigma_n = 0 \tag{4.64}$$

Then the solutions are:

$$(\sigma_n^{av})_1 = 0, \quad (\sigma_n^{av})_{2,3} = \frac{u_o}{\alpha}\left[-1 \pm \sqrt{A^* \alpha u_o}\right] \tag{4.65}$$

Since we are interested in the positive values of the normal stress, the selected solution for the normal stress is:

$$(\sigma_n^{av})_3 = \frac{u_o}{\alpha}\left[-1 + \sqrt{\frac{E}{1+v} \cdot \frac{\alpha u_o}{aL_b}}\right] \tag{4.66}$$

Parameters in Equation (4.66) are:

$$\tau_{tg} = 1.85 + \sigma_n^{av} \tan(58.78)$$

Parameters in Equation (4.60) are:

$$\alpha = 0.001 \quad u_o = 0.0017, \quad L_b = 0.0128, \quad v = 0.25$$

Using this concept, the measured peak shear strength in pull-out tests performed *in situ* as a function of the deformability of rock mass and predicted strength curves are plotted in Figure 4.38 and compared with each other.

4.8   PULL-OUT CAPACITY

The pull-out capacity of rockbolts are of paramount importance in association with the design of anchorages for transmission towers, suspension bridges, and the suspension of blocks of rock, etc. However, there are a few closed-form solutions, which are usually short of evaluating the bearing capacity of rockbolts/anchors. The closed-form solutions to be presented consider the possible shear failure along one of the interfaces, which are the most critical surfaces within the system of rockbolts/rockanchors. There are reports that a failure of rock in the shape of cones can also occur. However, this is only possible for rockanchors with an enlarged section at the bottom, where the load has to act to cause such a failure.

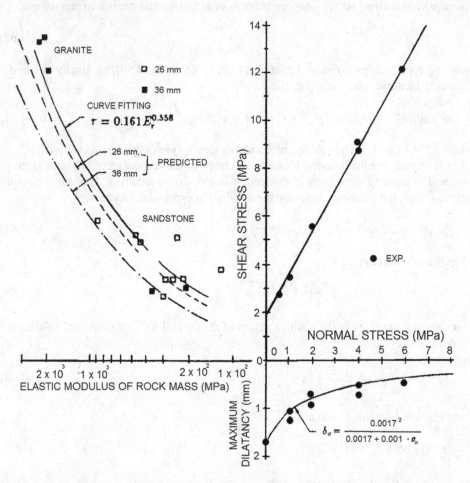

Figure 4.38 Estimated shear strength of tendon-grout (TG) interface in borehole.

Nevertheless, this type of failure can also be treated in a similar manner by just introducing the condition of failure and its direction in the procedure to be presented.

4.8.1   Constitutive equations

The constitutive laws of materials given below are somewhat simplified in order to obtain closed-form solutions.

1) *Constitutive law for steel*

The steel bar is assumed to show a bilinear elasto-plastic behavior, which is given by (Fig. 4.39a)

- Elastic behavior of bar:

$$\sigma_b = E_b \varepsilon_b, \quad \sigma_b \le \sigma_y$$

- Elasto-plastic behavior of bar:

$$\sigma_b = E_t\varepsilon_b + F \quad \sigma_b \geq \sigma_y \tag{4.67}$$

$$F = (1 - E_t/E_b)\sigma_y$$

where
E_b = elastic modulus of bar
E_t = deformation modulus after yielding
σ_b = axial stress in bar
σ_y = yielding strength of bar
ε_b = axial strain in bar

2) *Behavior of interfaces and grout annulus*

The behavior of interfaces and grout annulus during shearing is assumed to be elastic-softening–residual plastic flow type (Fig. 4.39b). This type of assumption will cover a wide range of behaviors, such as the elastic-brittle plastic behavior to the elastic-perfectly plastic behavior. As experimentally shown in the previous section, there is no difference in the behaviors of the grout annulus and the interfaces within the elastic range. Nevertheless, the yielding conditions and the post-failure characteristics differ somewhat. Accordingly, the behavior of the annulus and interfaces are written as:

- Elastic behavior:

$$\tau = G\gamma, \quad \tau \leq \tau_p$$

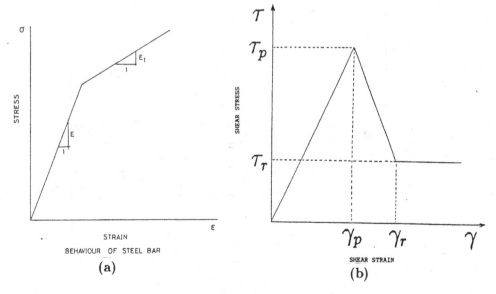

Figure 4.39 Idealized behavior of bar and interfaces.

- Softening behavior:

$$\tau = \tau_p - \frac{\gamma - \gamma_p}{\gamma_r - \gamma_p}(\tau_p - \tau_r), \quad \tau_r \leq \tau \leq \tau_p \tag{4.68}$$

- Residual plastic flow behavior:

$\tau = \tau_r$

where
G = elastic modulus of grout annulus
τ = shear stress at any point in grout annulus
τ_p = peak shear strength of the annulus or interfaces
τ_r = residual shear strength of the annulus or interfaces
γ = shear strain at any point in the annulus
γ_p = shear strain at peak shear strength of the annulus or interfaces
γ_r = shear strain at residual shear strength of the annulus or interfaces

4.8.2   Governing equations

Let us consider a rockbolt subjected to an axial load and assume that it is a one-dimensional member. Additionally, the applied load is also assumed to be transferred to the surrounding medium as shear stresses only. For a unit slice, the force equilibrium for the bolt-grout system can be written as (Fig. 4.40):

$$(\sigma_z + \Delta\sigma_z)A_b - \sigma_z A_b + \tau_b A_{pb}\Delta z = 0$$

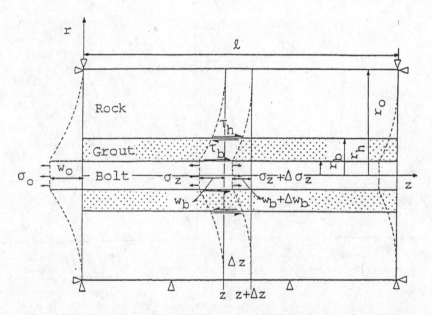

Figure 4.40 Idealized stress and deformation states in/along bolts and its vicinity.

where $A_b = \pi r_b^2, A_{pb} = 2\pi r_b$. Dividing the above relationship by $\pi r_b^2 \Delta z$ and taking the limit yields the governing equation for the bolt-grout system as:

$$\lim_{\Delta z \to 0} \frac{\Delta \sigma_z}{\Delta z} + \frac{2}{r_b}\tau_b = 0 \Rightarrow \frac{d\sigma_z}{dz} + \frac{2}{r_b}\tau_b = 0 \tag{4.69}$$

where
r_b = radius of bar
τ_b = shear stress at grout-bolt interface
σ_z = axial stress in bar
z = distance of a point from bolt head

If the stresses are assumed to be transferred into grout and rock as shear stresses, the force equilibrium condition for the elementary unit slice is written as:

$$2\pi(\tau_{rz} + \Delta \tau_{rz})(r + \Delta r)\Delta z - 2\pi \tau_{rz} r \Delta z = 0$$

Dividing above expression by $2\pi r \Delta r \Delta z$ and taking the limit yields the governing equation for the direction parallel to the bolt axis as:

$$\lim_{\Delta r \to 0} \frac{\Delta \tau_{rz}}{\Delta r} + + \frac{\tau_{rz}}{r} = 0 \Rightarrow \frac{d\tau_{rz}}{dr} + \frac{\tau_{rz}}{r} = 0 \tag{4.70}$$

where
τ_{rz} = shear stress in grout annulus or rock
r = radial distance of a point

With the use of yield conditions and boundary conditions, the above governing equations can be coupled and solved. By considering the failure at the bolt-grout interface, solutions for the following three cases are presented:

- Case I: Solutions for pure elastic behavior of the bar, interfaces, grout, and rock.
- Case II: Solutions for elastic-brittle plastic behavior of the bolt-grout interface and elastic behavior of bar, grout, and rock,
- Case III: Solutions for elastic-softening residual plastic behavior of the grout-rock interface and elasto-plastic behavior of bar.

To start with, the governing Equation (4.70) for the rock is solved together with boundary conditions and the constitutive law for the elastic behavior of rock. Writing the relationship between shear strain and shear stress as:

$$\tau_{rz} = G_r \gamma = G_r \frac{dw}{dr}$$

where G_r and w are the shear modulus of rock and displacement of rock in the direction of bolt axis, respectively, and inserting this relationship in Equation (4.70) yields the following differential equation:

$$\frac{d^2 w}{dr^2} + \frac{1}{r}\frac{dw}{dr} = \frac{1}{r}\frac{d}{dr}\left(r\frac{dw}{dr}\right) = 0. \tag{4.71}$$

The general solution of the above differential equation is:

$$w = C_1 \ln r + C_2 \tag{4.72}$$

where C_1 and C_2 are the integration constants. With the use of boundary conditions given as

$$w = -w_h \qquad \text{at } r = r_h,$$
$$w = 0 \qquad \text{at } r = r_0,$$

where r_h radius of borehole, the constants are obtained as:

$$C_1 = w_h \frac{1}{\ln(r_0 / r_h)}, \quad C_2 = -w_h \frac{\ln r_0}{\ln(r_0 / r_h)}.$$

Finally, the displacement in rock takes the following form:

$$w = w_h \frac{1}{\ln(r_0 / r_h)}(\ln r - \ln r_0). \tag{4.73}$$

The shear stress in the rock is then obtained as:

$$\tau_{rz} = G_r \frac{w_h}{r \ln(r_0 / r_h)}. \tag{4.74}$$

Together with the following boundary conditions

$$w = -w_h \qquad \text{at } r = r_h,$$
$$w = -w_b \qquad \text{at } r = r_b,$$

the displacement and shear stress in grout annulus can be obtained in a similar manner as

displacement:

$$w = \frac{1}{\ln(r_h / r_b)}((w_b - w_h)\ln r - w_b \ln r_h + w_h \ln r_b) \tag{4.75}$$

shear stress:

$$\tau_{rz} = G_g \frac{w_b - w_h}{r \ln(r_h / r_b)} \tag{4.76}$$

where G_g is the shear modulus of grout.

Requiring the continuity of shear stresses at grout-rock interface at $r = r_h$ yields the following relationship between w_b and w_h:

$$w_b = \chi w_h \tag{4.77}$$

where

$$\chi = \frac{G_r}{G_g} \frac{\ln(r_h / r_b)}{\ln(r_0 / r_h)} + 1.$$

Then, the shear stress at grout-bolt interface can be rewritten in the following form:

$$\tau_b = K_g w_b \qquad (4.78)$$

where

$$K_g = \frac{G_g}{r_b \ln(r_h / r_b)} \cdot \left(\frac{\chi - 1}{\chi} \right)$$

And the following relationship holds between the shear stress τ_b at the grout-bolt interface and the shear stress at a point $(r, z = constant)$, provided that rock and grout annulus behave elastically:

$$\tau_{r,z=constant} = \tau_b \frac{r_b}{r} \qquad (4.79)$$

4.8.2.1   Case I: Solutions for pure elastic behavior of the bar, interfaces, grout, and rock

The behavior of the whole system was assumed to be elastic. The governing equation (4.69) can be expressed in terms of the displacement with the use of the constitutive relationship between the axial stress and strain in the bolt:

$$\sigma_z = E_b \varepsilon_z = -E_b \frac{dw_b}{dz}, \qquad (4.80)$$

and the relationship given by Equation (4.68) as:

$$\frac{d^2 w_b}{dz^2} - \frac{2}{E_b r_b} \tau_b = \frac{d^2 w_b}{dz^2} - \alpha^2 w_b = 0 \qquad (4.81)$$

where

$$\alpha = \sqrt{\frac{2K_g}{E_b r_b}}.$$

The general solution of the above differential equation is:

$$w_b = A_1 e^{\alpha z} + A_2 e^{-\alpha z}. \qquad (4.82)$$

Integration constants A_1 and A_2 for the following boundary conditions

$$\sigma_z = 0 \qquad \text{at } z = L$$
$$\sigma_z = \sigma_0 \qquad \text{at } z = 0$$

where L is bolt length, take the following forms:

$$A_1 = \frac{\sigma_0}{E_b \alpha} \frac{e^{-\alpha L}}{e^{\alpha L} - e^{-\alpha L}}, \quad A_2 = \frac{\sigma_0}{E_b \alpha} \frac{e^{\alpha L}}{e^{\alpha L} - e^{-\alpha L}}$$

Then, axial displacement w_b and axial stress σ_z in bolt and shear stress τ_b at bolt-grout interface take the following forms:

- Axial displacement w_b

$$w_b = \frac{\sigma_0}{E_b \alpha} \frac{e^{\alpha(L-z)} + e^{-\alpha(L-z)}}{e^{\alpha L} - e^{-\alpha L}}$$

- Axial stress σ_z

$$\sigma_z = \sigma_0 \frac{e^{\alpha(L-z)} - e^{-\alpha(L-z)}}{e^{\alpha L} - e^{-\alpha L}}$$

(4.83)

- Shear stress τ_b at bolt-grout interface

$$\tau_b = \frac{\sigma_0 r_b \alpha}{2} \frac{e^{\alpha(L-z)} + e^{-\alpha(L-z)}}{e^{\alpha L} - e^{-\alpha L}}$$

In order to estimate the bolt-head stress at the yielding of the bolt-grout interface, the following boundary conditions can be used:

$$\sigma_z = 0 \qquad \text{at } z = L$$
$$\tau_b = \tau_p^{bg} \qquad \text{at } z = 0$$

where τ_p^{bg} is peak shear strength of the bolt-grout interface. Integration constants A_1 and A_2 take the following forms:

$$A_1 = \frac{\tau_p^{bg}}{K_g} \frac{e^{-\alpha L}}{e^{\alpha L} + e^{-\alpha L}}, \quad A_2 = \frac{\tau_p^{bg}}{K_g} \frac{e^{\alpha L}}{e^{\alpha L} + e^{-\alpha L}}$$

Then, axial displacement w_b and axial stress σ_z in bolt and shear stress τ_b at bolt-grout interface take the following forms:

- Axial displacement w_b

$$w_b = \frac{\tau_p^{bg}}{K_g} \frac{e^{\alpha(L-z)} + e^{-\alpha(L-z)}}{e^{\alpha L} + e^{-\alpha L}}$$

- Axial stress σ_z

$$\sigma_z = \frac{\tau_p^{bg} E_b \alpha}{K_g} \frac{e^{\alpha(L-z)} - e^{-\alpha(L-z)}}{e^{\alpha L} + e^{-\alpha L}}$$

(4.84)

- Shear stress τ_b at bolt-grout interface

$$\tau_b = \tau_p^{bg} \frac{e^{\alpha(L-z)} + e^{-\alpha(L-z)}}{e^{\alpha L} + e^{-\alpha L}}$$

The bolt-head stress σ_0 at the time yielding is obtained as:

$$\sigma_0 = \frac{\tau_p^{bg} E_b \alpha}{K_g} \frac{e^{\alpha L} - e^{-\alpha L}}{e^{\alpha L} + e^{-\alpha L}} = \frac{\tau_p^{bg} E_b \alpha}{K_g} \tanh(\alpha L) \tag{4.85}$$

4.8.2.2 Case II: Solutions for elastic-brittle plastic behavior of the bolt-grout interface, elastic behavior of bolt, grout, and rock

In this subsection, the elastic-brittle plastic behavior of interfaces is considered and the other materials in the system are assumed to behave elastically. The elastic-brittle plastic behavior is defined as:

$$\tau = G_g \gamma \qquad \text{when } \tau \leq \tau_p^{bg} \tag{4.86}$$

$$\tau = \eta \tau_p^{bg} \qquad \text{when } \gamma > \gamma_p \tag{4.87}$$

where $\eta = \tau_r^{pg} / \tau_p^{bg}$, τ_r^{bg} is the residual shear strength of the bolt-grout interface. The parameters in the above expression have the same meanings as those in the constitutive law given by Equation (4.68).

It is assumed that the bolt-grout interface is already yielded and there are two regions – elastic and plastic. Now, the expressions for the axial displacement and axial stress in the bolt and the shear stress at the bolt-grout interface will be derived for each respective region.

Plastic region: $0 \leq z \leq L_p$ (L_p: length of plastic zone measured from the bolt head)

As the plastification at the bolt-grout interface has already occurred, the differential equation (4.81) can be written in the following form, together with the use of identity $\tau_b = \eta \tau_p^{bg}$:

$$\frac{d^2 w_b}{dz^2} - \beta = 0 \tag{4.88}$$

where

$$\beta = \frac{2 \eta \tau_p^{bg}}{r_b E_b}.$$

Then, integrating the above differential equation twice yields:

$$w_b = \beta \frac{z^2}{2} + C_1 z + C_2 \tag{4.89}$$

With the use of the following boundary conditions:

$$w = w_0 \qquad \text{at } z = 0,$$
$$\sigma_z = \sigma_0 \qquad \text{at } z = 0,$$

the integration constants C_1 and C_2 are obtained as:

$$C_1 = -\frac{\sigma_0}{E_b}, \quad C_2 = w_0.$$

Then, axial displacement w_b and axial stress σ_z in bolt and shear stress τ_b at bolt-grout interface take the following forms:

- Axial displacement:

$$w_b = w_0 + \frac{\eta \tau_p^{bg}}{E_b r_b} z^2 - \frac{\sigma_0}{E_b}$$

- Axial stress:

$$\sigma_z = \sigma_0 - \frac{2\eta \tau_p^{bg}}{r_b} z \qquad (4.90)$$

- Shear stress:

$$\tau_b = \eta \tau_p^{bg}$$

Elastic region: $L_p \le z \le L$ (L: bolt length)

As the general solution is given by Equation (4.82), the problem is just to introduce the appropriate boundary conditions and yield condition to determine the integration constants A_1 and A_2. These conditions are:

$$\sigma_z = 0 \qquad \text{at } z = L,$$
$$\tau_b = \tau_p^{bg} \qquad \text{at } z = L_p.$$

Integration constants A_1 and A_2 take the following forms:

$$A_1 = \frac{\tau_p^{bg}}{K_g} \frac{1}{e^{\alpha(2L-L_p)} + e^{\alpha L_p}}, \quad A_2 = \frac{\tau_p^{bg}}{K_g} \frac{e^{2\alpha L}}{e^{\alpha(2L-L_p)} + e^{\alpha L_p}}.$$

Then, axial displacement w_b and axial stress σ_z in bolt and shear stress τ_b at bolt-grout interface take the following forms:

- Axial displacement w_b:

$$w_b = \frac{\tau_p^{bg}}{K_g} \frac{e^{\alpha z} + e^{\alpha(2L-z)}}{e^{\alpha(2L-L_p)} + e^{\alpha L_p}}$$

- Axial stress σ_z:

$$\sigma_z = \frac{\tau_p^{bg} E_b \alpha}{K_g} \frac{e^{\alpha(2L-z)} - e^{\alpha z}}{e^{\alpha(2L-L_p)} + e^{\alpha L_p}} \qquad (4.91)$$

- Shear stress τ_b at bolt-grout interface:

$$\tau_b = \tau_p^{bg} \frac{e^{\alpha z} + e^{\alpha(2L-z)}}{e^{\alpha(2L-L_p)} + e^{\alpha L_p}}$$

The bolt-head stress σ_0 at the time of yielding is obtained as:

$$\sigma_0 = \tau_p^{bg} \left[\frac{E_b \alpha}{K_g} \frac{e^{\alpha(2L-L_p)} - e^{\alpha L_p}}{e^{\alpha(2L-L_p)} + e^{\alpha L_p}} + \frac{2\eta}{r_b} L_p \right] \qquad (4.92)$$

4.8.2.3 Case III: Solutions for elastic-softening residual plastic behavior of the bolt-grout interface, elasto-plastic behavior of bolt, and the elastic behavior of grout and rock

In this subsection, it is assumed that the bolt-grout interface exhibits an elastic-softening–residual plastic flow behavior and the bar exhibits an elastic-strain hardening plastic behavior in the residual plastic behavior of the interface. The solutions for this particular case are as follows:

Region 1: $0 \leq z \leq L_{p1}$ (L_{p1}: length of plastic zone of bolt measured from the bolt head)

In this region, the bolt-grout interface and the bar are assumed to exhibit a residual plastic flow behavior and a strain-hardening plastic behavior, respectively. Inserting the constitutive law for the bar given by Equation (4.67) results in the following differential equation, together with $\tau_b = \tau_r^{bg} = \eta \tau_p^{bg}$:

$$\frac{d^2 w_b}{dz^2} - \beta = 0 \qquad (4.93)$$

where

$$\beta = \frac{2\eta \tau_p^{bg}}{r_b E_T}$$

and E_T modulus of deformation after yielding of bar. Then, integrating the above differential equation twice yields:

$$w_b = \beta \frac{z^2}{2} + C_1 z + C_2 \qquad (4.94)$$

With the use of the following boundary conditions:

$$w = w_0 \qquad \text{at } z = 0,$$
$$\sigma_z = \sigma_0 \qquad \text{at } z = 0,$$

the integration constants C_1 and C_2 are obtained as:

$$C_1 = -\frac{\sigma_0}{E_b}, \quad C_2 = w_0.$$

Then, axial displacement w_b and axial stress σ_z in bolt and shear stress τ_b at bolt-grout interface take the following forms:

- Axial displacement:

$$w_b = w_0 + \frac{\eta \tau_p^{bg}}{E_T r_b} z^2 - \frac{\sigma_0 - F}{E_T} z$$

(F is defined by Equation (4.67))

- Axial stress:

$$\sigma_z = \sigma_0 - \frac{2\eta \tau_p^{bg}}{r_b} z \tag{4.95}$$

- Shear stress:

$$\tau_b = \eta \tau_p^{bg}$$

where

$$\sigma_0 = \sigma_y + \frac{2\eta \tau_p^{bg}}{r_b} L_{p1}$$

$$w_0 = w_b^{ep} + \left(\frac{\sigma_y}{E_b} + \frac{2\eta \tau_p^{bg}}{E_b r_b} L_{p1} \right)(L_{p2} - L_{p1})$$

$$+ \frac{\eta \tau_p^{bg}}{r_b} \left(\frac{1}{E_b} - \frac{1}{E_t} \right) L_{p1}^2 - \frac{\eta \tau_p^{bg}}{E_b r_b} L_{p2}^2 + \frac{1}{E_t} (\sigma_0 - F) L_{p1}$$

Region 2: $L_{p1} \le z \le L_{p2}$ (L_{p2}: length of residual plastic zone of bolt-grout interface measured from the bolt head).

Bolt-grout interface at residual plastic state and bar is elastic. The form of the differential equation and its solution are the same as those in the previous case, except E_T is replaced by E_b in term β. Integration constants are determined from the following conditions:

$$\sigma_z = \sigma_y \qquad \text{at } z = L_{p1}$$
$$w_r = \zeta w_p \qquad \text{at } z = L_{p2}$$

The integration constants C_1 and C_2 are obtained together with the use of the relationship (4.67) as:

$$C_1 = -\left(\frac{\sigma_y}{E_b} + \frac{2\eta \tau_p^{bg}}{E_b r_b} L_{p1} \right), \quad C_2 = -\xi \frac{\tau_p^{bg}}{K_g} + \left(\frac{\sigma_y}{E_b} + \frac{2\eta \tau_p^{bg}}{E_b r_b} L_{p1} \right) L_{p2} - \frac{\eta \tau_p^{bg}}{E_b r_b} L_{p2}^2$$

Then, axial displacement w_b and axial stress σ_z in bolt and shear stress τ_b at bolt-grout interface take the following forms:

- Axial displacement:

$$w_b = \left(\frac{\sigma_y}{E_b} + \frac{2\eta\tau_p^{bg}}{E_b r_b} L_{p1} \right)(L_{p2} - z) - \frac{\eta\tau_p^{bg}}{E_b r_b}(L_{p2}^2 - z^2) + \xi \frac{\tau_p^{bg}}{K_g}$$

- Axial stress:

$$\sigma_z = \sigma_y - \frac{2\eta\tau_p^{bg}}{r_b}(z - L_{p1}) \tag{4.96}$$

- Shear stress:

$$\tau_b = \eta\tau_p^{bg}$$

Region 3: $L_{p2} \le z \le L_{p3}$ (L_{p3}: distance of the elastic-plastic boundary from the bolt head)

The bolt-grout interface at softening state and bar is elastic. The governing equation (4.81) together with the form of τ_b given by Equation (4.68) for the softening region and the resulting equation is transformed to the following form:

$$\frac{d^2 w_b}{dz^2} + \frac{2K_g}{E_b r_b}\left(\frac{1-\eta}{\xi-1} \right) w_b = \frac{2\tau_p^{bg}}{E_b r_b}\left(\frac{\xi-\eta}{\xi-1} \right) \tag{4.97}$$

The general solution of the above non-homogeneous differential equation is:

$$w_b = B_1 \cos(pz) + B_2 \sin(pz) + \frac{\tau_p^{bg}}{K_g}\left(\frac{\xi-\eta}{1-\eta} \right) \tag{4.98}$$

where

$$p = \sqrt{\frac{2K_g}{E_b r_b}\left(\frac{1-\eta}{\xi-1} \right)}.$$

Introducing the following conditions:

$$\tau_b = \tau_r^{bg} = \eta\tau_p^{bg} \qquad \text{at } z = L_{p2}$$
$$\tau_b = \tau_p^{bg} \qquad \text{at } z = L_{p3}$$

yields integration constants B_1 and B_2 as:

$$B_1 = \frac{\tau_p^{bg}}{K_g}\left(\frac{\xi-1}{1-\eta} \right)\frac{\sin(pL_{p2}) - \eta\sin(pL_{p3})}{\sin(p(L_{p3} - L_{p2}))}, B_2 = \frac{\tau_p^{bg}}{K_g}\left(\frac{\xi-1}{1-\eta} \right)\frac{\eta\cos(pL_{p3}) - \cos(pL_{p2})}{\sin(p(L_{p3} - L_{p2}))}.$$

Then, axial displacement w_b and axial stress σ_z in bolt and shear stress τ_b at bolt-grout interface take the following forms:

- Axial displacement:

$$w_b = \frac{\tau_p^{bg}}{k_g} \left\{ \frac{\xi-1}{1-\eta} \frac{(A-\eta B)\cos(pz)-(D-\eta C)\sin(pz)}{\sin(p(L_{p3}-L_{p2}))} + \frac{\xi-\eta}{1-\eta} \right\}$$

- Axial stress:

$$\sigma_z = \frac{p\tau_p^{bg} E_b}{K_g} \frac{\xi-1}{1-\eta} \left\{ \frac{(A-\eta B)\sin(pz)+(D-\eta C)\cos(pz)}{\sin(p(L_{p3}-L_{p2}))} \right\} \qquad (4.99)$$

- Shear stress:

$$\tau_b = \tau_p^{bg} \left\{ \frac{-(A-\eta B)\cos(pz)+(D-\eta C)\sin(pz)}{\sin(p(L_{p3}-L_{p2}))} \right\}$$

where

$$A = \sin(pL_{p2}), \quad B = \sin(pL_{p3}), \quad C = \cos(pL_{p3}), \quad D = \cos(pL_{p2})$$

Region 4: $L_{p3} \le z \le L$ (L: length of bolt)

Bolt-grout interface and bar are elastic. As the general solution is given by Equation (4.82), the problem is just to introduce the appropriate boundary condition and yield condition to determine the integration constant A_1 and A_2. These conditions are:

$$\sigma_z = 0 \qquad \text{at } z = L$$
$$\tau_b = \tau_p^{bg} \qquad \text{at } z = L_{p3}$$

Integration constants A_1 and A_2 take the following forms:

$$A_1 = \frac{\tau_p^{bg}}{K_g} \frac{1}{e^{\alpha(2L-L_{p3})}+e^{\alpha L_{p3}}}, \quad A_2 = \frac{\tau_p^{bg}}{K_g} \frac{e^{2\alpha L}}{e^{\alpha(2L-L_{p3})}+e^{\alpha L_{p3}}}.$$

Then, axial displacement w_b and axial stress σ_z in bolt and shear stress τ_b at bolt-grout interface takes the following forms:

- Axial displacement:

$$w_b = \frac{\tau_p^{bg}}{k_g} \frac{e^{\alpha(2L-L_{p3}-z)}+e^{\alpha(z-L_{p3})}}{e^{2\alpha(L-L_{p3})}+1}$$

- Axial stress:

$$\sigma_z = \frac{\tau_p^{bg} \alpha E_b}{k_g} \frac{e^{\alpha(2L-L_{p3}-z)} - e^{\alpha(z-L_{p3})}}{e^{2\alpha(L-L_{p3})}+1} \tag{4.100}$$

- Shear stress:

$$\tau_b = \tau_p^{bg} \frac{e^{\alpha(2L-L_{p3}-z)} + e^{\alpha(z-L_{p3})}}{e^{2\alpha(L-L_{p3})}+1}$$

4.9   SIMULATION OF PULL-OUT TESTS

In this subsection, the stress-strain state and the load-bearing capacity of a single fully grouted rockbolt under pull-out loads have been analyzed and discussed. It will be shown in later sections of this chapter that such a study has significant implications regarding the understanding on the mechanical performance and functions of rockbolts.

Two examples involve parametric studies on the effects of the ratio of elastic moduli of the bolt, grout, and rock and the ratio of the length of the bolt and borehole to their diameters by using the analytical model for the elastic behavior (Equation (4.83)). Figure 4.41 shows the effect of the ratio of the elastic modulus of the bolt to that of rock (no grout annulus in this specific case) on the axial and shear stress distributions in/along the bolt. As noted from the figure, the distributions are very much influenced by the ratio of the elastic moduli. The distributions become highly nonuniform as the elastic modulus of the embedment medium approaches the modulus of the bolt, and there is a high stress concentration near the bolt head where the load is applied. On the other hand, they tend to become uniformly distributed, and stress concentration tends to diminish as the ratio increases.

Next, the length of the bolt was varied. Figure 4.42 shows the axial stress σ_z distribution in bolt and shear stress τ_b distribution at bolt-grout interface along the bolt, normalized by the axial stress σ_0 and τ_0 at the bolt head for two ratios of the elastic moduli of the bolt and the embedment medium by using the analytical solution for the elastic behavior (Equations (4.83 and 4.84)). As noted from Figure 4.42, the stresses tend to concentrate near the bolt head where the load is applied, as the length of the bolt becomes longer. Besides the effect of the ratio of the elastic moduli, the length of bonding also influences the distributions even if the rock is very weak.

Figure 4.43 shows a simulation of a pull-out test, in which the bolt-grout interface fails by exhibiting an elastic-softening–residual plastic flow behavior. As seen from Figure 4.43, the load-displacement response of the bolt head rises linearly until the debonding process first occurs. At the same time, there is a stress concentration near the bolt head, which exponentially decays as the distance increases. By increasing the load furthermore, the debonded region comes into existence and the debonded region becomes larger. As the debonded region propagates further, there appears a point at which the maximum load-bearing capacity of the bolt is mobilized. From the axial load-displacement curve (Fig. 4.43a), it is possible to deduct three different regions in association with the shearing of the surface of weakness (the

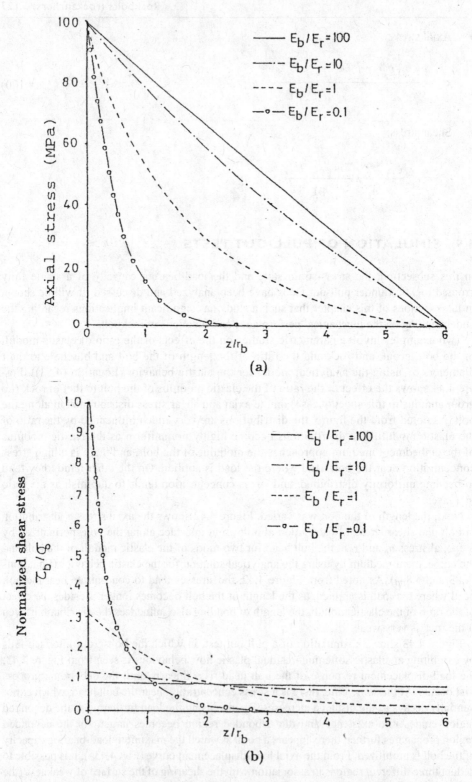

Figure 4.41 Variation of axial and shear stresses along the bolt for varying ratios of the elastic moduli of bolt and rock.

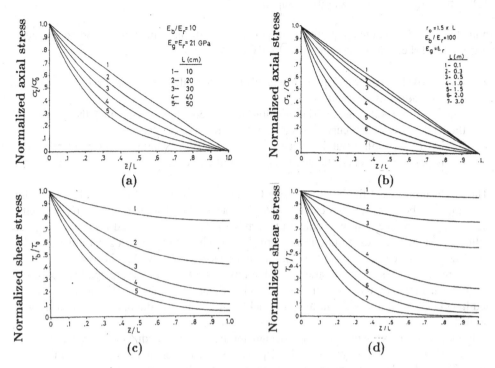

Figure 4.42 Variation of axial and shear stresses along the bolt for varying bonding length.

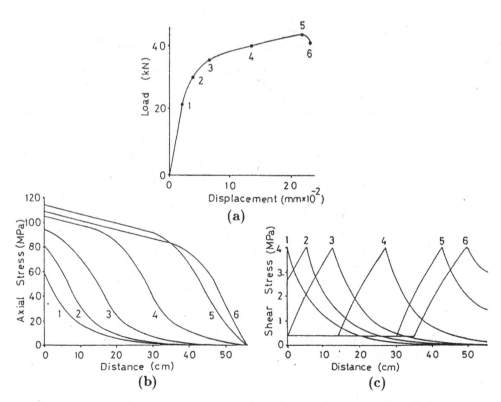

Figure 4.43 Load-displacement curve and distributions of stresses in/along the bar.

numbers 1,2,·,5,6 denote the points on the load versus displacement curve and corresponding axial and shear stress distributions along the bolt):

- Region (0–1) (system behaving elastically): the load-displacement curve rises linearly and its gradient is constant
- Region (1–3) (bolt-grout interface exhibiting elastic-softening behavior): the gradient of the load-displacement curve decreases with increase in displacement
- Region (3–6) (bolt-grout interface exhibiting elastic-softening–residual plastic flow behavior): the gradient of the load-displacement curve again becomes constant

The implication is that the bonding strength obtained from the pull-out tests by averaging over the total bolt surface area cannot represent the real bond strength, since the geometry of bars and borehole and mechanical properties of the bar, grout, and rock have a great influence on the stress distributions along the bar which, in turn, have a great impact on the mobilized intrinsic resistance of the material and interfaces involved in the system.

Figure 4.44 shows the axial strain distributions (solid lines) at various load levels along the bolt calculated by using the analytical model (Equations (4.90)–(4.92)) and those (dotted lines) measured in a pull-out test of a rockbolt of 13 mm in diameter, having a smooth surface.

Figure 4.45 shows the response of a rockanchor under cyclic loading performed *in situ*. As noted from the figure, an irrecoverable displacement of rockanchors occurs when the load exceeds a certain level. Furthermore, this irrecoverable displacement increases even if the

Table 4.1 Material properties and dimensions.

E_b	v_b	E_g	v_g	E_r	v_r	c_p	η	r_b	r_h	L	γ_r/γ_p
GPa		GPa		GPa		MPa		mm	mm	mm	
200	0.3	9.8	0.25	27	0.3	4	0.1	11	18	560	3.1

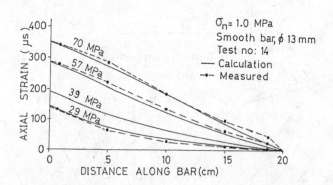

Figure 4.44 Comparison of calculated distributions of stresses in/along the bar with the experimental ones (analytical model).

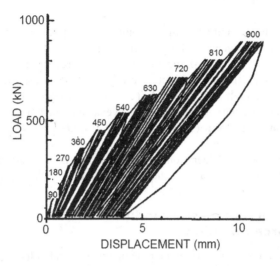

Figure 4.45 In situ cyclic pull-out tests.

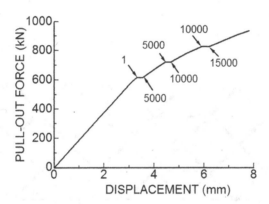

Figure 4.46 Simulation of in situ cyclic pull-out tests.

load level is kept the same. Therefore, the cyclic behavior of rockanchors can be evaluated by taking into account the shrinkage of the shear strength of interfaces as the number of cycles increases. This approach introduces a new concept, based on the principles of mechanics, to rockanchor design methodology. The approach is implemented in the earlier formulations for the estimation of pull-out response of rockanchors proposed by Aydan (1989) and Aydan et al. (1985, 1995). Figure 4.46 shows the predicted response of a cyclic pull-out tests. The general tendency is similar to the in situ performance, although there are some quantitative differences between calculations and observations. Nevertheless, this problem can be overcome by further experiments on the cyclic behavior of interfaces.

4.10   MESH BOLTING

There is an increasing demand for underground openings for energy storage projects (i.e. CAES and SMES projects), and it is lately an active field of research (Aydan *et al.*, 1995; Ebisu *et al.*, 1994). The shape of cavities is circular in cross-section with a diameter of 5–6 m. Cavities are subjected to high internal pressures varying between 4 MPa and 8 MPa. As such high internal pressures are expected to cause high tensile stresses in rock masses, which are generally weak against tensile stresses, the reinforcement of the opening by rockbolts is one of the options for resisting such high tensile stresses.

In this subsection, some considerations are presented on how to incorporate the effect of inclined bolting, which is different from the conventional rockbolting pattern.

4.10.1   Evaluation of elastic modulus of reinforced medium

Let us consider that a representative element contains two rockbolts having inclinations of α_1 and α_2 and densities η_1 and η_2 with respect to the global system (Fig. 4.47). Thus, the contributions of rockbolts in the respective directions are:

$$\Delta E_1 = (E_b - E)\left[\eta_1 a_{11}^2 + \eta_2 a_{21}^2\right]$$
$$\Delta E_2 = (E_b - E)\left[\eta_1 a_{12}^2 + \eta_2 a_{22}^2\right]$$

$$(4.101)$$

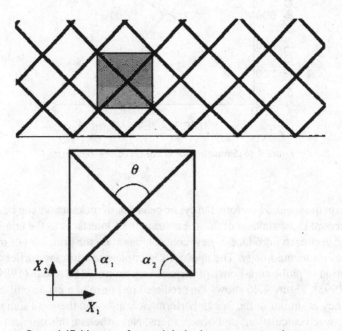

Figure 4.47 Notation for mesh-bolted representative element.

Let us assume that rockbolts are installed symmetrically with respect to the principal axes and the angle between bolts is θ. Then, the values of a_{ij} are:

$$
\begin{aligned}
a_{11} &= \cos(\alpha) \\
\alpha_{12} &= \cos(90 - \alpha) \\
\alpha_{21} &= \cos(\theta + \alpha) \\
\alpha_{22} &= \cos(\theta + \alpha - 90) \\
2\alpha + \theta &= 180, \quad \alpha_2 - \alpha_1 = \theta, \quad \alpha_1 = \alpha.
\end{aligned}
\tag{4.102}
$$

If we also assume that the density of bolts are the same, then Equation (4.101) becomes:

$$
\begin{aligned}
\Delta E_1 &= (E_b - E)\eta\left[\cos^2\alpha + \cos^2(\theta + \alpha)\right] \\
\Delta E_2 &= (E_b - E)\eta\left[\sin^2\alpha + \sin^2(\theta + \alpha)\right]
\end{aligned}
\tag{4.103}
$$

As noted from the above equations, the contributions are isotropic when the installation angle θ is 90°. If $\theta < 90°$, then $\Delta E_2 > \Delta E_1$. On the other hand, if $\theta > 90°$, then $\Delta E_2 < \Delta E_1$. A parametric study was carried out to see the variation of contributions as a function of installation angle, shown in Figure 4.48.

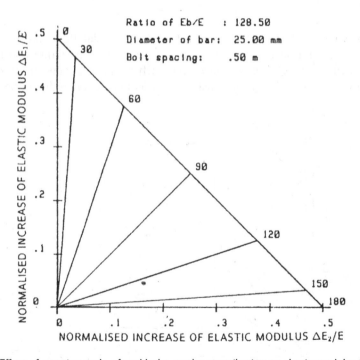

Figure 4.48 Effect of crossing angle of rockbolts on the contribution to elastic modulus in respective directions.

4.10.2   Evaluation of tensile strength of reinforced medium

As assumed in the evaluation of elastic modulus of reinforced medium, the averaged stress is a sum of stresses in reinforcing members and surrounding medium. In a uniaxial loading case, one can write the following by assuming that the uniaxial strain field is homogenous:

$$\sigma_{av} = \eta\sigma_b + \sigma_m(1 - \eta) \tag{4.103}$$

It is very well known that reinforcing materials, such as steel, behave elastic-perfectly plastic while rocks behave elastic-brittle plastic under a tensile stress field. It is most likely that the critical tensile strain of rocks is much less than that of steel and they will fracture before steel yields. The fracturing strength of reinforced rock may be given in the following form:

$$\sigma_{tp}^* = \left[1+\eta\left(\frac{E_b}{E}-1\right)\right]\sigma_t \tag{4.104}$$

where σ_t is the tensile strength of rock (medium). After the fracturing of rock, the reinforced medium can still take tensile load, and stress-strain behavior of the reinforced material purely depends upon the steel, provided that rock behaves elastic-brittle plastic. Thus the stress-strain relationship is:

$$\sigma_{av} = \eta E_b \varepsilon \tag{4.105}$$

If the yield strength of steel is exceeded, then the material flows until it fractures. An example of such a behavior is shown in Figure 4.49 by assuming that the tensile stress is created by using a servo-control high-stiff testing machine. If the medium is reinforced by two rockbolts symmetrically, then its tensile strength can be written in

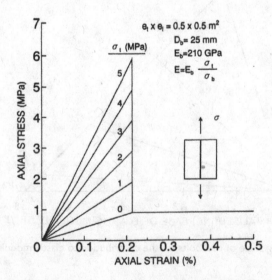

Figure 4.49 Uniaxial tensile response of bolted rock mass.

the following form, assuming that the reinforced medium is subjected to tensile stress parallel to axis x_1:

$$\sigma_{tp}^* = \left[1 + \eta \frac{\Delta E_1}{E_r}\right]\sigma_t \quad \cdot \tag{4.106}$$

Furthermore, its residual tensile strength may be given as:

$$\sigma_{tr}^* = \eta\sigma_b \tag{4.107}$$

where σ_b is the tensile strength of steel.

Chapter 5

Support members

5.1 INTRODUCTION

Support members fundamentally provide resistance against the movement of surrounding ground and/or intrusion of fluid into the open space, while reinforcement members such as rockbolts contribute to the mechanical properties of surrounding rock mass as well as restraining of ground movement into the open space. In some cases, support members may obstruct fluid flow into the surrounding medium. In other words, the support members are members separated from the surrounding ground, while the reinforcement members are an integral part of the rock mass on a macroscopic scale. The support members are conventionally shotcrete (with/without fibers), concrete liners, steel ribs/sets, and steel liners in underground excavations. Regarding surface structures, shotcrete, steel, or reinforced concrete piles and retaining walls may be visualized as support members. Fundamentally, support members convey the loads to the stable ground if they are unclosed or internally resist the forces directly if they are closed. In this chapter, the mechanical characteristics, constitutive modeling, and structural modeling of support members are presented and discussed.

5.2 SHOTCRETE

5.2.1 Historical background

Shotcrete first appeared as "gunit" in the world of tunneling, and it has become increasingly used as the NATM has become one of most popular tunneling methods since the early 1960s (Rabcewicz, 1964–1965, 1969). As will be discussed later on, shotcrete has no great resistance against rock loads or external fluid pressures, even under uniform loading conditions, when it is thin. However, engineers argue that it has a great supporting effect. According to Aydan et al. (1992), shotcrete has four important effects besides its internal pressure effect:

- Preventing rock at the excavation surface from becoming exposed to air and moisture changes directly. This is quite an important effect in some rocks, such as mudstone, shale, serpentine, and anhydrite, which weaken in strength as they lose their integrity.
- Preventing rock near the excavation surface from relaxation. This is particularly important in the case of some sedimentary rocks, having hard-soft intercalated layers which weaken in strength due to the interlayer sliding causing fractures in rock parallel to the excavation surface.

- Initiating an arch action within the rock mass through restraining interblock sliding and allowing interblock rotation.
- Initiating a wedging action by filling up the open discontinuities, which prevents rock mass from loosening.

Therefore, the supporting effect of shotcrete, although it is structurally small when thin, is an indirect one and manifests itself as the rock mass mobilizing the utmost available resistance.

5.2.2   Experiments on shotcrete

Despite its widespread use all over the world in various engineering works, ranging from tunneling to slope excavations, it is rare to find any fundamental study on the reinforcement function of shotcrete experimentally, theoretically, or numerically. Sezaki (1990) and his colleagues (Sezaki *et al.* 1989, 1992; Aydan *et al.*, 1992) were the first investigators to carry out laboratory and field tests on the mechanical behavior of shotcrete at various ages. These experimental studies have shown that the mechanical and physical properties of shotcrete vary with time, which may have important consequences on the resistance offered by shotcrete linings. These experimental studies on shotcrete with the consideration of the actual site practice are described herein.

5.2.2.1   Shotcreting procedure

There are two different kinds of shotcreting procedure – namely, dry or wet procedures. As the wet shotcreting procedure is commonly used in practice, wet shotcreting was chosen, and the composition of shotcrete used is given in Table 5.1. The accelerator for rapid hardening is chosen as 6%, 8%, and 10%. Table 5.1 gives the average values of the accelerator for each batch.

(a)   Materials

i)   cement: ordinary Portland cement
ii)   sand: Aichi prefecture Kiso river
iii) gravel: Mie prefecture, Kuwana, Tado-cho (max. size = 10 mm)
iv)   accelerator: Mikyou Coloid, Atack LQ-2

(b)   Size and shape of specimens

Cubic specimens with a size of 10 × 10 × 10 cm were prepared. Special molds fixed on a vertical wooden panel in place of a tunnel wall were used, and molds were shotcreted in a similar manner in the field.

Table 5.1 Shotcrete mixture.

Shotcreting Procedure	Maximum aggregate size (mm)	W/C (%)	Water (kg/m³)	Cement (kg/m³)	Sand (kg/m³)	Gravel (kg/m³)	Accelerator Concentration (Cx%)
wet	10	57	217	380	1115	633	6, 8, 10

(c) Shotcreting and curing

The volume of shotcrete at each batch was 0.315 m³. Considering some improper mixing, the first 1/3 of the batch was wasted. The velocity of shotcrete at the nozzle was 100 m/sec. The distance between the nozzle and the wall was 1 m, and shotcreting was perpendicular to the wall. The specimens were taken out of molds after 30 minutes following the shotcreting and shaped. They were kept in a dry room with a constant temperature of 20°C until they were tested.

5.2.2.2 Test items

(a) Uniaxial compression tests

Uniaxial compressive tests were done when the specimens were 1, 3, 6, and 12 hours old and 1, 3, 7, and 28 days old. A servo-control testing machine was used for uniaxial compression test (Fig. 5.1). Average strains were obtained from the readings from LVDT transducers. The strain rate was 0.2% per minute.

(b) Measurement of ultrasonic velocity (V_p)

The elastic wave velocity of each specimen was measured using an ultrasonic velocimeter (PUNDIT) and the relations among the uniaxial compressive strength, Young's modulus, and Poisson's ratio with the elastic wave velocity. The direction of wave velocity measurements is parallel to the loading direction.

Figure 5.1 Servo-control testing equipment at Nagoya University.

5.2.2.3 Test results

(a) The variation of uniaxial strength with time

Figure 5.2 shows strain-stress relation of shotcrete under uniaxial compression condition, while Figure 5.3 shows the variation of uniaxial strength of shotcrete with time. The strength is 0.3–1.3 MPa at 3 hours, 2–6 MPa at 12 hours, 4–8 MPa at 1 day, and 13–19 MPa at 28 days. The strength increases in proportion to the amount of accelerator used during the first 24 hours, but the increase thereafter seems not to depend upon the amount of accelerator.

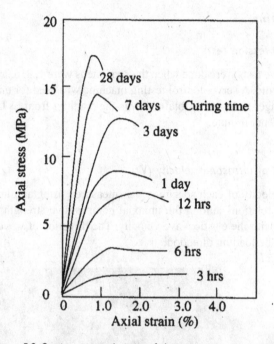

Figure 5.2 Strain-stress relations of shotcrete at various ages.

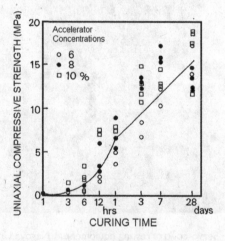

Figure 5.3 Variation of uniaxial compressive strength of shotcrete with curing time (modified from Sezaki, 1990).

To estimate the strength of shotcrete in relation to elapsed time, Kondo and Saka (1965) proposed a logarithmic empirical relation. However, this relation holds for specimens older than 1 day and is not applicable for younger shotcrete. Herein, we propose an exponential relation for the first 24 hours and a logarithmic expression for shotcrete at an age of 1 day or more. For averaged data of specimens having 6%, 8%, and 10% accelerator, an exponential function is determined by using the least-square technique. The relationships and its standard deviations r are given below:

younger than 24 hours:

$$\sigma_c = 0.238t^{1.050}, r = 0.995 \tag{5.1}$$

older than 24 hours:

$$\sigma_c = 6.64 + 6.165 \log t, r = 0.964 \tag{5.2}$$

where t is time in days. As seen from the standard deviation values, there is good correlation between experimental results and fitted expression. Nevertheless, it should be noted that average values are used for curve-fitting.

The strength of shotcrete is very much influenced by the curing process. If curing takes place with insufficient water content, the hydration of cement is not completed and the resulting strength of concrete decreases. Since the specimens were air-cured, the long-term strength of shotcrete was not very high. It is felt that further experimental work is necessary to investigate the effect of curing.

(b)   The variation of elastic modulus with time

There are very few data on the elastic modulus of shotcrete at early ages. The elastic modulus of shotcrete increases as time passes. The increase of the elastic modulus of shotcrete in relation to the advance of the tunnel face is an important phenomenon. The deformation of tunnels generally ceases when the face distance is twice the diameter of the tunnel, and a great proportion of the deformation takes place during the tunnel face advance. Therefore, information on the variation of properties of shotcrete with time is necessary for assessing tunnel stability.

Figure 5.4 shows the variation of elastic modulus of shotcrete with time. The elastic modulus is determined at a stress level of 1/3 of the strength. For data of specimens having 8% accelerator, the following functions are determined by using the least-square technique.

For averaged data of specimens having 6%, 8% and 10% accelerator, an exponential function is determined by using the least-square technique. The relations and its standard deviations r are given below:

younger than 24 hours:

$$E = 0.1656t^{1.232}; r = 0.996 \tag{5.3}$$

where the unit E is MPa, and t is time with a unit in hours.

older than 24 hours:

$$E = 0.702 + 0.991 \log t; r = 0.990 \tag{5.4}$$

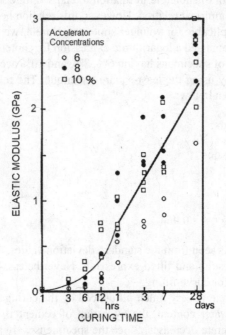

Figure 5.4 Variation of elastic modulus of shotcrete with curing time (modified from Sezaki, 1990).

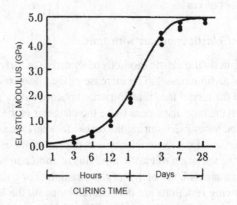

Figure 5.5 Variation of elastic modulus of shotcrete with curing time (from Aydan *et al.*, 1992).

where *t* is time in days. As seen from the standard deviation values, there is a good correlation between experimental results and fitted expression. Nevertheless, it should be noted that the average values used for curve-fitting and the real data has a wide scattering. In addition, Aydan *et al.* (1992) suggested the following function for time-dependent variation of the elastic modulus, which is given in the following form:

$$E = 5(1 - e^{-0.42t}) \tag{5.5}$$

Time in days and unit of the elastic modulus is in GPa. Figure 5.5 compares the experimental results with Equation (5.5). The estimations are in close agreement with experimental results.

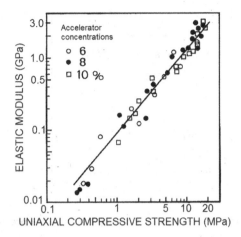

Figure 5.6 Relation between uniaxial compressive strength and elastic modulus of shotcrete (modified from Sezaki, 1990).

(c) The relation between uniaxial strength and elastic modulus

Figure 5.6 shows the relation between the uniaxial strength and elastic modulus. The elastic modulus of concrete is a function of the uniaxial strength in standards such as ACI and CEB/FIP, as follows:

$$E = a\rho^{3/2}\sqrt{\sigma_c} \tag{5.6}$$

where a is constant, ρ is unit weight of concrete.

However, the above relationship does not hold for shotcrete, and the following power function is proposed:

$$E = 0.1\sigma_c^{1.139}, \quad r = 0.990 \tag{5.7}$$

As seen from the figure, this expression fits very well with the experimental values. Instead of the power function, if a linear function is fitted, the following is obtained:

$$E = -0.01 + 0.15\sigma_c, \quad r = 0.951 \tag{5.8}$$

Though the standard deviation is a bit larger, it is still appropriate for engineering purposes.

(d) The relation between the elastic wave velocity and uniaxial strength

Figure 5.7 shows the relation between the elastic wave velocity V_p and uniaxial compressive strength. For the experimental data, two power functions, whose coefficients are determined by using the least square method, are proposed:

$$\sigma_c = 0.415V_p^{2.62}, \quad r = 0.925 \tag{5.9}$$
$$\sigma_c = 0.381V_p^{2.70}, \quad r = 0.924 \tag{5.10}$$

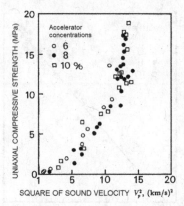

Figure 5.7 Relation between sound velocity and uniaxial compressive strength.

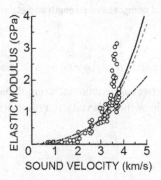

Figure 5.8 Relation between sound velocity and elastic modulus (modified from Sezaki, 1990).

As seen from the figure, the above functions fit well with the experimental data. Furthermore, the value of 2.7 seems to be appropriate for the power of the functions.

(e) The relation between the elastic wave velocity and elastic modulus

Figure 5.8 shows the relation between the elastic wave velocity V_p and elastic modulus. For the experimental data, we propose three power functions, whose coefficients are determined by using the least square method:

$$E = 0.0254V_p^{3.18}, \quad r = 0.826 \tag{5.11}$$

$$E = 0.0304V_p^{3.00}, \quad r = 0.825 \tag{5.12}$$

$$E = 845V_p^{2.00}, \quad r = 0.810 \tag{5.13}$$

As seen from the figure, when the velocity is about 3.5 km/sec or more, the Equation (5.11) fits better to experimental data. Furthermore, the expression with a power of 2, which results from the theory of wave propagation in elastic solids, also holds, but it does not fit the results. It is therefore suggested to use a value of 3.0 for the power of the function.

Figure 5.9 shows the time-dependent variation of Poisson's ratio of shotcrete. Aydan *et al.* (1992) proposed the following relationship for the variation of the Poisson's ratio.

$$E = 0.18 + 0.32e^{-5.6t} \tag{5.14}$$

where *t* in days. As noted from the figure, the experimental results are in good agreement with estimations from Equation (5.14).

5.2.2.4 Triaxial compression experiments

Figure 5.10 shows the strain-stress relations under a confining pressure of 0, 0.5 and 1.0 MPa. As noted from the figure, shotcrete becomes more brittle as the time elapses.

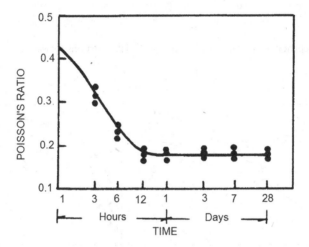

Figure 5.9 Variation of Poisson's ratio of shotcrete with curing time (from Aydan *et al.*, 1992).

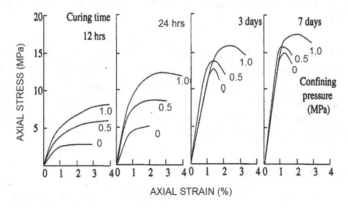

Figure 5.10 Triaxial strain-stress relations of shotcrete at different confining pressures and ages (from Aydan *et al.*, 1992).

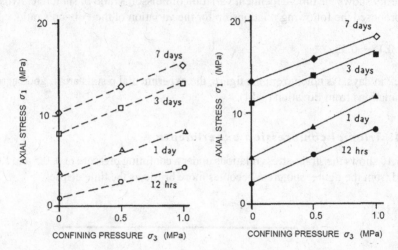

Figure 5.11 Time-dependent variation of initial and peak yield functions for various confining pressures.

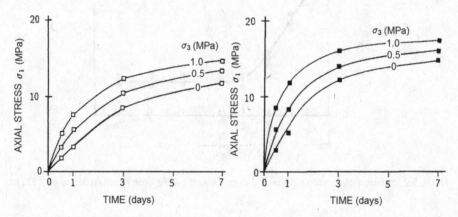

Figure 5.12 The initial and peak yield functions in the space of $\sigma_1 - \sigma_3$ for various ages (0.5, 1, 3, and 7 days).

Figure 5.11 shows the time-dependent variation of initial and peak yield functions for various confining pressures. Figure 5.12 shows the initial and peak yield functions in the space of $\sigma_1 - \sigma_3$ for various ages (0.5, 1, 3, and 7 days). It seems that the relations between σ_1 and σ_3 for the initial and peak yield functions do not depend on the age of shotcrete, and they are almost constant for four different ages. Furthermore, the elasto-plastic hardening of shotcrete does not seem to depend on the hydrostatic pressure or the age of shotcrete.

5.2.3 Constitutive modeling

The mechanical behavior of shotcrete is pseudo-time dependent during its hardening process, and the behavior following the hardening process becomes nearly time independent. Nevertheless, this time-dependent process may be very important, as the deformation of the excavated media mainly occurs during shotcrete's hardening process. Herein, we will propose a pseudo-time-dependent constitutive law based on the classical constitutive law, which utilizes the time-dependent nature of constitutive law parameters.

The incremental pseudo-time-dependent elasto-plastic constitutive law at a fixed time τ can be easily shown to be:

$$d\sigma = \left(\mathbf{D}^e - \frac{\mathbf{D}^e \dfrac{\partial G}{\partial \sigma} \otimes \dfrac{\partial F}{\partial \sigma} \mathbf{D}^e}{h + \dfrac{\partial F}{\partial \sigma} \cdot \left(D^e \dfrac{\partial G}{\partial \sigma} \right)} \right) d\varepsilon \tag{5.15}$$

where $F(\sigma,\kappa) = f(\sigma) - K(\kappa) = 0$: yield function; σ: stress tensor; $K(\kappa)$: hardening function; G: plastic potential; λ: proportionality coefficient; κ: hardening parameter; ε: strain tensor; ε^e: elastic strain tensor; ε^p: plastic strain tensor; $\mathbf{D}^e$: elasticity tensor; and ($\otimes$) and ($\cdot$) denote the tensor and dot products, respectively. It should be noted that the pseudo-time-dependent behavior is based on the concept; that is, the parameters of the elasto-plastic constitutive law can change depending upon the age of shotcrete, and stress-strain responses at loading and unloading at a given time τ are specified by the parameters at that time. Furthermore, the behavior of shotcrete is assumed to be a material exhibiting a non-dilatant and isotropically hardening plastic behavior.

5.2.4   Structural modeling of shotcrete

As discussed in the introduction, the structural resistance of shotcrete for surface structures is almost negligible. However, shotcrete is generally modeled as a thin-walled or thick-walled tube in literature (e.g. Aydan et al., 1992). As shown in Chapter 7, the relationship between the radial displacement of the tube with an outside pressure P_{is} and a zero internal pressure $P_i = 0$ at the adjacent side to the tunnel wall becomes:

$$u = \frac{1+\nu_s}{E_s} \frac{P_{is} a_o^2}{a_o^2 - a_i^2} \left[\frac{a_i^2}{a_o} + (1-2\nu_s)a_o \right] \tag{5.16}$$

The incremental form of the above expression is:

$$\Delta u = \frac{1+\nu_s}{E_s} \frac{\Delta P_{is} a_o^2}{a_o^2 - a_i^2} \left[\frac{a_i^2}{a_o} + (1-2\nu_s)a_o \right] \tag{5.17}$$

Inversely, we have:

$$\Delta P_{is} = K_s \Delta u \tag{5.18}$$

where

$$K_s = \frac{E_s}{1+\nu_s} \frac{a_o^2 - a_i^2}{a_o^2} \cdot \frac{a_o}{a_i^2 + (1-2\nu_s)a_o^2}$$

If the thickness of shotcrete is relatively small compared to the excavation radius, then the above expression can be further simplified to the following form:

$$\Delta P_{is} = K_s \Delta u \qquad (5.19)$$

where

$$K_s = \frac{E_s}{1-\nu_s^2} \frac{t}{a_o^2}$$

The above expression is equivalent to the expression for thin-walled tubes.

As the elastic modulus and Poisson's ratio of shotcrete vary with time, this expression should be integrated over time to yield the internal pressure offered by shotcrete as:

$$P_{is} = \int_{t=0}^{t} \Delta P_{is} dt \qquad (5.20)$$

5.3 CONCRETE LINERS

5.3.1 Historical background

Although the history of cement dates six thousand years before present (Sumer), modern concrete is associated with the development of cement from Portland stone by Joseph Aspdin of England, who is credited with the invention of modern Portland cement. Concrete is a mixture of cement, water, and aggregate (gravel, crushed rock, sand).

Concrete linings of various thicknesses are usually constructed for acquiring dry working conditions in structures such as shafts, tunnels, etc. throughout their service life, as well as for their stability. If such structures are excavated through water-bearing strata, the permeability of the concrete linings becomes important, as it governs the water flow into the openings. Although concrete itself may be regarded practically as an impermeable material from the engineering point of view, water inflow through linings would be mostly due to cracks. The cracks, which are often observed on sites, may result from various causes, such as ground and/or water pressures and thermal stresses developing during the hydration and hardening processes. As these cracks will cause various undesirable problems, such as water pumping, particularly in deep shafts, tunnels below the groundwater table, and even instability, counter measures against the occurrence of those should be undertaken.

Concrete liners can be cast in place or segmental with or without reinforcement. In rock engineering structures, liners are usually employed to resist fluid or gas pressures. They are regarded as a supplementary support member against rock loads, as their installation is usually delayed. They reduce the frictional resistance against water or air flow by smoothing the surface or they keep the excavations dry (e.g. Aydan, 1982). They are usually regarded as a good sealant when they have no cracks due to thermal stresses during the hydration of concrete or to other causes.

5.3.2 Mechanical behavior of concrete

Concrete's mechanical properties have been investigated for many years all over the world. Depending upon the mechanical characteristics of aggregates, cement, water/cement ratio, casting techniques, and admixtures, the uniaxial compressive strength may range from

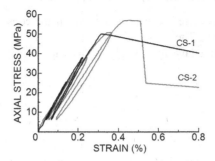

Figure 5.13 Mechanical behavior of concrete under uniaxial compression tests.

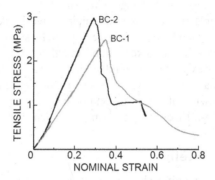

Figure 5.14 Mechanical behavior of concrete in Brazilian tests.

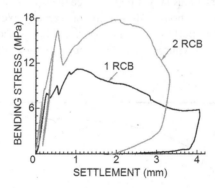

Figure 5.15 Mechanical behavior of reinforced concrete beam in 4-point bending tests.

10 MPa to 200 MPa. The uniaxial compressive strength of concrete commonly used in structures generally ranges between 20 MPa and 50 MPa. Figures 5.13 and 5.14 show the unconfined uniaxial compression, Brazilian tests of unreinforced concrete, while Figure 5.15 shows the mechanical behavior of concrete beam reinforced with one and two deformed bars. The height and diameter of concrete cylindrical samples were 200 mm and 100 mm in uniaxial compression tests. The length and diameter of cylindrical samples for Brazilian tests were 100 mm. The size of the beams was 150 × 150 × 450 mm.

5.3.3   Constitutive modeling of concrete

Mechanical behavior of concrete is generally modeled using constitutive laws by considering one-dimensional models. The constitutive models described for shotcrete are also used for concrete. The constitutive laws may be of the following types:

i)   Elastic
ii)   Visco-elastic
iii) Elasto-plastic
iv)   Elasto-visco-plastic

Nevertheless, the constitutive law of concrete is generally limited to elastic behavior.

The mechanical properties of concrete are generally standardized at the age of 28 days. However, their mechanical properties change with time. Figure 5.16 shows the variation of elastic modulus and Poisson's ratio of concrete utilizing rapid-hardening cement.

Concrete liners in underground excavations (e.g. tunnels, shafts, and large caverns) are generally constructed long after the excavation, and their interaction with surrounding ground is almost none. However, if the ground exhibits time-dependent behavior, the loads may act on the liners. Similarly, the concrete liners may experience water pressure following the construction and/or thawing of frozen ground under some special cases (Aydan, 1982; Aydan and Ersen, 1983, 1984; Ersen, 1983; Ersen and Aydan, 1984).

5.3.4   Structural modeling

When the loads act on the liners uniformly, they are usually designed as thick-walled cylinders. If the expected loads are nonuniform, they have to be reinforced to resist bending stresses. In such cases, segmented liners are usually used to reduce the bending stresses in the liners.

Concrete liners are generally non-reinforced in underground excavations, except for certain locations, such as tunnel portals, fracture zones, and heavy seepage zones. Reinforced concrete may be used for special purposes, such as for retaining walls, foundations of pylons, and suspension bridges.

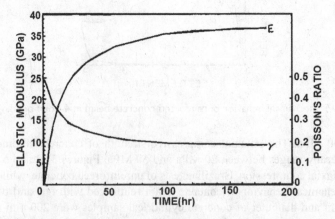

Figure 5.16 Variation of elastic modulus and Poisson's ratio of concrete with time (from Aydan and Ersen, 1984).

5.4   STEEL LINERS AND STEEL RIBS/SETS

5.4.1   Steel liners

Steel liners are generally used as sealants against water inflow-outflow and/or to reduce the frictional resistance to the flow of fluids or gases through the opening in many rock engineering projects. They are hardly used to resist rock pressures, as storing, transporting, and installing the liners is highly expensive. When the steel liners are used as a support member, the loading conditions should be uniform. Otherwise, they may be very weak under a nonuniform stress field and they may buckle under high compressive stress fields, as they are usually thin in relation to the dimensions of the openings. Steel liners are also used as shields for tunnel-boring machines.

5.4.2   Steel ribs/sets

Steel ribs are one of the most conventional support members used in geotechnical engineering structures and have been used for a long time, especially for resisting rock loads. They are generally employed together with wood lagging. However, the recent tendency is to use the ribs together with shotcrete and/or rockbolts in difficult rock conditions.

5.4.3   Constitutive modeling

The constitutive law of steel ribs/sets is fundamentally the same as for the reinforcing bar presented in Chapter 4. Readers should consult Chapter 4 for details.

5.4.4   Structural modeling

Steel ribs are structurally modeled as thin ribs subjected to axial and bending loads. In numerical analyses, they are modeled as beam elements.

Steel liners are structurally modeled as thin tubes (shells) in some simple theoretical analyses. In numerical analyses, they are represented by shell or plate elements.

Steel ribs can be modeled as a one-dimensional circular rib. The radial deformation of the rib can be shown to be:

$$u = \frac{1}{E_{rb}} \frac{P_{is} a_o^2 e_l}{A_{rb}} \tag{5.21}$$

The incremental form of the above expression is:

$$\Delta u = \frac{1}{E_{rb}} \frac{\Delta P_{is} a_o^2 e_l}{A_{rb}} \tag{5.22}$$

Inversely, we have:

$$\Delta P_{is} = K_{rb} \Delta u \tag{5.23}$$

where

$$K_{rb} = \frac{E_{rb} A_{rb}}{a^2 e_l}$$

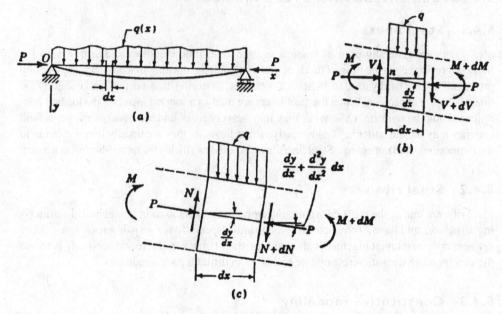

Figure 5.17 Notation.

For nonuniform load condition, the following equations based on the bending of beams are used for the structural modeling of steel ribs (Fig. 5.17):

$$EI\frac{d^4u}{dx^4} + P\frac{d^2u}{dx^2} = q$$

$$EI\frac{d^3u}{dx^3} + P\frac{du}{dx} = -V$$

$$EI\frac{d^2u}{dx^2} = -M \tag{5.24}$$

$$V = \frac{dM}{dx}$$

$$q = -\frac{dV}{dx}$$

where
u = deflection of beam
x = distance
q = intensity of load per unit width
P = axial force
V = shear force
M = moment
E = elastic modulus of beam
I = inertia modulus

The relationship between bending moments, shear forces, and stress and strain components are also given by (note that axial force P is omitted in relationships presented below):

$$\sigma_x = E\varepsilon_x$$

$$\sigma_x = \frac{My}{I}$$

$$\tau_{xy} = \frac{dM}{dx}\frac{1}{bI}\int_0^b\int_{y_1}^{t/2} ydydz \quad or \quad \frac{V}{bI}\int_0^b\int_{y_1}^{t/2} ydydz \qquad (5.25)$$

$$I = \int_0^b\int_{y_1}^{t/2} y^2 dydz$$

where
σ_x = axial stress in x-direction
ε_x = axial strain in x-direction
L = span
t = thickness
b = width
y_1 = distance from the neutral axis of beam

In some structural analyses of steel ribs, even concrete liners are analyzed using moment resistant frames (e.g. Kovari, 1977). In such analyses, steel ribs/concrete liners may be assumed to be supported by the ground using Winkler and Pasternak support. For such modeling, the first equation of Equation (5.26) may be rewritten as follows:

$$EI\frac{d^4u}{dx^4} + P\frac{d^2u}{dx^2} + k_n u - k_G\frac{d^2u}{dx^2} = q \qquad (5.26)$$

where k_n and k_G are Winkler and Pasternak–type ground reaction stiffnesses. The loads resulting from rock mass are applied to steel ribs as distributed loads or point-loads.

Chapter 6

Finite element modeling of reinforcement/support system

6.1 INTRODUCTION

Among all numerical analysis techniques, the finite element method is superior to other numerical methods when the flexibility of mesh generation and size, ease of dealing with material nonlinearity and formulation are taken into account. Some special elements developed particularly for finite element analyses are explained herein.

6.2 MODELING REINFORCEMENT SYSTEMS: ROCKBOLTS

The effect of rockbolts was represented as an equivalent surface load or as equal and opposite point forces located at either end of the rockbolt in earlier numerical analyses (Barla and Cravero, 1972; Zienkiewicz, 1977). Following this, bar elements were used to represent rockbolts in numerical analyses to account for the axial stiffness of bar. Though the bar element was initially used to represent the pre-stressed mechanically anchored rockbolts, it was also used to represent grouted rockbolts by increasing the points of fixation to the surrounding medium. To account for the phenomenon of slippage of the interface between steel bar and the grout, the linkage element proposed by Ngo and Scordelis (1967) for reinforcing bars in concrete started to be employed in numerical analyses.

To represent the reinforcement effect of rockbolts at the discontinuities, Heuze and Goodman (1973) have proposed using a short bar element attached to both sides of the discontinuity. This concept was also used in the representation of rockbolts by Lorig (1985). Nevertheless, determining parameters with a physical meaning in these representations is extremely difficult. The model suggested by John and Van Dillen (1983) was the most realistic and appropriate model among all models suggested previously. This model also had some shortcomings such as the proper modeling of interfaces, which has a great impact on the overall response and resistance of the bolt system.

In this section, a rockbolt element developed by Aydan (Aydan, 1989; Aydan et al., 1986c, 1986d, 1987c, 1987d, 1988a) to fully account for the mechanism involved in the rockbolt system (steel bar in axial and shear loading, interfaces and grout annulus in shear and normal loading) is described in detail. The mechanical modeling of the rockbolt system is presented, and the governing equations for the bar and grout annulus are derived in consideration of responses to the finite element model for the rockbolt system.

6.2.1   Mechanical modeling of steel bar

The steel bar is assumed to be a cylindrical one-dimensional axisymmetric object with a finite volume. The mechanical responses of the bar against applied loads were assumed to consist of an axial response parallel to the longitudinal axis of the bar and a shear response perpendicular to that axis. Let us consider an elementary unit slice in a local axisymmetric coordinate system $r\theta z$; as shown in Figure 6.1.

Equilibrium equation for axial loading: The force equilibrium condition for the elementary unit slice is written as:

$$(\sigma_{zz}^{b} + \Delta\sigma_{zz}^{b})A_{b} - \sigma_{zz}^{b}A_{b} = 0 \tag{6.1}$$

where
σ_{zz}^{b} = axial stress in bolt
$A_{b} = \pi r_{b}^{2}$
r_{b} = radius of bar

Rearranging the above expression, dividing by $\Delta z A_{b}$, and taking the limit give the equilibrium equation for the axial direction of the bar as:

$$\lim_{\Delta z \to 0} \frac{\Delta\sigma_{zz}^{b}}{\Delta z} = 0 \Rightarrow \frac{d\sigma_{zz}^{b}}{dz} = 0 \tag{6.2}$$

Equilibrium equation for shear loading: Similarly, the force equilibrium condition for the elementary unit slice yields the following for shear loading:

$$(\tau_{zr}^{b} + \Delta\tau_{zr}^{b})A_{b} - \tau_{zr}^{b}A_{b} = 0 \tag{6.3}$$

where τ_{zr}^{b} is the shear stress in the bolt.

Rearranging the above expression, dividing by $\Delta z A_{b}$, and taking the limit give the equilibrium equation for the radial direction of the bar as:

$$\lim_{\Delta z \to 0} \frac{\Delta\tau_{zr}^{b}}{\Delta z} = 0 \Rightarrow \frac{d\tau_{zr}^{b}}{dz} = 0 \tag{6.4}$$

6.2.2 Mechanical modeling of grout annulus

The grout annulus, together with two interfaces between bolt and grout, and grout and rock, is an important element for the stress transfer between the steel bar and the surrounding medium. This transfer is made mainly through the shear response of the grout annulus. The transverse response of the grout annulus is also an important factor in evaluating the form of failure somewhere within the grout annulus and the dilatancy, which may arise during the debonding process. In the following, mechanical models are suggested for the above responses.

The grout annulus is also assumed to be a cylindrical axisymmetric object. The mechanical responses of the annulus against applied loads were assumed to consist of a shear response parallel to the longitudinal axis of the bar and a normal response perpendicular to that axis. Let us consider an elementary unit slice of the annulus in a local axisymmetric coordinate system $(r\theta z)$, as shown in Figure 6.1.

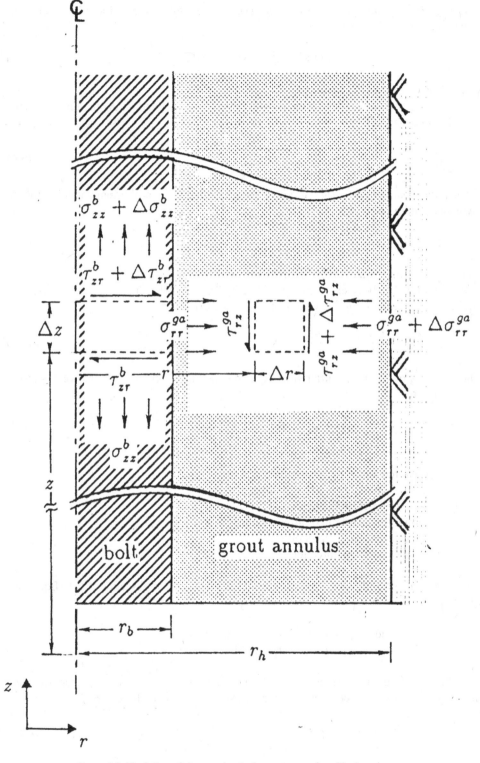

Figure 6.1 Modeling of the mechanical responses of rockbolt system.

Equilibrium equation for shear loading: The force equilibrium condition for the elementary unit slice is written as:

$$(\tau_{rz}^{ga} + \Delta\tau_{rz}^{ga})2\pi\Delta z(r+\Delta r)\Delta\theta - \tau_{rz}^{ga}2\pi\Delta zr\Delta\theta = 0 \tag{6.5}$$

where τ_{rz} is shear stress in annulus.

Rearranging the above expression, dividing by $2\pi\Delta zr\Delta r\Delta\theta$, and taking the limit gives the equilibrium equation for the axial direction of the bar as:

$$\lim_{\Delta r \to 0}\frac{\Delta\tau_{rz}^{ga}}{\Delta r} + \frac{\tau_{rz}^{ga}}{r} = 0 \Rightarrow \frac{d\tau_{rz}^{ga}}{dr} + \frac{\tau_{rz}^{ga}}{r} = 0 \tag{6.6}$$

Equilibrium equation for normal loading: Similarly, the force equilibrium condition for the elementary unit slice is written as:

$$(\sigma_{rr}^{ga} + \Delta\sigma_{rr}^{ga})2\pi\Delta z(r+\Delta r)\Delta\theta - \sigma_{rr}^{ga}2\pi\Delta zr\Delta\theta = 0 \tag{6.7}$$

where σ_{rr} is normal (radial) stress in grout annulus.

Rearranging the above expression, dividing by $2\pi\Delta zr\Delta r\Delta\theta$, and taking the limit give the equilibrium equation for the direction normal to the bar axis as:

$$\lim_{\Delta r \to 0}\frac{\Delta\sigma_{rz}^{ga}}{\Delta r} + \frac{\sigma_{rz}^{ga}}{r} = 0 \Rightarrow \frac{d\sigma_{rr}^{ga}}{dr} + \frac{\sigma_{rr}^{ga}}{r} = 0 \tag{6.8}$$

The solutions of the above differential equations for linear relationships between stresses and strains are:

$$\tau_{rz} = G_{ga}\gamma_{ga} = G_{ga}\frac{dw^{ax}}{dr} \qquad \sigma_{rz} = E_{ga}\varepsilon_{ga} = E_{ga}\frac{du^{ax}}{dr} \tag{6.9}$$

where
G_{ga} = shear modulus of grout annulus
E_{ga} = Young's modulus of grout annulus
γ_{ga} = shear strain of grout annulus
ε_{ga} = normal (radial) strain of grout annulus
w^{ax} = displacement in z-direction
u^{ax} = displacement in r-direction

Together with the following boundary conditions (see also Subsection 4.6.2 (Equations (4.60)–(4.64)):

$w^{ax} = -w_b^{ax}$ and $u^{ax} = -u_b^{ax}$ at $r = r_b$; r_b: radius of bar
$w^{ax} = -w_h^{ax}$ and $u^{ax} = -u_h^{ax}$ at $r = r_h$; r_h: radius of hole

yield the following relationships for shear strain and normal strain in terms of displacements at bolt-grout and grout-rock interfaces and dimensions of bar and borehole:

$$\gamma_{ga}^a = A_{sr}\frac{(w_b^{ax} - w_h^{ax})}{r} = A_{sr}\frac{\Delta w^{ax}}{r} \qquad \varepsilon_{ga}^r = A_{nr}\frac{(u_b^{ax} - u_h^{ax})}{r} = A_{nr}\frac{\Delta u^{ax}}{r} \tag{6.10}$$

where

$A_{sr} = 1/\ln(r_h/r_b)$

$A_{nr} = 1/\ln(r_h/r_b)$

u_b^{ax} = radial displacement at bolt-grout interface

u_h^{ax} = radial displacement at rock-grout interface

w_b^{ax} = axial displacement at bolt-grout interface

w_h^{ax} = axial displacement at rock-grout interface

Δw^{ax} = relative axial displacement of grout annulus

Δw^{ax} = relative normal (radial) displacement of grout annulus

These solutions are utilized in the finite element model derivations for the rockbolt system, regarding the definitions of shear strain and normal strain of grout annulus in terms of relative displacement between bolt-grout interface and grout-rock interface.

6.2.3   Finite element formulation of rockbolt element

In this section, an explicit finite element modeling of the bolt system is described based on the rockbolt element of Aydan (1989). The element is of a coupled form of a bar element and interface elements of Ghaboussi-type. The presented description is a general representation of rockbolts, which can be simplified depending upon the required effects of rockbolts and the reduction of cost and labor of computations. The element is assumed to be consisting of eight nodes, two of which are attached to the steel bar and the rest to the surrounding rock mass in the three-dimensional case (Fig. 6.2).

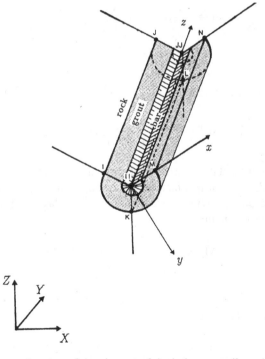

Figure 6.2 Perspective view of the element of the bolt-system (from Aydan, 1989).

Let us take a body in which rockbolts together with their grout annuli are embedded. Provided that the stress field σ is statically admissible and the displacement field $\mathbf{u}$ is kinematically admissible for this body with its traction and displacement boundaries, the virtual work equations for bar and grout annulus of a rockbolt system in the body can be written in the following forms, respectively, by taking variations on the displacement $\delta\mathbf{u}$ in the bar and on the relative displacement $\delta\Delta\mathbf{u}$ in the grout annulus:

bar:

$$\int_a^b \int_0^{2\pi} \int_0^{r_b} \left\{ \delta u^{ax} \frac{d\tau_{zr}^b}{dz} + \delta w^{ax} \frac{d\sigma_{zz}^b}{dz} \right\} r\,dr\,d\theta\,dz \tag{6.11}$$

grout annulus:

$$\int_a^b \int_0^{2\pi} \int_{r_b}^{r_h} \left\{ \delta(\Delta u^{ax}) \frac{1}{r} \frac{d}{dr}(r\sigma_{rr}^{ga}) + \delta(\Delta w^{ax}) \frac{1}{r} \frac{d}{dr}(r\tau_{rz}^{ga}) \right\} r\,dr\,d\theta\,dz \tag{6.12}$$

Where u^{ax} and w^{ax} are displacement components associated with r − direction and z − direction respectively.[1] Integrating the above expressions by parts yields the following expressions:

bar:

$$\left[\int_0^{2\pi} \int_0^{r_b} \left(\delta u^{ax} \tau_{zr}^b + \delta w^{ax} \sigma_{zz}^b \right) r\,dr\,d\theta \right]_{z=a}^{z=b}$$

$$- \int_a^b \int_0^{2\pi} \int_0^{r_b} \left\{ \frac{d\delta u^{ax}}{dz} \tau_{zr}^b + \frac{d\delta w^{ax}}{dz} \sigma_{zz}^b \right\} r\,dr\,d\theta\,dz \tag{6.13}$$

grout annulus:

$$\left[\int_a^b \int_0^{2\pi} \left(\delta(\Delta u^{ax})\sigma_{rr}^{ga} + \delta(\Delta w^{ax})\tau_{rz}^{ga} \right) d\theta\,dz \quad r \right]_{r=r_b}^{r=r_h}$$

$$- \int_a^b \int_0^{2\pi} \int_{r_b}^{r_h} \left\{ \frac{1}{r} \frac{d}{dr}(r\delta(\Delta u^{ax}))\sigma_{rr}^{ga} + \frac{1}{r} \frac{d}{dr}(r\delta(\Delta w^{ax}))\tau_{rz}^{ga} \right\} r\,dr\,d\theta\,dz \tag{6.14}$$

The displacements in bar and the relative displacements of grout annulus are dependent of z only. With the consideration of this fact and choosing the linear type interpolation (shape) functions, let us take the following approximations to displacement $\mathbf{u}$ of bar and relative displacement $\Delta\mathbf{u}$ of grout annulus:

$$\mathbf{u} = N(x)U_b^{ax} \quad \Delta\mathbf{u} = N(x)\Delta U_{ga}^{ax} \tag{6.15}$$

where

$$\mathbf{u} = \{u^{ax}, w^{ax}\}^T; \Delta\mathbf{u} = \{\Delta u^{ax}, \Delta w^{ax}\}^T$$

---

1 Superscript ax is used to denote the displacement components in the axisymmetric state.

$$\mathbf{N} = \begin{bmatrix} N_{II} & 0 & N_{JJ} & 0 \\ 0 & N_{II} & 0 & N_{JJ} \end{bmatrix}$$

$$N_{II} = \frac{z_{JJ} - z}{L}, \quad N_{JJ} = \frac{z - z_{II}}{L}, \quad L = z_{JJ} - z_{II}$$

$$\mathbf{U}_b^{ax} = \left\{ U_{II}^{ax}, W_{II}^{ax}, U_{JJ}^{ax}, W_{JJ}^{ax} \right\}^T; \quad \Delta\mathbf{U}_{ga}^{ax} = \left\{ \Delta U_{II}^{ax}, \Delta W_{II}^{ax}, \Delta U_{JJ}^{ax}, \Delta W_{JJ}^{ax} \right\}^T$$

The relative displacement of grout annulus in terms of displacements of nodes (II, JJ), attached to bar and averaged displacements of nodes (I, J, K, L, M, N) attached to borehole are explicitly defined as:

$$\Delta U_{II}^{ax} = U_{II}^{ax} - U_{IKM}^{ax}, \quad \Delta W_{II}^{ax} = W_{II}^{ax} - W_{IKM}^{ax},$$
$$\Delta U_{JJ}^{ax} = U_{JJ}^{ax} - U_{JLN}^{ax}, \quad \Delta W_{JJ}^{ax} = W_{JJ}^{ax} - W_{JLN}^{ax}.$$

The displacements of nodes attached to borehole in respective directions is averaged as given below to define the displacement field of the borehole:

$$U_{IKM}^{ax} = \frac{1}{3}(U_I^{ax} + U_K^{ax} + U_M^{ax}), \quad W_{IKM}^{ax} = \frac{1}{3}(W_I^{ax} + W_K^{ax} + W_M^{ax}),$$

$$U_{JLN}^{ax} = \frac{1}{3}(U_J^{ax} + U_L^{ax} + U_N^{ax}), \quad W_{JLN}^{ax} = \frac{1}{3}(W_J^{ax} + W_L^{ax} + W_N^{ax}).$$

In averaging displacements of nodes, the displacements of three nodes attached to the borehole about each node of the bar are employed. To define the true displacement field of the borehole, more nodes are necessary. Nevertheless, the present *three nodes approach* has been found to be satisfactory in numerical analysis.

The relationships for strains of bar in terms of nodal displacements are of the following form:

bar:

$$\varepsilon^b = \mathbf{B}_b \mathbf{U}_b^{ax}$$

or explicitly:

$$\begin{Bmatrix} \gamma_{zr}^b = du^r / dz \\ \varepsilon_{zz}^b = dw / dz \end{Bmatrix} = \frac{1}{L} \begin{bmatrix} -1 & 0 & 1 & 0 \\ 0 & -1 & 0 & 1 \end{bmatrix} \begin{Bmatrix} U_{II}^{ax} \\ W_{II}^{ax} \\ U_{JJ}^{ax} \\ W_{JJ}^{ax} \end{Bmatrix} \qquad (6.16)$$

The relationships for strains of grout annulus in terms of nodal displacements are of the following form with the use of relationship (6.10):

grout annulus:

$$\varepsilon^{ga} = \mathbf{B}_{ga} \Delta\mathbf{U}_{ga}^{ax}$$

or explicitly:

$$\begin{Bmatrix} \varepsilon_{rr}^{ga} = A_{nr}\Delta u^{ax}/r \\ \gamma_{rz}^{ga} = A_{sr}\Delta w^{ax}/r \end{Bmatrix} = \frac{1}{r\ln(r_h/r_b)}\begin{bmatrix} N_{II} & 0 & N_{JJ} & 0 \\ 0 & N_{II} & 0 & N_{JJ} \end{bmatrix}\begin{Bmatrix} \Delta U_{II}^{ax} \\ \Delta W_{II}^{ax} \\ \Delta U_{JJ}^{ax} \\ \Delta W_{JJ}^{ax} \end{Bmatrix} \tag{6.17}$$

The use of theoretical expressions given by Equation (6.10) saves one from the necessity of discretization for radial direction of the rockbolt system.

The constitutive law for a linear elastic behavior of bar and grout annulus may be written as:

$$\boldsymbol{\sigma}^b = \mathbf{D}^b\boldsymbol{\varepsilon}^b, \boldsymbol{\sigma}^{ga} = \mathbf{D}^{ga}\boldsymbol{\varepsilon}^{ga}$$

or explicitly:

$$\begin{Bmatrix} \tau_{zr}^b \\ \sigma_{zz}^b \end{Bmatrix} = \begin{bmatrix} G_b & 0 \\ 0 & E_b \end{bmatrix}\begin{Bmatrix} \gamma_{zr}^b \\ \varepsilon_{zz}^b \end{Bmatrix} \quad \begin{Bmatrix} \sigma_{rr}^{ga} \\ \tau_{rz}^{ga} \end{Bmatrix} = \begin{bmatrix} E_{ga} & 0 \\ 0 & G_{ga} \end{bmatrix}\begin{Bmatrix} \varepsilon_{rr}^{ga} \\ \gamma_{rz}^b \end{Bmatrix} \tag{6.18}$$

where
E_b = Young's modulus of bar
G_b = shear modulus of bar
E_{ga} = Young's modulus of grout annulus
G_{ga} = shear modulus of grout annulus

Finally, the finite element discretization of Equations (6.13) and (6.14), together with relationships given above for an element numbered e, yield the stiffness matrices for bar and grout annulus in the following forms:

bar:

$$\mathbf{K}_b^e = \int_{z_{II}}^{z_{JJ}}\int_0^{2\pi}\int_0^{r_b}\mathbf{B}_b^T\mathbf{D}^b\mathbf{B}_b\, rdrd\theta dz \tag{6.19}$$

grout annulus:

$$\mathbf{K}_{ga} = \int_{z_{II}}^{z_{JJ}}\int_0^{2\pi}\int_{r_b}^{r_h}\mathbf{B}_{ga}^T\mathbf{D}^{ga}\mathbf{B}_{ga}\, rdrd\theta dz \tag{6.20}$$

Carrying out the integration yields the stiffness matrices $\mathbf{K}_b^e$ and $\mathbf{K}_{ga}^e$ for bar and grout annulus of the bolt element explicitly as:

bar

$$\mathbf{K}_b^e = \begin{bmatrix} K_b^r & 0 & -K_b^r & 0 \\ 0 & K_b^z & 0 & -K_b^z \\ -K_b^r & 0 & K_b^r & 0 \\ 0 & -K_b^z & 0 & K_b^z \end{bmatrix} \tag{6.21}$$

where

$$K_b^r = \frac{G_b A}{L} \quad K_b^z = \frac{E_b A}{L} \quad A = \pi r_b^2, \quad L = z_{JJ} - z_{II}.$$

grout annulus:

$$\mathbf{K}_{ga} = \begin{bmatrix} 2K_{ga}^r & 0 & K_{ga}^r & 0 \\ 0 & 2K_{ga}^z & 0 & K_{ga}^z \\ K_{ga}^r & 0 & 2K_{ga}^r & 0 \\ 0 & K_{ga}^z & 0 & 2K_{ga}^z \end{bmatrix} \tag{6.22}$$

where

$$K_{ga}^r = \pi E_{ga} \frac{L}{3 \ln(r_h / r_b)} \quad K_{ga}^z = \pi G_{ga} \frac{L}{3 \ln(r_h / r_b)}$$

As finite element analyses are generally carried out using a Cartesian coordinate system, the above stiffness matrices for such analysis should be converted to its equivalent Cartesian representations. For this purpose, let us choose a local Cartesian coordinate system (xyz) whose z-axis coincides with the z-axis of the polar coordinate system $(r\theta z)$. The following relationships should hold among the displacement components $\{u^{ax}, w^{ax}\}$ in axisymmetric representation and the components $\{u, v, w\}$ in Cartesian representation, since the rockbolt system under consideration is assumed to be an axisymmetric body:

$$\begin{aligned} u^{ax} &= u, & u^{ax} &= v, & w^{ax} &= w, \\ \Delta u^{ax} &= \Delta u, & \Delta u^{ax} &= \Delta v, & \Delta w^{ax} &= \Delta w. \end{aligned} \tag{6.23}$$

By virtue of the above relationship (6.23), the stiffness matrices derived above can be transformed to their equivalent Cartesian representations as given below:

bar:

$$\mathbf{K}_b^e = \begin{bmatrix} K_b^r & 0 & 0 & -K_b^r & 0 & 0 \\ 0 & K_b^r & 0 & 0 & -K_b^r & 0 \\ 0 & 0 & K_b^z & 0 & 0 & -K_b^z \\ -K_b^r & 0 & 0 & K_b^r & 0 & 0 \\ 0 & -K_b^r & 0 & 0 & K_b^r & 0 \\ 0 & 0 & -K_b^z & 0 & 0 & K_b^z \end{bmatrix} \tag{6.24}$$

grout annulus:

$$\mathbf{K}_{ga} = \begin{bmatrix} 2K_{ga}^r & 0 & 0 & K_{ga}^r & 0 & 0 \\ 0 & 2K_{ga}^r & 0 & 0 & K_{ga}^r & 0 \\ 0 & 0 & 2K_{ga}^z & 0 & 0 & K_{ga}^z \\ K_{ga}^r & 0 & 0 & 2K_{ga}^r & 0 & 0 \\ 0 & K_{ga}^r & 0 & 0 & 2K_{ga}^r & 0 \\ 0 & 0 & K_{ga}^z & 0 & 0 & 2K_{ga}^z \end{bmatrix} \tag{6.25}$$

The above stiffness matrices will be now converted to obtain the stiffness matrix of the bolt-element in terms of nodal displacement of the element in local coordinates. The relationship among nodal bolt displacements, annulus-relative displacements, and nodal displacement of the element can be written as:

$$\mathbf{U}_{u_b, \Delta u_{ga}} = \mathbf{A}\mathbf{U}_{nodal}$$

or explicitly:

$$\begin{Bmatrix} \mathbf{U}_{II} \\ \mathbf{U}_{JJ} \\ \Delta\mathbf{U}_{II} \\ \Delta\mathbf{U}_{JJ} \end{Bmatrix} = \begin{bmatrix} [0] & [0] & [0] & [0] & [0] & [0] & [T]_1 & [0] \\ [0] & [0] & [0] & [0] & [0] & [0] & [0] & [T]_1 \\ [T]_{1/3} & [0] & [T]_{1/3} & [0] & [T]_{1/3} & [0] & [T]_1 & [0] \\ [0] & [T]_{1/3} & [0] & [T]_{1/3} & [0] & [T]_{1/3} & [0] & [T]_1 \end{bmatrix} \begin{Bmatrix} \mathbf{U}_I \\ \mathbf{U}_J \\ \mathbf{U}_K \\ \mathbf{U}_L \\ \mathbf{U}_M \\ \mathbf{U}_N \\ \mathbf{U}_{II} \\ \mathbf{U}_{JJ} \end{Bmatrix} \tag{6.26}$$

where

$$[0] = \begin{bmatrix} 0 & 0 & 0 \\ 0 & 0 & 0 \\ 0 & 0 & 0 \end{bmatrix}, \quad [T]_{1/3} = -\frac{1}{3}\begin{bmatrix} 1 & 0 & 0 \\ 0 & 1 & 0 \\ 0 & 0 & 1 \end{bmatrix}, \quad [T]_1 = \begin{bmatrix} 1 & 0 & 0 \\ 0 & 1 & 0 \\ 0 & 0 & 1 \end{bmatrix},$$

$$\mathbf{U}_I = \{U_I, V_I, W_I\}^T \quad \mathbf{U}_J = \{U_J, V_J, W_J\}^T \quad \mathbf{U}_K = \{U_K, V_K, W_K\}^T$$
$$\mathbf{U}_L = \{U_L, V_L, W_L\}^T \quad \mathbf{U}_M = \{U_M, V_M, W_M\}^T \quad \mathbf{U}_N = \{U_N, V_N, W_N\}^T$$
$$\mathbf{U}_{II} = \{U_{II}, V_{II}, W_{II}\}^T \quad \mathbf{U}_{JJ} = \{U_{JJ}, V_{JJ}, W_{JJ}\}^T$$

With the use of the above relationships and the transformation law given by:

$$\mathbf{K}_{local}^e = \mathbf{A}^T \mathbf{K}_{b,ga}^e \mathbf{A} \tag{6.27}$$

where

$$\mathbf{K}^e_{b,ga} = \begin{bmatrix} [K]_b & [0] \\ [0] & [K]_{ga} \end{bmatrix},$$

The stiffness matrix of the bolt element in terms of the nodal displacements of the element takes the following form:

$$\mathbf{K}^e_{local} = \begin{bmatrix} \dfrac{1}{9}[K]_{ga} & \dfrac{1}{9}[K]_{ga} & \dfrac{1}{9}[K]_{ga} & -\dfrac{1}{3}[K]_{ga} \\[2mm] \dfrac{1}{9}[K]_{ga} & \dfrac{1}{9}[K]_{ga} & \dfrac{1}{9}[K]_{ga} & -\dfrac{1}{3}[K]_{ga} \\[2mm] \dfrac{1}{9}[K]_{ga} & \dfrac{1}{9}[K]_{ga} & \dfrac{1}{9}[K]_{ga} & -\dfrac{1}{3}[K]_{ga} \\[2mm] -\dfrac{1}{3}[K]_{ga} & -\dfrac{1}{3}[K]_{ga} & -\dfrac{1}{3}[K]_{ga} & [K]_{ga}+[K]_b \end{bmatrix}.$$

The stiffness matrix in local coordinates can be transformed to the global stiffness matrix by using the following relationship:

$$\mathbf{K}^e_{global} = \mathbf{T}^T \mathbf{K}^e_{local} \mathbf{T} \tag{6.28}$$

where

$$\mathbf{T} = \begin{bmatrix} [T]^* & [0] & [0] & [0] & [0] & [0] & [0] & [0] \\ [0] & [T]^* & [0] & [0] & [0] & [0] & [0] & [0] \\ [0] & [0] & [T]^* & [0] & [0] & [0] & [0] & [0] \\ [0] & [0] & [0] & [T]^* & [0] & [0] & [0] & [0] \\ [0] & [0] & [0] & [0] & [T]^* & [0] & [0] & [0] \\ [0] & [0] & [0] & [0] & [0] & [T]^* & [0] & [0] \\ [0] & [0] & [0] & [0] & [0] & [0] & [T]^* & [0] \\ [0] & [0] & [0] & [0] & [0] & [0] & [0] & [T]^* \end{bmatrix}$$

$$[0] = \begin{bmatrix} 0 & 0 & 0 \\ 0 & 0 & 0 \\ 0 & 0 & 0 \end{bmatrix}, \quad [T]^* = \begin{bmatrix} l_1 & m_1 & n_1 \\ l_2 & m_2 & n_2 \\ l_3 & m_3 & n_3 \end{bmatrix}$$

The relationships for direction cosines l_i, m_i, n_i; $i = 1,2,3$ between the local coordinate system (xyz) and global coordinate system (XYZ), are specified by using the spherical coordinates (r, θ, ψ) as:

$$\begin{aligned} l_1 &= \cos\theta\cos\psi, & m_1 &= \sin\theta\cos\psi, & n_1 &= -\sin\psi, \\ l_2 &= -\sin\theta, & m_2 &= \cos\theta, & n_2 &= 0, \\ l_3 &= \cos\theta\sin\psi, & m_3 &= \sin\theta\sin\psi, & n_3 &= \cos\psi. \end{aligned}$$

θ is the angle of the trace of the position vector $\mathbf{r}$ of bolt element on Z-plane, measured from axis X of the global coordinate system, and ψ is the angle between the position vector $\mathbf{r}$ of the bolt element and axis Z of the global coordinate system. Determination of these angles from the coordinates $(X_i^e, Y_i^e, Z_i^e; i = II, JJ)$ of nodes of the bolt element in global system are as follows:

$$\theta = \tan^{-1}\frac{\Delta Y}{\Delta X} \quad \psi = \cos^{-1}\frac{\Delta Z}{\|\mathbf{r}\|} \tag{6.29}$$

where

$$\|\mathbf{r}\| = \sqrt{\Delta X^2 + \Delta Y^2 + \Delta Z^2}$$
$$\Delta X = X_{JJ} - X_{II} \quad \Delta Y = Y_{JJ} - Y_{II} \quad \Delta Z = Z_{JJ} - Z_{II}$$

Although the 3D representation is desirable for rockbolts/rockanchors, in the 2D case, the node number can be reduced to six. In some extreme cases, the node number may be reduced to four nodes. If the interactions between rockbolts and surrounding ground are neglected, the node number can be reduced to two, which would correspond to the bar element.

6.3   FINITE ELEMENT MODELING OF SHOTCRETE

Shotcrete in numerical simulations can be modeled in various ways. As the thickness of shotcrete is small, it can be modeled as shell elements. Aydan and Kawamoto (1991) and Aydan *et al.* (1992) described a two-noded element for shotcrete linings and named as shotcrete element. The details of this element are as follows. Let us now consider a two-noded element (I, J) in a two-dimensional space and take two coordinate systems (oxy) and $(o'x'y')$ as shown in Figure 6.3. As shotcrete linings are generally thin, it may be reasonable to assume that the variation of displacement v' and u' along the thickness y' of shotcrete lining in the local coordinate system $(o'x'y')$ is negligible. By the virtue of this assumption, the strain component $\varepsilon_{y'y'}$ vanishes and the remaining strain components take the following form:

$$\varepsilon_{x'x'} = \frac{\partial u'}{\partial x'}, \quad \gamma_{x'y'} = \frac{\partial v'}{\partial x'} \tag{6.30}$$

Let us assume that the shape functions are linear such that:

$$N_I = \frac{1}{2}(1-\xi), \quad N_J = \frac{1}{2}(1+\xi) \tag{6.31}$$

where $\xi = (2x' + x_I' + x_J')/L$, $L = (x_J' - x_I')$. Then, the relationship between the strains and nodal displacements becomes

$$\begin{Bmatrix} \varepsilon_{x'x'} \\ \gamma_{x'y'} \end{Bmatrix} = \frac{1}{L}\begin{bmatrix} -1 & 0 & 1 & 0 \\ 0 & -1 & 0 & 1 \end{bmatrix}\begin{Bmatrix} U_I' \\ V_I' \\ U_J' \\ V_J' \end{Bmatrix} \quad or \quad \boldsymbol{\varepsilon}' = \mathbf{B}\mathbf{U}' \tag{6.32}$$

Thus, the stiffness matrix in the local coordinate system is obtained in an explicit form by using the above relations as given below:

$$K' = \begin{vmatrix} K_a' & 0 & -K_a' & 0 \\ 0 & K_s' & 0 & -K_s' \\ -K_a' & 0 & K_a' & 0 \\ 0 & -K_s' & 0 & K_s' \end{vmatrix} \tag{6.33}$$

where

$$K_a' = \frac{E}{1-v^2} \cdot \frac{t_{y'}t_{z'}}{x_J' - x_I'} \quad K_s' = \frac{E}{2(1+v)} \cdot \frac{t_{y'}t_{z'}}{x_J' - x_I'}$$

The stiffness matrix in the local coordinate system is then transformed to the stiffness matrix in the global coordinate system by the following relationship:

$$\mathbf{K} = \mathbf{T}^T \mathbf{K}' \mathbf{T} \tag{6.34}$$

where

$$\mathbf{T} = \begin{vmatrix} c & s & 0 & 0 \\ -s & c & 0 & 0 \\ 0 & 0 & c & s \\ 0 & 0 & -s & c \end{vmatrix}, \quad \theta = \tan^{-1}\left(\frac{y_J - y_I}{x_J - x_I}\right); c = \cos\theta; s = \sin\theta$$

This shotcrete element can be easily extended to 3D analyses by utilizing a plane element through the addition of stiffness components with the consideration of plane-stress condition.

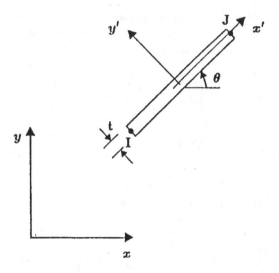

Figure 6.3 Shotcrete element. (from Aydan *et al.*, 1992)

6.4   FINITE ELEMENT MODELING OF STEEL RIBS/SETS OR SHIELDS

The element described above can be also used to simulate steel ribs/sets in finite element analysis by replacing with the appropriate stiffness properties. This procedure is adopted by Aydan and Kawamoto (1991) and Aydan et al. (1992). However, the bending resistance resulting from their 3D geometry may be counted through the introduction of bending element. The finite element form of governing equation for the bending of steel ribs/sets may be given in the following form in a local coordinate system $(o'x'y'z')$ attached to the a representative element having two nodes (I, J) as

$$\int_{x_i'}^{x_j'} \delta w' \frac{d^2}{dx'^2}\left(EI\frac{d^2 w'}{dx'^2}\right)dx' + \int_{x_i'}^{x_j'} \delta w' P\frac{d^2 w'}{dx'^2}dx' = \int_{x_i'}^{x_j'} \delta w' q(x)dx' \qquad (6.35)$$

Replacing the integrand of the first component in Equation (6.35) by its equivalence with the use of derivation by parts yields the following:

$$\int_{x_i'}^{x_j'} \frac{d^2\delta w'}{dx'^2}\left(EI\frac{d^2 w'}{dx'^2}\right)dx' = -\int_{x_i'}^{x_j'} \delta w' q(x)dx' + \delta w' \frac{d}{dx'}\left(EI\frac{d^2 w'}{dx'^2}\right)\Bigg|_{x_i'}^{x_j'} \qquad (6.36)$$

The interpolation function of flexure w' is generally assumed to be of the following form in the local coordinate system attached to nodes (i, j):

$$w' = \beta_1 + \beta_2\xi + \beta_3\xi^2 + \beta_4\xi^3 \qquad (6.37)$$

where

$$\xi = \frac{x' - x_i'}{x_j' - x_i'}$$

Coefficients β_k can be evaluated using the knowing values of nodal flexure and flexure angle at nodes (i, j). Thus, flexure w' can be represented in terms of nodal flexure and flexure angle at nodes (i, j) in the following form:

$$w' = [N_i \quad N*_i \quad N_j \quad N*_j]\begin{Bmatrix} w_i' \\ \theta_i' \\ w_j' \\ \theta_j' \end{Bmatrix} \quad \text{or} \quad w' = [N]\{W\} \qquad (6.38)$$

where

$$N_i = 1 - 3\left(\frac{\xi}{L}\right)^2 + 2\left(\frac{\xi}{L}\right)^3 ; N_j = 3\left(\frac{\xi}{L}\right)^2 - 2\left(\frac{\xi}{L}\right)^3 \qquad (6.39a)$$

$$N*_i = L\left[\left(\frac{\xi}{L}\right)2\left(\frac{\xi}{L}\right)^2 + 2\left(\frac{\xi}{L}\right)^3\right] ; N*_j = L\left[-\left(\frac{\xi}{L}\right)^2 + \left(\frac{\xi}{L}\right)^3\right] \qquad (6.39b)$$

Introducing Equation (6.38) into Equation (6.36) yields the following relationship, element-wise:

$$\int_{x_i'}^{x_j'} EI[B]^T [B]dx'\{W'\} = -\int_{x_i'}^{x_j'} [N]^T q(x)dx' + [\bar{N}]^T Q\Big|_{x_i'}^{x_j'} \quad \text{or} \quad [K']_b\{W'\} = \{F'\} \qquad (6.40)$$

Matrix $[B]$ is given in the following form:

$$\cdot \quad [B] = [B_i \quad B^*_i \quad B_j \quad B^*_j] \qquad (6.41)$$

where

$$B_i = \frac{1}{L^2}[6-12\xi]; B^*_i = \frac{1}{L}(4-6\xi); N_j = \frac{1}{L^2}[-6+12\xi]; B^*_i = \frac{1}{L}(2-6\xi) \qquad (6.42)$$

Specifically $[K']_b$ is obtained as follows:

$$[K']_b = \frac{EI}{L^3} \begin{bmatrix} 12 & 6L & -12 & 6L \\ 6L & 4L^2 & -6L & 2L^2 \\ -12 & -6L & 12 & -6L \\ 6L & 2L^2 & -6L & 4L^2 \end{bmatrix} \quad \text{or} \quad [K']_b = \begin{bmatrix} K_{11}^b & K_{12}^b & K_{13}^b & K_{14}^b \\ & K_{22}^b & K_{23}^b & K_{24}^b \\ & & K_{33}^b & K_{34}^b \\ sym & & & K_{44}^b \end{bmatrix} \qquad (6.43)$$

If the axial stiffness of the steel ribs/sets is taken into account, the local stiffness matrix for axial response is given in the following form:

$$[K']_a = \frac{EA}{L} \begin{bmatrix} 1 & -1 \\ -1 & 1 \end{bmatrix} \quad \text{or} \quad [K']_a = \begin{bmatrix} K_{11}^a & K_{12}^a \\ sym & K_{22}^a \end{bmatrix} \qquad (6.44)$$

The final form of stiffness matrix can be rewritten using 6.43 and 6.44 in 6.40 as:

$$[K]_{ab}^l = \begin{bmatrix} K_{11}^a & 0 & 0 & K_{12}^a & 0 & 0 \\ 0 & K_{11}^b & K_{12}^b & 0 & K_{13}^b & K_{14}^b \\ 0 & K_{21}^b & K_{22}^b & 0 & K_{23}^b & K_{24}^b \\ K_{21}^a & 0 & 0 & K_{22}^a & 0 & 0 \\ 0 & K_{31}^b & K_{32}^b & 0 & K_{33}^b & K_{34}^b \\ 0 & K_{41}^b & K_{42}^b & 0 & K_{34}^b & K_{44}^b \end{bmatrix} \qquad (6.45)$$

If Winkler and Pasternak supports are considered, the contribution to the stiffness matrices as follows:

Winkler Support:

$$[K]_W^l = \frac{kL}{420} \begin{bmatrix} 0 & 0 & 0 & 0 & 0 & 0 \\ 0 & 156 & 22L & 0 & 54 & -13L \\ 0 & 22L & 4L^2 & 0 & 13L^2 & -3L^2 \\ 0 & 0 & 0 & 0 & 0 & 0 \\ 0 & 54 & 13L & 0 & 156 & 22L \\ 0 & -13L & -3L^2 & 0 & 22L & 4L^2 \end{bmatrix} \qquad (6.46)$$

Pasternak Support:

$$[K]_P^l = \frac{k_G}{30L} \begin{bmatrix} 0 & 0 & 0 & 0 & 0 & 0 \\ 0 & 36 & 3L & 0 & -36 & -3L \\ 0 & 3L & 4L^2 & 0 & -3L & -L^2 \\ 0 & 0 & 0 & 0 & 0 & 0 \\ 0 & -36 & -3L & 0 & 36 & 3L \\ 0 & -3L & -L^2 & 0 & 3L & 4L^2 \end{bmatrix} \qquad (6.47)$$

The total stiffness matrix is a sum of bending and Winkler and Pasternak support stiffness matrices:

$$[K]_l = [K]_{ab}^l + [K]_W^l + [K]_P^l \qquad (6.48)$$

Equation (6.48) can be transformed to a global stiffness matrix using the following relationship:

$$[K]_{ab}^g = [T^*]^T [K]_l [T^*] \qquad (6.49)$$

where

$$[T^*] = \begin{bmatrix} c & s & 0 & 0 & 0 & 0 \\ -s & c & 0 & 0 & 0 & 0 \\ 0 & 0 & 1 & 0 & 0 & 0 \\ 0 & 0 & 0 & c & s & 0 \\ 0 & 0 & 0 & -s & c & 0 \\ 0 & 0 & 0 & 0 & 0 & 1 \end{bmatrix}$$

6.5   FINITE ELEMENT ANALYSIS OF SUPPORT AND REINFORCEMENT SYSTEMS

The finite element analysis is carried out by considering either elastic behavior or elasto-plastic behavior of the rockbolt system and the medium. For the elasto-plastic analysis, an iteration scheme is necessary when a part of the domain starts to exhibit plastic behavior. The iteration schemes adopted in analysis are (Owen and Hinton, 1980) (Fig. 6.4):

1) Direct iteration scheme
2) Newton-Raphson iteration scheme
3) Tangential stiffness iteration scheme
4) Initial stiffness iteration scheme
5) Hybrid iteration schemes

Each iteration scheme has its own merits and demerits. Among these iteration schemes, the initial stiffness iteration scheme proposed by Zienkiewicz et al. (1969) are generally employed in the finite element analyses, as the debonding phenomena in the rockbolt system cannot be treated by any of the other iteration schemes. As the details of this iteration

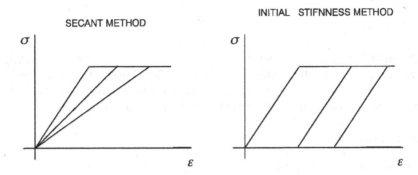

Figure 6.4 Illustration of some of iteration schemes.

scheme is well presented in the article by Zienkiewicz *et al.* (1969) and the textbooks by Zienkiewicz (1977) and Owen and Hinton (1980), no presentation about the scheme is be made herein.

6.6   DISCRETE FINITE ELEMENT METHOD (DFEM-BOLT) FOR THE ANALYSIS OF SUPPORT AND REINFORCEMENT SYSTEMS

Aydan and Mamaghani (Aydan, 1997; Aydan *et al.*, 1996; Mamaghani *et al.*, 1999) have jointly developed the discrete finite element method (DFEM) for assessing the response and stability of rock block systems, based on the finite element method together with the utilization of updated Lagrangian scheme. It consists of a mechanical model to represent the deformable blocks and contact models that specify the interaction among them. Small displacement theory is applied to the intact blocks, while blocks can take finite displacement. Blocks are polygons with an arbitrary number of sides that are in contact with the neighboring blocks, and they are idealized as a single or multiple finite elements. Block contacts are represented by a contact element. The essentials of DFEM is described in the following.

6.6.1   Mechanical modeling

The general equation of motion is given by:

$$\nabla \cdot \boldsymbol{\sigma} + \mathbf{b} = \rho \ddot{\mathbf{u}} \tag{6.50}$$

where $\boldsymbol{\sigma}$, $\mathbf{b}$, ρ, $\ddot{\mathbf{u}}$ are stress tensor, body force, density, and acceleration, respectively.

The following presentation is restricted to the framework of the small-strain theory. The strain-displacement relationships are:

$$\boldsymbol{\varepsilon} = \frac{1}{2}(\nabla \mathbf{u} + (\nabla \mathbf{u})^T) \tag{6.51}$$

The strain rate-velocity relationships are:

$$\dot{\boldsymbol{\varepsilon}} = \frac{1}{2}(\nabla \mathbf{v} + (\nabla \mathbf{v})^T) \tag{6.52}$$

where $\mathbf{v} = \dot{\mathbf{u}}$

The following constitutive relationship among stresses and strains and strain rates holds:

$$\sigma = D_e \varepsilon + D_v \dot{\varepsilon} \tag{6.53}$$

where D_e and Dv are elasticity and viscosity tensors (see Aydan *et al.*, 1995). However, they can be replaced by elasto-plastic and visco-plastic tensors if necessary (Aydan and Naw-rocki, 1998). This type constitutive law allows us to model intact blocks as well as contacts, interfaces, or rock discontinuities. The boundary and initial conditions are as follows:

Boundary conditions:

$$\hat{t} = \sigma \cdot n \quad on \quad \Gamma_t \text{ and } u = \hat{u} \quad on \quad \Gamma_u$$

where $\hat{t}$ is the surface traction in the n direction on Γ_t, while initial conditions are:

$$u_0, \quad \dot{u}_0 \text{ at } t = 0$$

6.6.2 Finite element modeling

In the following, the finite element form of the equation of motion is derived. Taking a variation on δu, the following integral form of the Equation (6.50) can be written as:

$$\int_\Omega (\nabla \cdot \sigma) \cdot \delta u d\Omega + \int_\Omega b \cdot \delta u d\Omega = \int_\Omega \rho \ddot{u} \cdot \delta u d\Omega \tag{6.54}$$

With the use of the Gauss divergence theorem and the boundary conditions, the weak form of the governing equation takes the following form:

$$\int_{\Gamma_t} \hat{t} \cdot \delta u d\Gamma + \int_\Omega b \cdot \delta u d\Omega = \int_\Omega \sigma \cdot (\nabla \delta u) d\Omega + \int_\Omega \rho \ddot{u} \cdot \delta u d\Omega \tag{6.55}$$

i) Discretization in space domain

Let us assume that displacements are approximated by the following expression:

$$u = NU(t) \tag{6.56}$$

Using the above approximate form and the constitutive law, the following expressions in a condensed form are obtained for a typical finite element:

$$M\ddot{U} + C\dot{U} + KU = F \tag{6.57}$$

where

$$M = \int_{\Omega_e} \rho N^T N d\Omega, C = \int_{\Omega_e} B^T D_v B d\Omega, \quad K = \int_{\Omega_e} B^T D_e B d\Omega, \quad F = \int_{\Omega_e} N^T b d\Omega + \int_{\Gamma_e} N^T t d\Gamma$$

ii) Discretization in time domain

The time domain in the finite element method is usually discretized using the finite difference technique. There are a number of procedures to treat the problem of time discretization. One of the most commonly used procedures is the central difference method. Employing the

Taylor expansion of series at steps $n - 1$, n and $n + 1$, and neglecting the third order terms, Equation (6.57) can be transformed to:

$$\bar{K}U_{n+1} = \bar{F}_{n+1} \tag{6.58}$$

where

$$\bar{K} = \frac{1}{\Delta t^2}M + \frac{1}{2\Delta t}C$$

$$\bar{F}_{n+1} = \left(\frac{2}{\Delta t^2}M - K\right)U_n - \left(\frac{1}{\Delta t^2}M - \frac{1}{2\Delta t}C\right)U_{n-1} + F_n$$

6.6.3   Finite element modeling of block contacts

The contact element is used to model block contacts. Let us now consider a two-noded element (l, m) in a two-dimensional space and take two coordinate systems (oxy) and $(o'x'y')$ as shown in Figure 6.4. Assuming that the strain component $\varepsilon_{y'y'}$ is negligible, the remaining strain components take the following form:

$$\varepsilon_{x'x'} = \frac{\partial u'}{\partial x'}, \quad \gamma_{x'y'} = \frac{\partial v'}{\partial x'} \tag{6.59}$$

Let us assume that the shape functions are linear such that:

$$N_l = \frac{1}{2}(1 - \xi), \quad N_m = \frac{1}{2}(1 + \xi) \tag{6.60}$$

where $\xi = (2x' + x'_l + x'_m)/L$, $L = (x'_l - x'_m)$. Then, the relationship between the strains and nodal displacements becomes:

$$\begin{Bmatrix} \varepsilon_{x'x'} \\ \gamma_{x'y'} \end{Bmatrix} = \frac{1}{L}\begin{bmatrix} -1 & 0 & 1 & 0 \\ 0 & -1 & 0 & 1 \end{bmatrix}\begin{Bmatrix} U'_l \\ V'_l \\ U'_m \\ V'_m \end{Bmatrix} \tag{6.61}$$

Thus, the stiffness matrix in the local coordinate system is explicitly obtained as:

$$K' = \begin{bmatrix} k'_n & 0 & -k'_n & 0 \\ 0 & k'_s & 0 & -k'_s \\ -k'_n & 0 & k'_n & 0 \\ 0 & -k'_s & 0 & k'_s \end{bmatrix} \tag{6.62}$$

where

$$k'_n = E_n \cdot \frac{A_c}{x'_m - x'_l}, \quad k'_s = G_s \cdot \frac{A_c}{x'_m - x'_l}$$

A_c is the contact area, E_n, G_s are normal and shear elastic moduli of discontinuity, respectively.

The stiffness matrix in the local coordinate system is then transformed to the stiffness matrix in the global coordinate system by the following relationship:

$$\mathbf{K} = \mathbf{T}^T \mathbf{K}' \mathbf{T}$$ (6.63)

where

$$\mathbf{T} = \begin{bmatrix} \cos\theta & \sin\theta & 0 & 0 \\ -\sin\theta & \cos\theta & 0 & 0 \\ 0 & 0 & \cos\theta & \sin\theta \\ 0 & 0 & -\sin\theta & \cos\theta \end{bmatrix}, \quad \theta = \tan^{-1}\left(\frac{y_m - y_l}{x_m - x_l}\right)$$

The viscosity (damping) matrix in the local coordinate system can be also obtained in a similar manner as given below:

$$\mathbf{C}' = \begin{bmatrix} c_n' & 0 & -c_n' & 0 \\ 0 & c_s' & 0 & -c_s' \\ -c_n' & 0 & c_n' & 0 \\ 0 & -c_s' & 0 & c_s' \end{bmatrix}$$ (6.64)

where

$$c_n' = V_n \cdot \frac{A_c}{x_m' - x_l'} \quad c_s' = V_s \cdot \frac{A_c}{x_m' - x_l'}$$

V_n, V_s are normal and shear viscosity moduli of discontinuity, respectively.

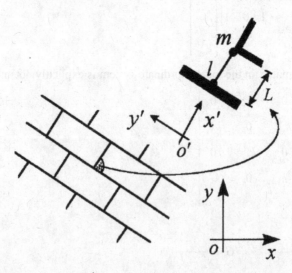

Figure 6.5 Contact element for the representation of block contacts.

6.6.4   Considerations of support and reinforcement system

DFEM is fundamentally a finite element method and is very easy to implement the elements used for rock reinforcement and support systems. The original DFEM code developed by the author, named DFEM-BOLT, is extended to include the support and reinforcement members. Most of the computations reported in the related chapters are based on a pseudo-dynamic version of the method.

Chapter 7

Applications to underground structures

7.1 INTRODUCTION

Analytical and numerical methods to evaluate the support and reinforcement effects of reinforcement and support system for underground excavations are described in this chapter. Ground-response-support reaction incorporating various support members and rockbolts and rockanchors and the face effect is formulated and several examples of applications are given. Furthermore, the effect mesh bolting for compressed air energy storage schemes is formulated and several examples of excavations are presented. The effects of various conditions for the effective utilization of reinforcement and support system for underground structures are analyzed through finite element simulations. The response of rockbolts in discontinuum is analyzed and their implications on the interpretation of the field measurements on rockbolt performances are discussed. The presently available proposals on the suspension effect, the beam building effect, and the arch formation effect of rockbolts are re-examined and more generalized solutions are presented. Furthermore, solutions for the reinforcement effect of bolts against the flexural and columnar type of toppling failure are given in addition to reinforcement effect of rockbolts against sliding type of failure.

7.2 ANALYTICAL APPROACH

Analytical approaches to tunneling are based on the closed-form solutions of the equation of motion without inertia for static case. In some cases, time dependency of surrounding rock may also be taken into account. The simplest condition for deriving analytical solutions is a circular opening subjected to hydrostatic *in situ* stress. However, there are several solutions for non-hydrostatic conditions (Kastner, 1961; Gerçek, 1993, 1996, 1997; Gerçek and Geniş, 1999). In this section, analytical solutions developed for a circular tunnel under hydrostatic stress condition and for tunnels subjected to non-hydrostatic stress state are described.

7.2.1 Solutions for hydrostatic in situ stress state for support system and fully grouted rockbolts

A vast number of studies were performed for the determination of stress and strain fields about cylindrical (circular) and spherical openings excavated in a hydrostatic far-field stress field, as it is fairly easy to obtain analytical solutions for this particular situation. An

analytical solution for elasto-plastic behavior of rocks is important for providing fundamental information for assessing the stability of openings, as well as for the support design
and excavation stage of underground openings. The first analytical solution, developed by
Fenner (1938), assumes that rock mass exhibits an elastic-perfectly plastic behavior. Talobre
(1957) also developed his own solution. Since these earlier studies, a numerous solutions
have been proposed and used for the design. The major differences among these methods are
associated with assumed elasto-plastic behavior, yield function, and the modeling of support
members. One can find a summary of most solutions in the article by Brown *et al*. (1983),
including their own solutions. Egger (1973b,c) also discussed the utilization of analytical
solutions for spherical openings to infer the stress distributions and displacement of rock
mass in the vicinity of tunnel faces.

Aydan (1989) developed analytical solutions for tunnels supported by rockbolts, shotcrete, steel ribs, and concrete linings, and the interaction between rockbolts and surrounding ground are explicitly considered. In his solutions, the modification of the deformation
modulus of rock mass were taken into account.

Aydan *et al*. (1993, 1996) developed solutions for determining stress and strain fields
around cylindrical tunnels in squeezing ground, which were extended to spherical openings
to obtain a unified solution for the radially symmetric problem by Aydan and Geniş (2010).
The derivation of a unified solution for radially symmetric openings developed by Aydan
and Geniş (2010) are presented herein. It should also be noted that the improvement of the
deformation modulus in a bolted zone, which was considered in the original solution by
Aydan (1989), is neglected in the solution given here.

Rock mass around an opening is assumed to obey the Mohr-Coulomb yield criterion, and
the solution presented in study is developed for the elastic-perfect-residual plastic material
behavior. Although it is possible to develop solutions for the Hoek-Brown criterion, it is not
done on purpose, as the generalized Hoek-Brown criterion violates the Euler theorem used
in the classical theory of plasticity for constitutive modeling of rocks. In the proposed unified
solution, a shape coefficient (*n*) is defined, which is 1 for a cylindrical opening and 2 for a
spherical opening. Following the presentation of the fundamentals of the unified analytical
method and the material behavior, the method of derivation has been given. In addition,
some procedures are presented for the consideration of effects of the support system its and
long-term properties.

i)   Unified analytical solutions for deformation, strains, and stress around a circular tunnel supported by rockbolts, shotcrete, and steel ribs

Constitutive laws

A generalized form of the constitutive law between stresses and strains of rock in an
elastic region for a radially symmetric problem (cylindrical and spherical openings) can
be given as:

$$\begin{Bmatrix} \sigma_r \\ \sigma_\theta \end{Bmatrix} = \begin{bmatrix} \lambda+2\mu & n\lambda \\ \lambda & n\lambda+2\mu \end{bmatrix} \begin{Bmatrix} \varepsilon_r \\ \varepsilon_\theta \end{Bmatrix} \tag{7.1}$$

where n is shape coefficient and has a value of 1 for a cylindrical opening and 2 for a spherical opening; σ_r, radial stress; σ_θ, tangential stress; ε_r, radial strain; ε_θ, tangential strain. λ and μ, are Lamé constants and given as:

$$\lambda = \frac{E\nu}{(1+\nu)\,(1-2\nu)}\,; \quad \mu = \frac{E}{2(1+\nu)} \tag{7.3}$$

where E is elastic modulus of rock and ν is Poisson's ratio of rock.

Equilibrium equation

When the problem is radially symmetric, the momentum law for static case takes the following form:

$$\frac{d\sigma_r}{dr} + n\frac{\sigma_r - \sigma_\theta}{r} = 0 \tag{7.4}$$

where r is the distance from opening center.

Compatibility condition

Compatibility condition between strain components for radially symmetric openings is given as:

$$\frac{d\varepsilon_\theta}{dr} + \frac{\varepsilon_\theta - \varepsilon_r}{r} = 0 \tag{7.5}$$

Relationships between strain components and radial displacement (u) are:

$$\varepsilon_r = \frac{du}{dr}, \quad \varepsilon_\theta = \frac{u}{r} \tag{7.6}$$

Behavior of rock material

An elastic-perfect-residual plastic model, as shown in Figure 7.1, approximates the behavior of rock. Although it is possible to consider the strain-softening behavior, it is extremely difficult to obtain closed-form solutions, and numerical techniques would be necessary. Rock was assumed to obey the Mohr-Coulomb yield criterion. Although it is possible to derive solutions for Hoek-Brown criterion, it is not intentionally done, as the generalized Hoek-Brown criterion violates the Euler theorem used in the classical theory of plasticity for constitutive modeling of rocks. Failure zones about radially symmetric openings excavated in rock mass for elastic-perfect-residual plastic behavior and yield functions for each region are illustrated in Figure 7.2 and given by:

$$\sigma_1 = q\sigma_3 + \sigma_c, \quad q = \frac{1+\sin\phi}{1-\sin\phi} \qquad \text{perfectly plastic region,} \tag{7.7a}$$

$$\sigma_1 = q^*\sigma_3 + \sigma_c^*, \quad q^* = \frac{1+\sin\phi^*}{1-\sin\phi^*} \qquad \text{residual plastic region,} \tag{7.7b}$$

where, σ_1: maximum principal stress; σ_3: minimum principal stress; σ_c: uniaxial compressive strength of intact rock; σ_c^*: uniaxial compressive strength of broken rock; ϕ: internal friction

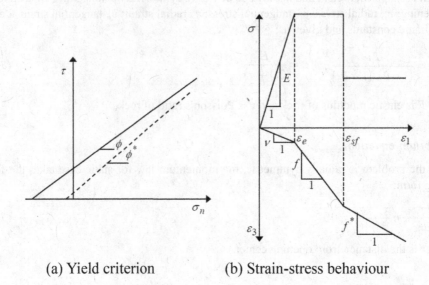

(a) Yield criterion   (b) Strain-stress behaviour

Figure 7.1 Mechanical models for rock mass.

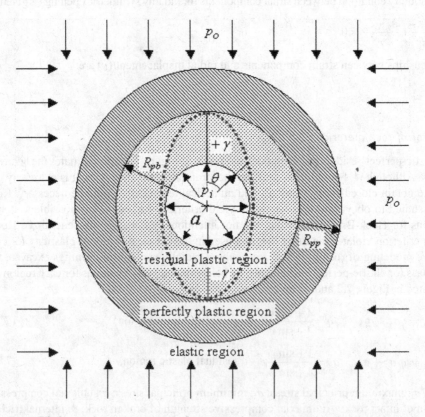

Figure 7.2 States about an opening and notations (gravity is considered in dotted zone).

angle of intact rock; and ϕ^*: internal friction angle of broken rock. Relationships between total radial and tangential strains in plastic regimes are assumed to be of the following form

$$\varepsilon_r = -f \, \varepsilon_\theta \qquad \text{for perfectly plastic region,} \tag{7.8a}$$

$$\varepsilon_r = -f^* \varepsilon_\theta \qquad \text{for residual plastic region,} \tag{7.8b}$$

where f and f^* are physical constants obtained from tests. These constants may be interpreted as plastic Poisson's ratios (Aydan et al., 1995b).

1)   STRESS AND STRAIN FIELD AROUND OPENING

1)   Residual plastic zone ($a \leq r \leq R_{pb}$)

Inserting the yield criterion Equation (7.7b) into the governing Equation (7.4) with $\sigma_3 = \sigma_r$ and $\sigma_1 = \sigma_\theta$ yields

$$\frac{d\sigma_r}{dr} + n\,(1-q^*)\frac{\sigma_r}{r} = n\frac{\sigma_c^*}{r} \tag{7.9}$$

The solution of the above differential equation is:

$$\sigma_r = C\, r^{n(q^*-1)} - \frac{\sigma_c^*}{q^*-1}. \tag{7.10}$$

The integration constant C is obtained from the boundary condition $\sigma_r = p_i$ at $r = a$ as:

$$C = \left(p_i + \frac{\sigma_c^*}{q^*-1} \right) \frac{1}{a^{n(q^*-1)}}, \tag{7.11}$$

where p_i is internal or support pressure. Thus, the stresses now take the following forms:

$$\sigma_r = \left(p_i + \frac{\sigma_c^*}{q^*-1} \right) \left(\frac{r}{a} \right)^{n(q^*-1)} - \frac{\sigma_c^*}{q^*-1}, \tag{7.12}$$

$$\sigma_\theta = q^* \left(p_i + \frac{\sigma_c^*}{q^*-1} \right) \left(\frac{r}{a} \right)^{n(q^*-1)} - \frac{\sigma_c^*}{q^*-1} \tag{7.13}$$

Solving the differential equation obtained by inserting the relationship given by Equation (7.8b) in Equation (7.5) yields:

$$\varepsilon_\theta = \frac{A}{r^{f^*+1}} \tag{7.14}$$

The integration constant A is determined from the continuity of the tangential strain at perfect-residual plastic boundary $r = R_{pb}$ as:

$$A = \varepsilon_\theta^{pb} R_{pb}^{f^*+1}, \tag{7.15}$$

ε_θ^{pb} in Equation (7.15) is the tangential strain at perfect-residual plastic boundary ($r = R_{pb}$) and it is specifically given by:

$$\varepsilon_\theta^{pb} = \eta_{sf} \varepsilon_\theta^{ep} \qquad \eta_{sf} = \frac{\varepsilon_{sf}}{\varepsilon_e}, \tag{7.16}$$

where η_{sf} is tangential strain level at perfect-residual plastic boundary (Fig. 8.3); ε_θ^{ep} is tangential strain at elastic-perfect plastic boundary as:

$$\varepsilon_\theta^{ep} = \frac{1+\nu}{nE}\left(p_0 - \sigma_{rp}\right) \tag{7.17}$$

σ_{rp} in Equation (7.17) is radial stress at elastic-perfect plastic boundary. As a result, the tangential strain in surrounding rock becomes:

$$\varepsilon_\theta = \frac{1+\nu}{nE}\left(p_0 - \sigma_{rp}\right) \eta_{sf} \left(\frac{R_{pb}}{r}\right)^{f^*+1}. \tag{7.18}$$

2) Perfectly-plastic zone ($R_{pb} \leq r \leq R_{pp}$)

Inserting the yield criterion Equation (7.7a) into the governing Equation (7.4) with $\sigma_3 = \sigma_r$ and $\sigma_1 = \sigma_\theta$ gives:

$$\frac{d\sigma_r}{dr} + n\,(1-q)\frac{\sigma_r}{r} = n\frac{\sigma_c}{r} \tag{7.19}$$

The solution of the above differential equation is:

$$\sigma_r = Cr^{n(q-1)} - \frac{\sigma_c}{q-1} \tag{7.20}$$

The integration constant C is obtained from the boundary condition $\sigma_r = \sigma_{rp}$ at $r = R_{pp}$ as:

$$C = \left(\sigma_{rp} + \frac{\sigma_c}{q-1}\right)\frac{1}{R_{pp}^{n(q-1)}} \tag{7.21}$$

Thus, the stresses now take the following forms:

$$\sigma_r = \left(\sigma_{rp} + \frac{\sigma_c}{q-1}\right)\left(\frac{r}{R_{pp}}\right)^{n(q-1)} - \frac{\sigma_c}{q-1}, \tag{7.22}$$

$$\sigma_\theta = q\left(\sigma_{rp} + \frac{\sigma_c}{q-1}\right)\left(\frac{r}{R_{pp}}\right)^{n(q-1)} - \frac{\sigma_c}{q-1} \tag{7.23}$$

Since the derivation of the tangential strain is similar to the previous case, the final expression takes the following form:

$$\varepsilon_\theta = \frac{1+\nu}{nE}\left(p_0 - \sigma_{rp}\right)\left(\frac{R_{pp}}{r}\right)^{f^*+1} \tag{7.24}$$

The relationship between the plastic zone radii is also found from the requirement of the continuity of tangential strain at $r = R_{pb}$ and relationship Equation (7.18) as:

$$\frac{R_{pp}}{R_{pb}} = \eta_{sf}^{\frac{1}{f+1}} \tag{7.25}$$

3) Elastic zone ($R_{pp} \leq r$)

The derivation of stresses and displacement expressions for cylindrical opening was previously given in detail by Aydan and Ersen (1985) and Aydan (1989) with the consideration of initially stressed elastic medium by a far-field hydrostatic *in situ* stress (p_0). The final forms of the expressions for radially symmetric openings are of the following forms:

$$\sigma_r = p_0 - \left(p_0 - \sigma_{rp}\right)\left(\frac{R_{pp}}{r}\right)^{n+1}, \tag{7.26}$$

$$\sigma_\theta = p_0 + \frac{1}{n}\left(p_0 - \sigma_{rp}\right)\left(\frac{R_{pp}}{r}\right)^{n+1}, \tag{7.27}$$

$$\varepsilon_\theta = \frac{1+\nu}{nE}\left(p_0 - \sigma_{rp}\right)\left(\frac{R_{pp}}{r}\right)^{n+1} \tag{7.28}$$

The specific form for σ_{rp} is obtained from the continuity condition of tangential stresses at $r = R_{pp}$ by equality Equation (7.23) and Equation (7.27) as:

$$\sigma_{rp} = \frac{p_0 + n\left(p_0 - \sigma_c\right)}{1 + nq} \tag{7.29}$$

II) PLASTIC ZONES RADIUS AROUND OPENING

1) Perfectly plastic-residual plastic zone boundary radius (R_{pb})

The perfectly plastic-residual plastic zone boundary radius is found from the requirement of the continuity of radial stresses, i.e. by equality of Equations (7.12) and (7.22), at $r = R_{pb}$ as:

$$\frac{R_{pb}}{a} = \left[\frac{\frac{(1+n)\left[(q-1)+\alpha\right]}{(1+nq)(q-1)}(\eta_{sf})^{\frac{n(1-q)}{f+1}} - \frac{\alpha}{q-1} + \frac{\alpha^*}{q^*-1}}{\beta + \frac{\alpha^*}{q^*-1}}\right]^{\frac{1}{n(q^*-1)}}, \tag{7.30}$$

where β is the support pressure normalized by overburden pressure and given as:

$$\beta = \frac{p_i}{p_0}, \tag{7.31}$$

and α is also competency factor as:

$$\alpha = \frac{\sigma_c}{p_0} \tag{7.32}$$

2)   Perfectly plastic and elastic zone boundary radius (R_{pp})

The perfectly plastic and elastic zone boundary radius is also found by inserting σ_{rp} given by Equation (7.29) in the radial stress Equation (7.22) with $\sigma_r = p_i$ at $r = a$ as:

$$\frac{R_{pp}}{a} = \left\{ \frac{(1+n)\left[(q-1)+\alpha\right]}{(1+nq)\left[(q-1)\beta+\alpha\right]} \right\}^{\frac{1}{n(q-1)}} \tag{7.33}$$

III)   NORMALIZED OPENING WALL STRAINS

1)   Elastic state

Tangential strain at opening wall can be obtained as:

$$\varepsilon_\theta^a = \frac{1+\nu}{nE}(p_0 - p_i), \tag{7.34}$$

$$\sigma_\theta^a = \frac{n+1}{n}p_0 - \frac{1}{n}p_i \tag{7.35}$$

If the opening is strained to its elastic limit, then $\sigma_\theta^a = \sigma_c$ for $p_i = 0$. Thus, we have the elastic strain limit as:

$$\varepsilon_\theta^e = \frac{1+\nu}{E} \cdot \frac{\sigma_c}{n+1} \tag{7.36}$$

Using the above relationship in Equation (7.34), one obtains the normalized opening wall strain (ξ) as:

$$\xi = \frac{\varepsilon_\theta^a}{\varepsilon_\theta^e} = \frac{n+1}{n}\left(\frac{1-\beta}{\alpha}\right) \le 1 \tag{7.37}$$

2)   Perfectly-plastic state

Tangential strain at opening wall can be obtained as:

$$\varepsilon_\theta^a = \frac{1+\nu}{nE}(p_0 - \sigma_{rp})\left(\frac{R_{pp}}{a}\right)^{f+1}, \tag{7.38}$$

elastic limit is given as:

$$\varepsilon_\theta^e = \frac{1+\nu}{nE}(p_0 - \sigma_{rp}) \tag{7.39}$$

Using the above relationship in Equation (7.38), one obtains the normalized opening wall strain as:

$$\xi = \frac{\varepsilon_\theta^a}{\varepsilon_\theta^e} = \left\{ \frac{(1+n)\left[(q-1)+\alpha\right]}{(1+nq)\left[(q-1)\beta+\alpha\right]} \right\}^{\frac{f+1}{n(q-1)}} \tag{7.40}$$

3)   Residual plastic state

Tangential strain at opening wall can be obtained as:

$$\varepsilon_\theta^a = \frac{1+\nu}{nE}\left(p_0 - \sigma_{rp}\right)\eta_{sf}\left(\frac{R_{pb}}{a}\right)^{f^*+1}$$

(7.41)

Using Equations (7.39) and (7.41), one obtains the normalized opening wall strain as:

$$\xi = \frac{\varepsilon_\theta^a}{\varepsilon_\theta^e} = \eta_{sf}\left\{\frac{\dfrac{(1+n)\left[(q-1)+\alpha\right]}{(1+nq)\,(q-1)}\,(\eta_{sf})^{\frac{n\,(1-q)}{f+1}} - \dfrac{\alpha}{q-1} + \dfrac{\alpha^*}{q^*-1}}{\beta + \dfrac{\alpha^*}{q^*-1}}\right\}^{\frac{f^*+1}{n\,(q^*-1)}},$$

(7.42)

where

$$\alpha^* = \frac{\sigma_c^*}{p_0}$$

(7.43)

ii)   Consideration of support system

Although shotcrete, rockbolts, and steel ribs are principal support members and are widely used, it is very rare to find a fundamental study on the proper design method for the reinforcement effect of support systems consisting of shotcrete, rockbolts, and steel ribs. As a certain displacement of ground takes place before the installation of supports, the resulting internal pressure provided by the support members must be evaluated in terms of relative displacement. Depending upon the state of rock before the installation of supports, there may be several combinations. In this section, it is assumed that rock is in elastic state at the time of installation of rockbolts, shotcrete, and steel ribs (Fig. 7.3), and relative displacement at each region is evaluated for this special case as below (refer to Aydan (1989) for other combinations).

At the time of installation of support members, if rock behaves elastically, the radial deformation of the tunnel is given by:

$$u_{in} = \frac{1+\nu}{nE}r\left(p_0 - \sigma_{rp}\right)\left(\frac{a}{r}\right)^{n+1}$$

(7.44)

Subtracting this deformation from equations 7.18, 7.24, and 7.28, we obtain the following relationships:

1)   Flow region ($a \le r \le R_{pb}$):

$$\Delta u = \frac{1+\nu}{nE}r\left[\left(p_0 - \sigma_{rp}\right)\eta_{sf}\left(\frac{R_{pb}}{r}\right)^{f^*+1} - \left(p_o - p_{in}\right)\left(\frac{a}{r}\right)^{n+1}\right]$$

(7.45)

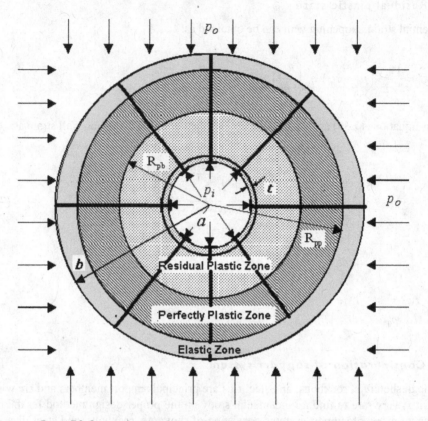

Figure 7.3 Support system for a radially symmetric opening and notations.

2) Perfectly plastic region ($R_{pb} \leq r \leq R_{pp}$):

$$\Delta u = \frac{1+\nu}{nE} r \left[(p_0 - \sigma_{rp}) \left(\frac{R_{pp}}{r} \right)^{f+1} - (p_o - p_{in}) \left(\frac{a}{r} \right)^{n+1} \right] \qquad (7.46)$$

3) Elastic region ($R_{pp} \leq r$):

$$\Delta u = \frac{1+\nu}{nE} r \left[(p_0 - \sigma_{rp}) \left(\frac{R_{pp}}{r} \right)^{n+1} - (p_o - p_{in}) \left(\frac{a}{r} \right)^{n+1} \right] \qquad (7.47)$$

(i) Modeling rockbolts

For a prescribed displacement (Δu_h) at grout-rock interface, the governing equation of rock-bolts takes the following form (Aydan, 1989):

$$\frac{d^2 u_{ax}}{d\xi^2} - \alpha^2 (u_{ax} - \Delta u_h) = 0; \quad \alpha^2 = \frac{2}{r_b^2 \ln(r_h / r_b)} \frac{G_g}{E_b} \qquad (7.48)$$

where $\xi = r$, r_h: borehole radius; r_b: bar radius; G_g: shear modulus of grout; E_b: elastic modulus of rockbolt. If the relative displacements of rock are introduced in the above non-homogenous second order differential equation, the solution of the above equation by elementary methods of integration becomes impossible, and the use of a numerical integration becomes necessary. For the sake of simplification, we introduce the following form for differential displacement:

$$f(\xi) = C_o^* e^{-D\xi} \tag{7.49}$$

Constants C_o^* and D are determined from the values of Δu_h at each respective interface of regions. Specifically, they are as follows:

1) Flow region ($a \leq r \leq R_{pb}$):

$$D_{pb} = \frac{\ln(\Delta u_a / \Delta u_{pb})}{(R_{pb} - a)}; \quad C_o^{*pb} = \Delta u_{pb} e^{D_{pb} R_{pb}} \tag{7.50a}$$

$$\Delta u_a = \frac{1+\nu}{nE} a \left[(p_0 - \sigma_{rp}) \eta_{sf} \left(\frac{R_{pb}}{a} \right)^{f^*+1} - (p_o - p_{in}^b) \right] \tag{7.50b}$$

$$\Delta u_{pb} = \frac{1+\nu}{nE} R_{pb} \left[(p_0 - \sigma_{rp}) \eta_{sf} - (p_o - p_{in}^b) \left(\frac{a}{R_{pb}} \right)^{n+1} \right] \tag{7.50c}$$

2) Perfectly plastic region ($R_{pb} \leq r \leq R_{pp}$):

$$D_{pp} = \frac{\ln(\Delta u_{pb} / \Delta u_{pp})}{(R_{pp} - R_{pb})}; \quad C_o^{*pp} = \Delta u_{pp} e^{D_{pp} R_{pp}}; \tag{7.51a}$$

$$\Delta u_{pp} = \frac{1+\nu}{nE} \left[(p_0 - \sigma_{rp}) R_{bp} \eta_{sf} \left(\frac{R_{pp}}{R_{pb}} \right)^{f^*+1} - (p_o - p_{in}^b) \frac{a^{n+1}}{R_{pp}^n} \right] \tag{7.51b}$$

3) Elastic region ($R_{pp} = r$):

$$D_{pe} = \frac{\ln(\Delta u_b / \Delta u_{pe})}{(b - R_{pp})}; \quad C_o^{*pe} = \Delta u_b e^{D_{pe} b} \tag{7.52a}$$

$$\Delta u_{pe} = \frac{1+\nu}{nE} R_{pe} \left[(p_0 - \sigma_{rp}) \eta_{sf} - (p_o - p_{in}^b) \left(\frac{a}{R_{pe}} \right)^{n+1} \right] \tag{7.52b}$$

$$\Delta u_b = \frac{1+\nu}{nE} b \left[(p_0 - \sigma_{rp}) \left(\frac{R_{pe}}{b} \right)^{n+1} \frac{R_{pb}}{b} \eta_{sf} - (p_o - p_{in}^b) \left(\frac{a}{b} \right)^{n+1} \right] \tag{7.52c}$$

The axial displacement and axial stress of a rockbolts takes the following form its solution:

$$u_{ax} = A_1 e^{-\alpha\xi} + A_2 e^{\alpha\xi} + C_o e^{-D\xi} \tag{7.53a}$$

$$\sigma_b = E_b \alpha \left[A_1 e^{-\alpha\xi} - A_2 e^{\alpha\xi} + \frac{D}{\alpha} C_o e^{-D\xi} \right] \tag{7.53b}$$

Integration constants (A_1 and A_2) at each region are determined from the continuity condition and boundary conditions. Accordingly, the internal pressure provided by rockbolts can be obtained from the well-known formula:

$$\Delta p_i^b = \sigma_b(r=a) \frac{A_b}{e_t e_l} \tag{7.54}$$

where A_b, e_t, and e_l are cross-section area and spacing of a typical rockbolt.

(ii)   Modeling shotcrete

The thin-wall cylinder or thick-wall cylinder approach is commonly used to assess the internal pressure effect of shotcrete in tunneling. For radially symmetric situations, a similar approach can be adopted. One can easily derive the relationship for the displacement and outer pressure p_{io}^s acting on the shotcrete with internal pressure $p_{ii}^s = 0$:

$$u = \frac{1+\nu_s}{nE_s} p_{io}^s \frac{a_o^{n+2}}{a_o^{n+1} - a_i^{n+1}} \left[\frac{1-2\nu_s}{1-\nu_s+n\nu_s} + \frac{1}{n}\left(\frac{a_i}{a_o}\right)^{n+1} \right] \tag{7.55}$$

The incremental form of the above equation is:

$$\Delta u = \frac{1+\nu_s}{nE_s} \Delta p_{io}^s \frac{a_o^{n+2}}{a_o^{n+1} - a_i^{n+1}} \left[\frac{1-2\nu_s}{1-\nu_s+n\nu_s} + \frac{1}{n}\left(\frac{a_i}{a_o}\right)^{n+1} \right] \tag{7.56}$$

and its inverse is:

$$\Delta p_{io}^s = K_s \Delta u \tag{7.57}$$

where

$$K_s = \frac{nE_s}{1+\nu_s} \frac{a_o^{n+1} - a_i^{n+1}}{a_o^{n+2}} \left[\frac{(1-\nu_s+n\nu_s)a_o^{n+1}}{n(1-2\nu_s)a_o^{n+1} + (1-\nu_s+n\nu_s)a_i^{n+1}} \right] \tag{7.58}$$

If the thickness of shotcrete is negligible as compared with the radius of opening, then we have the following:

$$K_s = \frac{nE_s}{1+\nu_s} \frac{t}{a_o^2} \frac{(1-\nu_s+n\nu_s)}{(1-\nu_s)} \tag{7.59}$$

(iii) Modeling steel ribs

If the steel rib is modeled as a one-dimensional rib, its radial deformation is given by:

$$u = \frac{1}{nE_{rb}} p_i^{rb} \frac{a_o^2 e_l}{A_{rb}} \tag{7.60}$$

Its incremental form becomes:

$$\Delta u = \frac{1}{nE_{rb}} \Delta p_i^{rb} \frac{a_o^2 e_l}{A_{rb}} \tag{7.61}$$

Or inversely, we have:

$$\Delta p_i^{rb} = K_{rb} \Delta u \text{ and } K_{rb} = nE_{rb} \frac{A_{rb}}{a_o^2 e_l} \tag{7.62}$$

The total internal pressure of the support system may be given as:

$$p_i^{ss} = \Delta p_i^b + \Delta p_{io}^s + \Delta p_i^{rb} \tag{7.63}$$

If support members yield during the deformation of surrounding rock, their behaviors are assumed to be elastic-perfectly plastic. This will particularly require further formulation of axial stress evaluation of rockbolts.

iii) Consideration of body forces in plastic zone

In a general sense, the consideration of body forces violates the radial symmetry of the governing equation. However, there are some proposals in literature for this purpose (i.e. Fenner, 1938; Hoek and Brown, 1980; Aydan, 1989; Sezaki et al., 1994). Some slight modifications to the developed solutions are described in order to consider the effect of body forces in residual-plastic zones. As the perfectly plastic region sustains its original strength, the effect of body forces should be negligible in this region. However, the effect of body forces may be important in residual plastic (flow) region. Using a similar approach proposed by Aydan (1989), the maximum body force on the support system in the residual plastic region may be approximately obtained as follows (Fig. 7.2):

$$p_{ib} = \left(1 - \left(\frac{a}{R_{pb}}\right)^{n(q^*-1)}\right) \frac{\gamma a \cos\theta}{n(q^*-1)-1} \tag{7.64}$$

Where γ is unit weight of residual plastic zone. The values of angle (θ) for crown, sidewall, and invert are 0, $\pi/2$, and π, respectively. It should be noted that this assumption is only valid provided that the radial symmetry of the problem is not violated.

iv) Consideration of creep failure

It is well known that deformation and strength characteristics of rocks depend upon the stress rate or strain rate used in tests (i.e. Bieniawski, 1970; Lama and Vutukuri, 1978; Aydan et al., 1994, 2010). Creep tests and relaxation tests are commonly used to determine the time-dependent characteristics of rocks. There is a common conception that the creep of rocks does not occur unless the applied stress is greater than a threshold stress value, which is called the creep threshold (Aydan *et al.*, 1995a, 1995b; Aydan and Nawrocki, 1998). This threshold stress level is generally related to the stress level at which fractures are initiated. The so-called transient creep is likely to be as a result of actual visco-elastic behavior of rock. Secondary creep, on the other hand, is due to stable crack propagation, and tertiary creep is due to unstable crack propagation. Therefore, secondary and tertiary creep are visco-plastic phenomena rather than visco-elastic phenomena, as they involve energy dissipation by fracturing. Figure 7.4 shows the normalized uniaxial creep stress by the short-term uni-axial strength for various rocks as a function of the failure time (time elapsed until the failure). The shrinkage of the uniaxial strength of rock may be represented by the following functional form:

$$\frac{\sigma_c}{\sigma_c^o} = F(t, t_s)$$ (7.65)

where t and t_s are time and short-term test duration. σ_c^o is the short-term uniaxial strength of rock mass. The first authors explored several specific forms of function (*F*) to evaluate the experimental results. Figure 7.4 compares the experimental results with the following functional form:

$$\frac{\sigma_c}{\sigma_c^o} = 1 - b \ln\left(\frac{t}{t_s}\right)$$ (7.64)

This simple concept may be also extended to a multi-axial stress state, in which the yield surface of rocks under multi-axial loading conditions is modeled as the shrinkage of the yield-ing surface. An illustration of this concept on the Mohr-Coulomb yield criterion is shown in Figure 7.5.

If the time-dependent characteristics of the uniaxial compressive strength of rocks is known, it may be possible to determine the time-dependent variation of various mechani-cal properties of rocks from the relationships obtained by Aydan *et al.* (1993, 1995a, 1995b). Using the approach originally proposed by Ladanyi (1974), together with the time-dependent variation of mechanical properties involved in the equations given in the previous

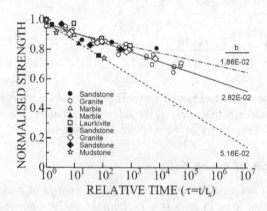

Figure 7.4 Creep strength of various rocks (from Aydan and Nawrocki, 1998).

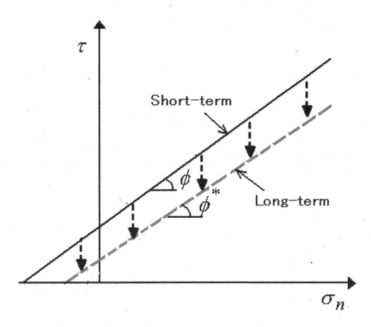

Figure 7.5 Illustration of triaxial long-term strength (modified from Aydan et al., 1995b).

subsection, it is possible to determine the time-dependent deformation of tunnels. Under multi-axial initial stress conditions and complex tunnel geometry, excavation scheme, and boundary conditions, the use of numerical techniques is necessary. For numerical analyses, the time-dependent behavior of rocks may be modeled by an approach proposed by Aydan et al. (1995b, 1996).

v) Consideration of face effect of tunnels in analytical solutions

Advancing tunnels utilizing support systems consisting of rockbolts, shotcrete, steel ribs, and concrete lining are three-dimensional complex structures, which is a dynamic process. However, tunnels are often modeled as a one-dimensional axisymmetric structure subjected to hydrostatic initial stress state as a static problem. The effect of tunnel face advance on the response and design of support systems is often replaced through an excavation stress release factor determined from pseudo three-dimensional (axisymmetric) or pure three-dimensional analyses as given below:

$$f_{ae} = \frac{e^{-bx/D}}{1/B + e^{-bx/D}}$$ (7.67)

where x is distance from tunnel face and the values for coefficients B and b suggested by Aydan (2011) are 2.33 and 1.7, respectively.

Figure 7.6a illustrates an unsupported circular tunnel subjected to an axisymmetric initial stress state. The variation of displacement and stresses along the tunnel axis were computed using the elastic finite element method. The radial displacement at the tunnel wall is

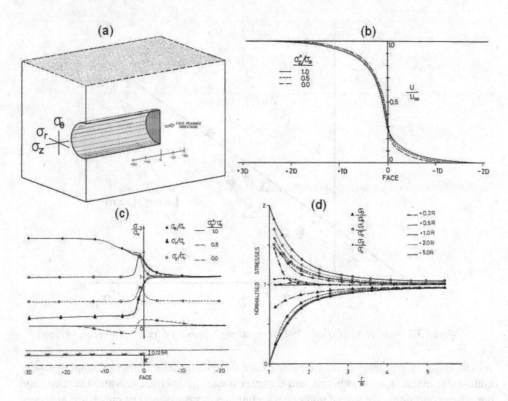

Figure 7.6 (a) Computational model for elastic finite element analysis; (b) Normalized radial displacement of the tunnel surface; (c) Normalized stress components along tunnel axis at a distance of 0.125 R; (d) The variation of stresses along *r*-direction at various distances from tunnel face.

normalized by the largest displacement and is shown in Figure 7.6b. As seen from the figure, the radial displacement takes place in front of the tunnel face. The displacement is about 28–30% of the final displacement. Its variation terminates when the face advance is about +2D. Almost 80% of the total displacement takes place when the tunnel face is about +1D. The effect of the initial axial stress on the radial displacement is almost negligible.

Figure 7.6c shows the variation of radial, tangential, and axial stress around the tunnel at a depth of 0.125 R. As noted from the figure, the tangential stress gradually increases as the distance increases from the tunnel face. The effect of the initial axial stress on the tangential stress is almost negligible. The radial stress rapidly decreases in the close vicinity of the tunnel face and the effect of the initial axial stress on the radial stress is also negligible. The most interesting variation is associated with the axial stress distribution. The axial stress increases as the face approaches, and then it gradually decreases to its initial value as the face effect disappears. This variation is limited to a length of 1 R (0.5 D) from the tunnel face. It is also interesting to note that if the initial axial stress is nil, even some tensile axial stresses may occur in the vicinity of tunnel face.

Figure 7.6d shows the stress distributions along the r-axis of the tunnel at various distances from the face when the initial axial stress is equal to initial radial and tangential stresses.

As noted from the figure, the maximum tangential stress is 1.5 times the initial hydrostatic stress and becomes twice the distance from the tunnel face is +5 R, which is almost equal to theoretical estimations for tunnels subjected to hydrostatic initial stress state. The stress state near the tunnel face is also close to that of the spherical opening subjected to hydrostatic stress state. The stress state seems to change from spherical state to the cylindrical state (Aydan, 2011). It should be noted that it would be almost impossible to simulate exactly the same displacement and stress changes of 3D analyses in the vicinity of tunnels by 2D simulations using the stress-release approach, irrespective of constitutive law of surrounding rock as a function of distance from tunnel (Aydan *et al.*, 1988; Aydan and Geniş, 2010).

vi) Applications

Elasto-plastic responses of cylindrical and spherical openings were obtained by using an analytical solution, which is described in this study. It is assumed that a cylindrical and spherical opening with a diameter of 6 m is excavated at a depth of below 1000 m from ground surface in rock mass having $\sigma_c = 20$ MPa and $\gamma = 25$ kN/m³.

Empirical equations given by Aydan *et al.* (1993, 1996) were used for determining properties of rock mass. In these equations, the properties of rock mass were related to the uniaxial compressive strength of rock material. The selected parameters used in the analyses are given in Table 7.1.

First, the ground reaction curves were obtained for the cylindrical and spherical openings. The opening wall strain was related to the normalized ground pressure and normalized plastic radius about the opening (Fig. 7.7). At the same internal support pressure, plastic zone

Table 7.1 The properties of strength and deformation of rock material and rock mass used in the parametric study.

σ_c	σ_c^*	E	U	γ	ϕ	ϕ^*	η_{sf}	f	f^*	p_0
MPa	MPa	GPa		kN/m³	(°)	(°)				MPa
20	0.05	5.5	0.25	25	42.3	54.8	1.311	2.60	3.96	25

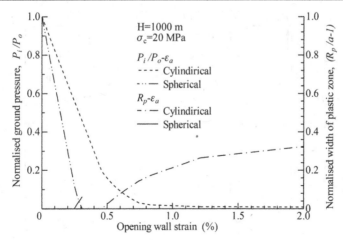

Figure 7.7 Strain, plastic zone radius around cylindrical and spherical openings, and ground reaction curves.

radius and wall strain around the cylindrical openings are greater than those for the spherical opening (Fig. 7.7 and 7.8). In the present example, elastic-perfect plastic zone radius
and residual-perfect plastic zone radius about cylindrical openings are 5.84 m and 5.07 m,
respectively. On the other hand, no residual plastic zone was observed, but elastic-perfect
plastic zone radius with 3.19 m was found about the spherical opening.

Then, the distribution of tangential and radial stresses is obtained around cylindrical and
spherical openings for the condition; that is, internal pressure is zero. It is clear that tangential stresses around the cylindrical openings are greater than the spherical openings. In addition, the radial and tangential stresses around spherical openings approach faster to the *in
situ* stress as a function distance from the ground surface. As a result, the radius of the plastic
zone about a spherical opening is smaller than that for a cylindrical one (Fig. 7.8).

The next comparison was concerned with the comparison of an elastic-perfectly plastic-
residual plastic model with a strain-softening model. The tunnel was assumed to be 200 m
below the ground surface in a rock mass having a uniaxial strength of 2.5 MPa. Empirical
equations given by Aydan *et al.* (1993, 1996) were used for determining other properties of
rock mass for analyses. Figures 7.9 and 7.10 show the computed ground reaction curve and
stress distributions for two constitutive models. As expected, there is almost no difference
regarding ground response curves if the consideration of the softening part of the constitutive model is taken into account by the proposed scheme. The only difference is associated
with the distribution of tangential stress in rock mass as seen in Figure 7.10.

Some comparisons on the modeling of support systems described in Chapter 6 are made
with finite element analyses reported by Aydan *et al.* (1986, 1987). The material properties
used in the analysis are given in Table 7.2 and the behavior of rock mass was assumed to be
elastic-perfectly plastic, which can be easily considered by the analytical solution presented
in this article. Figure 7.11 shows the axial stress distribution in the bolt at the equilibrium
state. As seen from the figure, the calculated axial stress distribution of a typical bolt is
almost the same as that calculated by the finite element model proposed by Aydan *et al.*
(1986). In this particular case, the rockbolts behave elastically.

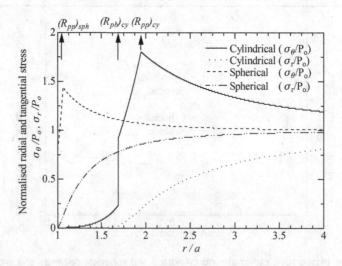

Figure 7.8 Tangential and radial stresses distribution around cylindrical and spherical openings ($P_i = 0$).

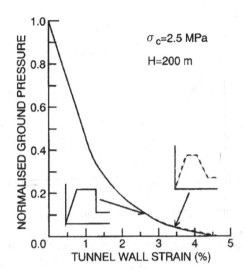

Figure 7.9 Comparison of ground response curves for different constitutive relations.

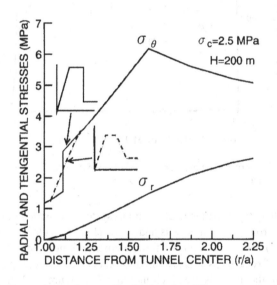

Figure 7.10 Comparison of stress distributions for different constitutive relations.

Table 7.2 Parameters used in computations.

Rock Mass				Bolt		Shotcrete			In situ
E (MPa)	v	σ_c (MPa)	Φ (o)	E(MPa)	σ_t (MPa)	E (MPa)	v	σ_c (MPa)	p_0 (MPa)
250	0.4	2	35	210	450	10	0.2	35	10

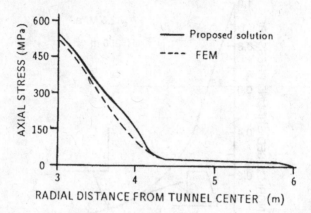

Figure 7.11 Comparison of computed axial stress distribution of rockbolts with FEM results.

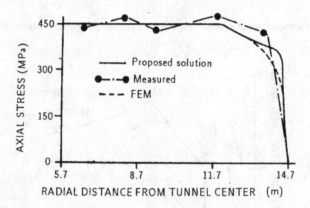

Figure 7.12 Comparison of computed axial stress distribution of rockbolts with FEM results and *in situ* measurements.

In the next example, the performance of a support system consisting of rockbolts and shotcrete lining, used in a roadway tunnel, was analyzed. This particular case was also analyzed by Aydan *et al.* (1987) using elasto-plastic finite element method with Aydan's rockbolt element. The behavior of rock mass was assumed to be elastic-perfectly plastic and the material properties used in the analysis are given in Table 7.2. Figure 7.12 shows the axial stress distribution in the bolt at the equilibrium state. As seen from the figure, the calculated axial stress distribution of a typical bolt is almost the same as that calculated by the finite element model. In this particular case, the bolts exhibit an elasto-perfectly plastic behavior.

The third example illustrates the performance of a support system consisting of rockbolts, shotcrete, and steel ribs. The material properties used in the analysis were also the same as those given in Table 7.3. Figure 7.13 shows ground response and support reaction curves and axial stress development in each support member as a function of the displacement of the tunnel wall. From these examples, it can be said that the presented models for a support system consisting of rockbolts, shotcrete, and steel ribs are appropriate and can be used for preliminary assessment of support systems of underground openings.

Table 7.3 Parameters used in computations.

Material	E (GPa)	v	σ_c (MPa)	σ_t (MPa)	ϕ (o)	Geometry	
Rock Mass	0.289	0.25	2.5	–	25	$D = 6$	m
Rockbolt	210	0.30	450	450	–	$D_b = 25$	mm
Grout	3.4	0.30	–	–	–	$D_h = 36$	mm
Shotcrete	5.0	0.20	10	–	–	$t = 100$	mm
Steel rib	210	0.30	450	450	–	$A_{rb} = 2000$	mm²

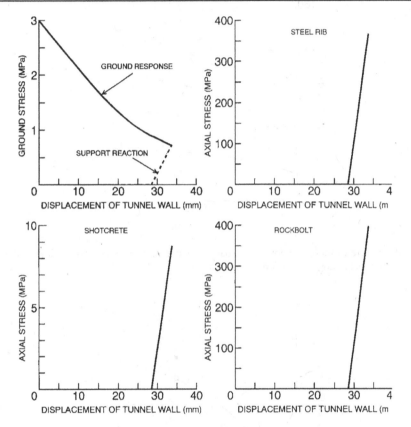

Figure 7.13 Computed ground response-support reaction curves and axial stress development in support members.

The final example is concerned with the time-dependent deformation of an actual tunnel in Nagasaki Prefecture, Japan. The name of tunnel is Tawarazaka and the deformation of the tunnel in the squeezing section continued for more than 1500 days after the completion of excavation (Aydan et al., 1995a, 1995b). Therefore, some investigations were necessary for the causes of time-dependent deformations. Additional laboratory tests and in situ pressuremeter tests and borings were conducted at certain locations along the section. Some swelling tests were also performed. The swelling tests indicated that rocks in this section had no swelling potential. Some attempts have been made to gather information on long-term tests on soft rocks available in literature and observations on the time-history of deformations, so that the degradation models for mechanical characteristics could be developed. The degradation model described in d) Consideration of creep failure and its details are given in the article by Aydan *et al.* (1995a, 1995b) have been incorporated in the theoretical model, in which each parameter changes its value with time.

Figure 7.14 shows predicted deformation responses for the tunnel without any support system for different values of short-term uniaxial strength of rock in the squeezing section. Figure 7.15 shows measured displacement responses at the center of the floor, together with

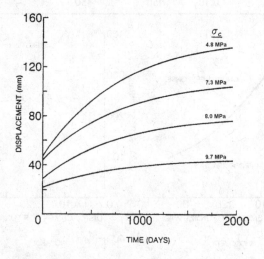

Figure 7.14 Predicted time-dependent deformation of the squeezing section of Tawarazaka tunnel (uniaxial strength of rock mass is variable).

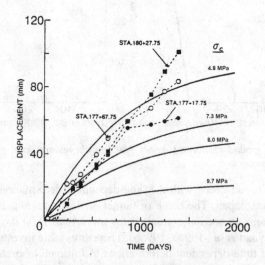

Figure 7.15 Comparison of predicted time-dependent deformation response with observed floor heave for the squeezing section of Tawarazaka tunnel.

those of some parametric analyses (note that the initial deformation response due to excavation is subtracted from the calculated displacement responses). In the parametric study, the
short-term uniaxial strength of surrounding rock mass varied between 4.8 to 9.7 MPa and the
retardation failure time for the degradation models was selected as 500 days. The calculated
displacement-time response curves are similar to those observed.

7.2.2   Solutions for hydrostatic in situ stress state for pre-stressed rockanchors

Aydan (1993) developed an analytical solution for the reinforcement of underground openings with tensioned grouted rockanchors, which corresponds a special form of fully grouted
rockbolts. As shown in Chapters 4 and 6, the governing equations of rockanchors may be
written as given below (Fig. 7.16):

Grout annulus and rock mass:

$$\frac{d\tau_{\xi\eta}}{d\xi} + \frac{\tau_{\xi\eta}}{\xi} = 0 \tag{7.65}$$

Rockanchor:

$$\frac{d\sigma_{\eta}}{d\eta} + \frac{2}{r_b}\tau_b = 0 \tag{7.66}$$

where
  $\tau_{\xi\eta}$ = shear stress
  σ_{η} = axial stress in rockanchor
  τ_b = shear stress at anchor-grout interface
  r_b = radius of rockanchor

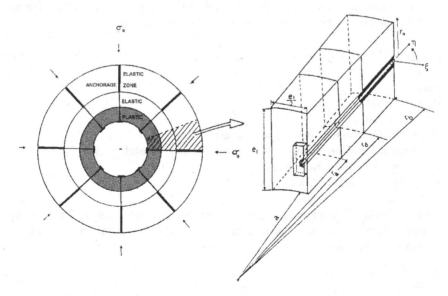

Figure 7.16 Illustration of a circular opening supported by pre-stressed rockanchors.

As shown previously, the above equations can be combined as:

$$\frac{d^2 u_b}{d\eta^2} - \alpha^2 u_b = -\alpha^2 \Delta u_{r0} \tag{7.67}$$

where

$$\alpha^2 = \frac{2K_g}{E_b r_b}, \quad K_g = \frac{G_g}{r_b \log(r_h / r_b)} \cdot \frac{1}{K_a + 1}, \quad K_a = \frac{G_g}{G_r} \cdot \frac{\log(r_0 / r_h)}{\log(r_h / r_b)} + 1$$

u_b = axial deformation of rockanchor
Δu_{r0} = deformation at radius r_0 after installation of rockanchors
r_h = hole radius of rockanchor
r_0 = radius where the deformation defined
E_b = elastic modulus of rockanchor
G_g = shear modulus of grout
G_r = shear modulus of rock mass

The most important aspect in this type modeling is how to evaluate the deformation increment Δu_{r0}. Provided that, the system is axisymmetric, the radial deformation of surrounding rock can be given as:

$$u_r = \frac{A}{r} \tag{7.68}$$

Parameter A depends upon the elastic properties of rock mass and *in situ* stress. The deformation increment following the installation of rockanchors may be given as:

$$\Delta u_{r0} = (A^e - A^i)\frac{1}{r} \tag{7.69}$$

Indices e and i correspond to equilibrium and installation of rockanchors. Thus, the increment of deformation may be easily given by the following relation:

$$\Delta u_{r0} = \frac{1 + \mu_r}{E_r}\left[(\sigma_0' - \sigma_{rd})r_d^2 - (\sigma_0' - P_{in})a^2\right]\frac{1}{r} \tag{7.70}$$

where
$\sigma_0' = \sigma_0 + \Delta\sigma_0$
σ_0 = *In situ* stress
$\Delta\sigma_0$ = Stress increment due to rockanchors
P_{in} = Internal pressure acting at the time of rockanchor installation
μ_r = Poisson's ratio of rock mass
E_r = Elastic modulus of rock mass

As discussed previously (see Equation 7.49), the exact solution of Equation (7.67) is not possible and numerical integration would be necessary. It is also approximated by the following relationship as given below:

$$\Delta u_{r0} = C_0' e^{-Dr} \tag{7.71}$$

where

$$C_0' = \frac{1+\mu_r}{E_r}\left[(\sigma_0' - \sigma_{rd})r_d^2 - (\sigma_0' - P_{in})a^2\right]\frac{e^{Dr_a}}{r_a}; \quad D = \frac{\ln(r_b/r_a)}{r_b - r_a}$$

Deformation of rockanchors can be obtained by inserting Equation (7.71) into Equation (7.67) as given below:

$$u_b = C_1 e^{\alpha\eta} + C_2 e^{-\alpha\eta} + C_0 e^{-D\eta} \tag{7.72}$$

where

$$C_0 = C_0' \frac{\alpha^2}{\alpha^2 - D^2}$$

Integration coefficients C_1 and C_2 are obtained from the following boundary conditions:

$\sigma_\eta = \sigma_b$ at $\eta = r_a$
$\sigma_\eta = 0$ at $\eta = r_b$

as given below:

$$C_1 = \frac{1}{e^{\alpha(r_b - r_a)} - e^{-\alpha(r_b - r_a)}}\left[\frac{\sigma_b}{E_b\alpha}e^{-\alpha r_b} - C_0 D\left(e^{-\alpha r_b - Dr_a} - e^{-\alpha r_a - Dr_b}\right)\right] \tag{7.73a}$$

$$C_2 = \frac{1}{e^{\alpha(r_b - r_a)} - e^{-\alpha(r_b - r_a)}}\left[\frac{\sigma_b}{E_b\alpha}e^{\alpha r_b} - C_0 D\left(e^{-\alpha r_b - Dr_a} - e^{-\alpha r_a - Dr_b}\right)\right] \tag{7.73b}$$

Thus, the axial stress of rockanchor is obtained as:

$$\sigma_\eta = -\alpha\left(C_1 e^{\alpha\eta} - C_2 e^{-\alpha\eta} - C_0 D e^{-D\eta}\right) \tag{7.74}$$

If pre-stress is applied to rockanchors, the deformation of a rockanchor is given by the following relationship (see 4.8 Pull-out capacity for details):

$$u_b^{pr} = \frac{\sigma_{pr}}{E_b\alpha} \cdot \frac{e^{\alpha(r_b - \eta)} + e^{-\alpha(r_b - \eta)}}{e^{\alpha(r_b - r_a)} - e^{-\alpha(r_b - r_a)}} \tag{7.75}$$

where σ_{pr} is pre-stress value. It should be noted that it is assumed that the radial direction coincides with the longitudinal axial η of the rockanchor.

The effect of rockanchors on the deformation and stress state of circular openings would follow the approach proposed by Egger (1973b,c). In this approach two cocentric cylinders are considered and the axial stress of rockanchors with distributed uniform pressure P_b over the surface of cylinders are r_d, r_a, and ∞ (see Figure 7.16). It should be also noted that internal pressure P_{a1} at $r = r_a$ will act on the inner cylinder and internal pressure P_{a2} will act on the outer cylinder. Thus, there will be internal pressure jump P_a at $r = r_a$ as given below:

$$P_a = P_{a1} - P_{a2} \tag{7.76}$$

$$P_a = P_b \frac{a}{r_a} \quad P_b = \frac{n \cdot A_b \cdot \sigma_b}{2\pi \cdot a \cdot t} \tag{7.77}$$

where t, n, A_b, and σ_b are spacing, number, cross-section area, and axial stress of representative anchor at $r = r_a$.

If the continuity of radial deformation is required at $r = r_a$, the increment of *in situ* stress as a function of P_b can be given as:

$$\Delta\sigma_0 = P_b \frac{a}{r_a} \tag{7.78}$$

Thus, the effective *in situ* stress acting in the inner cylinder may be given as:

$$\sigma_0' = \sigma_0 + \Delta\sigma_0 \tag{7.79}$$

Assuming the brittle plastic behavior of yielded rock (Mohr-Coulomb type), the stress components may be given as (Aydan and Ersen, 1985):

- Radial stress:

$$\sigma_r = (P_i + \frac{\sigma_c^*}{q^*-1})\left(\frac{r}{a}\right)^{q^*-1} - \frac{\sigma_c^*}{q^*-1}) \tag{7.80a}$$

- Tangential stress:

$$\sigma_\theta = (P_i + \frac{\sigma_c^*}{q^*-1})\left(\frac{r}{a}\right)^{q^*-1} + \frac{\sigma_c^*}{q^*-1}) \tag{7.80b}$$

- Elastic-plastic zone radius:

$$r_d = a\left[\frac{\sigma_{rd} + \frac{\sigma_c^*}{q^*+1}}{P_i + \frac{\sigma_c^*}{q^*-1}}\right]^{\frac{1}{q^*-1}} \tag{7.80c}$$

σ_{rd} is at $r = r_d$ and given specifically as:

$$\sigma_{rd} = \frac{2\sigma_0 - \sigma_c}{q+1} \tag{7.80d}$$

Where
P_i = internal pressure
σ_c = uniaxial compressive strength of non-yielded rock mass
σ_c^* = uniaxial compressive strength of yielded rock mass
ϕ = friction angle of non-yielded rock mass
ϕ^* = friction angle of yielded rock mass

$$q = \frac{1+\sin\phi}{1-\sin\phi} \quad q^* = \frac{1+\sin\phi^*}{1-\sin\phi^*}$$

If volumetric strain is assumed to be zero, the radial deformation at elastic-plastic deformation boundary can be given as (e.g. Brown et al., 1983; Aydan and Ersen, 1985):

$$u = \frac{r}{f+1}\frac{1+\mu_r}{E_r}(\sigma_0' - \sigma_{rd})\left[2\left(\frac{r_d}{r}\right)^{f+1} + (f-1)\right] \tag{7.81}$$

where f is interpreted as Plastic Poisson's ratio.

Stress and deformation in elastic zones may be given as:

Inner Cylinder $(r_d - r_a)$

- Radial stress:

$$\sigma_r = \sigma_0' - (\sigma_0' - \sigma_{rd})\left(\frac{r_d}{r}\right)^2 \tag{7.82a}$$

- Tangential stress:

$$\sigma_\theta = \sigma_0' + (\sigma_0' - \sigma_{rd})\left(\frac{r_d}{r}\right)^2 \tag{7.82b}$$

- Radial deformation:

$$u = \frac{1+\mu_r}{E_r}(\sigma_0' - \sigma_{rd})\left(\frac{r_d}{r}\right)^2 \tag{7.82c}$$

Outer Cylinder $(r_a - \infty)$

- Radial stress:

$$\sigma_r = \sigma_0 - (\sigma_0 - \sigma_{ra})\left(\frac{r_d}{r}\right)^2 \tag{7.83a}$$

- Tangential stress:

$$\sigma_\theta = \sigma_0 + (\sigma_0 - \sigma_{ra})\left(\frac{r_a}{r}\right)^2 \tag{7.83b}$$

- Radial deformation:

$$u = \frac{1+\mu_r}{E_r}(\sigma_0 - \sigma_{ra})\left(\frac{r_a}{r}\right)^2 \tag{7.83c}$$

where

$$\sigma_{ra} = \sigma_0' - (\sigma_0' - \sigma_{rd})\left(\frac{r_d}{r}\right)^2 - P_a$$

To evaluate P_b and ground-support reaction responses, the internal pressure is gradually reduced. This internal pressure involves internal pressure (P_s) provided by other support members, internal pressure (P_{bd}) resulting from rockanchor as a result of deformation of ground, and internal pressure (P_{pr}) resulting from pre-stress applied to rockanchors.

Deformation of surrounding rock following the installation of rockanchors may be given as:

$$\Delta u_{ra} = \frac{a}{f+1}\frac{1+\mu_r}{E_r}(\sigma_0' - \sigma_{rd})\left[2\left(\frac{r_d}{a}\right)^{f+1} + (f-1)\right] - \frac{1+\mu_r}{E_r}(\sigma_0 - P_{in})a \qquad (7.84)$$

Deformation of rockanchors may be also given as:

$$u_b^a = u_f + u_b^{ra} \qquad (7.85)$$

where u_b^{ra} and u_f deformation of grout annulus at $r = r_a$ and deformation of unbonded part of rockanchor, which is given by:

$$u_f = \frac{\sigma_b}{E_b}(r_a - a)$$

Requiring Equation (7.85) and (7.84) to become equal to each other, which is given as:

$$F = \Delta u_{ra} - u_b^a = 0 \qquad (7.86)$$

the internal pressure provided rockanchors can be obtained from an iteration technique, such as the Newton-Raphson iteration method, as given below:

$$P_b^k = P_b^{k-1} - \frac{F(P_b^{k-1})}{F'(P_b^{k-1})} \qquad (7.87)$$

Specifically for each value of P_i, P_b and P_s subsequently are computed. If there are no other support members, the value of P_s is zero. On the other hand, if pre-stress is applied to rockanchors, P_{pr} is added to P_{in} and the deformation (u_b^{pr}) of rockanchors due to pre-stress given by Equation (7.75) is added to Equation (7.85), and it is solved through the procedure similar to the previous situation.

The solution given above is applied to investigate the following four different situations:

- Effect of installation time
- Effect of rock mass
- Spacing of rockanchors
- Effect of pre-stressing and anchorage length

Table 7.4 gives the material properties and geometrical parameters used in computations. Figure 7.17 shows the axial stress in rockanchors for perfectly plastic behavior for different installation times. As expected, the axial stress in rockanchors is higher, as the installation time is quicker. This would indicate that the deformation of surrounding ground stabilizes at earlier stages.

Table 7.4 Material properties used in analyses.

σ_0 (MPa)	E_r (GPa)	v	σ_c (MPa)	σ_c^* (MPa)	ϕ (°)	ϕ^* (°)	E_b (GPa)	a (m)	b (m)	r_b (mm)	r_h (mm)
15	5	0.25	10	10	5 32	32	180	3	3	12.5	18

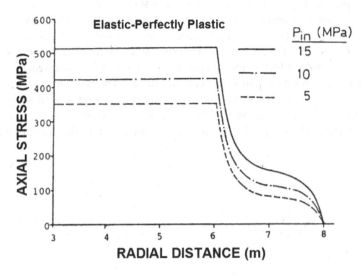

Figure 7.17 Effect of installation time on the axial stress of rockanchors.

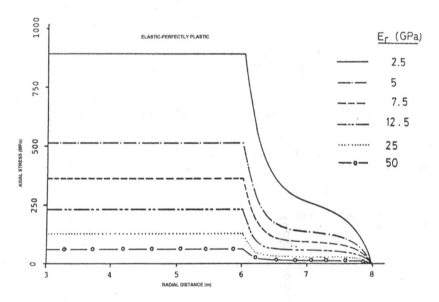

Figure 7.18 Axial stress in rockanchors as a function of rock mass with varying elastic modulus.

Next, the elastic modulus of rock mass is varied from 2.5 GPa to 50 GPa. Computed results are shown in Figure 7.18 for elastic-perfectly plastic behavior. As noted from the figure, the axial stress in weak rock mass is much larger than it is for hard rock mass. Furthermore, the radius of the plastic zone is much smaller in weak rocks compared with that in hard rock mass.

In order to investigate the effect of rock mass behavior on the response of rockanchors, the post-failure deformation of rock mass is assumed to be elastic-perfectly plastic and elastic-brittle plastic. Figure 7.19 shows the stress state around the tunnel. As expected, the effect of rockanchors becomes larger in the case of rock mass with elastic-brittle plastic behavior. It should be noted that the material properties for elastic-perfectly plastic and elastic-brittle plastic were the same except the post-failure strength of rock mass.

Figure 7.20 shows the effect of anchor spacing on the ground-support reaction curves. As expected, the denser rockanchor spacing decreases the axial stresses in rockanchors.

Figure 7.21 shows the effect of pre-stressing of rockanchors on ground-support reaction curves. As noted from the figure, the final support reaction offered by non-pre-stressed rockanchors are basically the same. This results may also imply that the pre-stressing may be unnecessary.

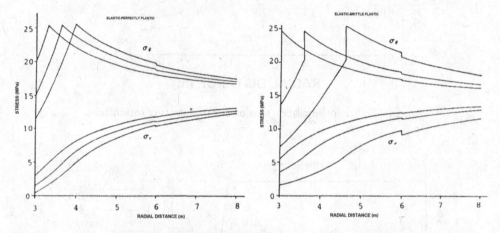

Figure 7.19 Effect of post-failure behavior on the stress state of tunnel.

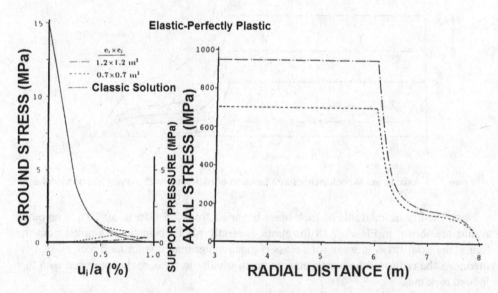

Figure 7.20 Effect of rockanchor spacing on axial stress of rockanchors and ground-support reaction curves.

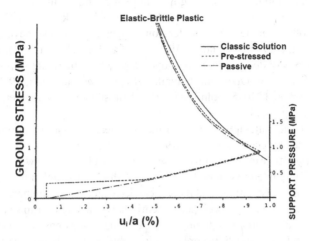

Figure 7.21 The effect of pre-stressing on ground-support reaction curves.

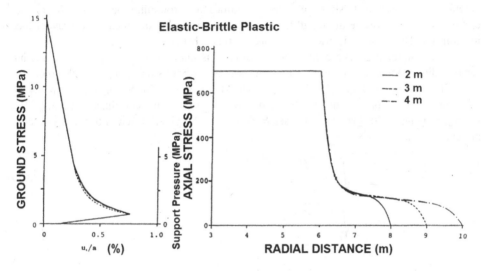

Figure 7.22 Effect of anchorage length on axial stress of rockanchors and ground-support reaction curves.

Next the effect of anchorage length is investigated. Results are shown in Figure 7.22 for elastic-perfectly plastic behavior. Fundamentally, the effect of anchorage length of rockanchors is not pronounced.

7.2.3 Analytical solutions for non-hydrostatic in situ stress state

The analytical solution developed by Kirsch (1898) for stress and deformation distribution in elastic medium about a circular opening in a biaxial far-field stress state was widely used by engineers to understand stress concentration and estimation of possible yield zones.

Kastner (1961) was a pioneer in applying it to tunneling, and his method has been also now used to infer the stress state from borehole breakouts (Zoback *et al.*, 1980). The analytical solutions for other geometries of underground openings are also developed by several pioneers such as Inglis (1913); Mindlin (1940) and Muskhelishvili (1953). These solutions and their applications to deep underground openings in rock mass are summarized in some textbooks and articles (e.g. Obert and Duvall, 1967; Jaeger and Cook, 1979; Terzaghi and Richart, 1952; Mindlin, 1940; Sokolnikoff, 1956; Timoshenko and Goodier, 1951; Muskhelishvili, 1953; Verruijt, 1997).

There were also some attempts to consider the nonlinear behavior of surrounding medium. The solution obtained by Galin (see Savin, 1961 for English description) for a medium behaving elastic-perfectly plastic manner with the use of Tresca yield criterion was the first of its kind. Detournay (1983) attempted to extend his solution to the Mohr-Coulomb material. However, the solutions are not always unique and theoretical derivations are extremely cumbersome.

Gerçek (1988, 1996, 1997) developed a semi-analytical method to evaluate the stress distribution around underground openings with various shapes based on complex variable method. Gerçek and Geniş (1999) applied this technique to obtain possible yield zones around underground openings. Aydan and Geniş (2010) extended this method to obtain potential yield zone due to slippage of discontinuities in surrounding rock mass. This method can be used to determine the length and number of rockbolts/rockanchors to prevent local instabilities. This method is briefly presented in this subsection.

The formula for stresses around openings of various shapes in an isotropic elastic medium for two-dimensional cases can be found in various textbooks and papers on the theory of elasticity (Timoshenko and Goodier, 1970; Muskhelishvili, 1953; Savin, 1961; Greenspan, 1944, etc.). Stress components in a two-dimensional space with a coordinate system ξ, η are $\sigma_{\xi\xi}$, $\sigma_{\eta\eta}$, $\tau_{\xi\eta}$ (Fig. 7.23). Principal stresses σ_1 and σ_3 can be given in terms of the above components as:

$$\sigma_1 = \frac{1}{2}(\sigma_{\xi\xi} + \sigma_{\eta\eta}) + \sqrt{(\frac{\sigma_{\xi\xi} - \sigma_{\eta\eta}}{2})^2 + \tau_{\xi\eta}^2} \qquad (7.87a)$$

$$\sigma_3 = \frac{1}{2}(\sigma_{\xi\xi} + \sigma_{\eta\eta}) - \sqrt{(\frac{\sigma_{\xi\xi} - \sigma_{\eta\eta}}{2})^2 + \tau_{\xi\eta}^2} \qquad (7.87b)$$

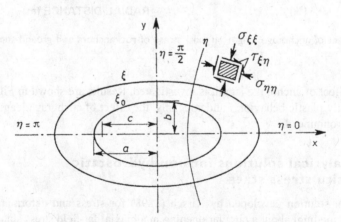

Figure 7.23 Coordinate system and notations.

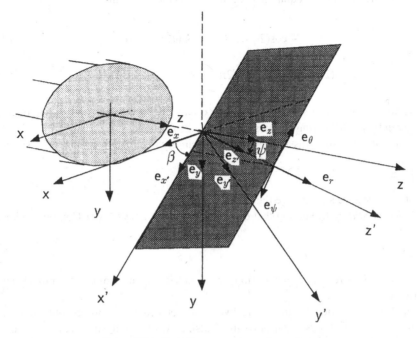

Figure 7.24 Coordinate system and notations.

Let us assume a Cartesian coordinate system, of which axes x, y, and z coincide with the horizontal and vertical directions and the axis of the opening, respectively, and also consider a discontinuity plane which makes angles β and ψ with the axes x and z respectively as shown in Figure 7.24.

The components of the transformation matrix take the following values for this particular case:

$$
\begin{aligned}
l_1 &= -\sin\beta\sin\psi & m_1 &= \cos\beta\sin\psi & n_1 &= \cos\psi \\
l_2 &= -\cos\beta & m_2 &= -\sin\beta & n_1 &= 0 \\
l_3 &= -\sin\beta\cos\psi & m_3 &= \cos\beta\cos\psi & n_3 &= -\sin\psi
\end{aligned} \tag{7.88}
$$

Using this transformation matrix, it can be shown that the normal stress and shear stress in terms of stresses $\sigma_{\xi\xi}$, $\sigma_{\eta\eta}$, $\tau_{\xi\eta}$ take the following forms:

$$
\sigma_n = \sigma_{eq}\cos^2\psi + \sigma_{zz}\sin^2\psi \; ; \; \tau_s = \sqrt{\tau_{x'z'}^2 + \tau_{z'y'}^2} \tag{7.89}
$$

where

$$
\sigma_{eq} = \frac{\sigma_{\xi\xi} + \sigma_{\eta\eta}}{2} - \frac{\sigma_{\xi\xi} - \sigma_{\eta\eta}}{2}\cos(2\eta - 2\beta) + \tau_{\xi\eta}\sin(2\eta - 2\beta)
$$

$$
\tau_{x'z'} = [\sigma_{eq} - \sigma_{zz}]\sin 2\psi
$$

$$
\tau_{x'y'} = \left[\frac{\sigma_{\xi\xi} - \sigma_{\eta\eta}}{2}\sin(2\eta - 2\beta) - \tau_{\xi\eta}\cos(2\eta - 2\beta)\right]\cos\psi
$$

When $\psi = 0$, then the above equations are reduced to the following forms:

$$\sigma_n = \frac{\sigma_{\xi\xi}+\sigma_{\eta\eta}}{2} - \frac{\sigma_{\xi\xi}-\sigma_{\eta\eta}}{2}\cos(2\eta-2\beta) - \tau_{\xi\eta}\sin(2\eta-2\beta) \qquad (7.90a)$$

$$\tau_s = \frac{\sigma_{\xi\xi}-\sigma_{\eta\eta}}{2}\sin(2\eta-2\beta) + \tau_{\xi\eta}\cos(2\eta-2\beta) \qquad (7.90b)$$

Once the normal and shear stresses are known on a plane, separation and sliding conditions can be given by the following expressions (Fig. 7.25):

Condition of Separation (assuming that discontinuity plane has a finite tensile strength):

$$-\sigma_t^j \le \sigma_n \qquad (7.91)$$

note that compressive stresses are regarded as positive.

Condition of Sliding (assuming that discontinuity plane obeys Mohr-Coulomb criterion):

$$\tau_s \ge c_j + \sigma_n \tan\phi_j \qquad (7.92)$$

where σ_t^j is tensile strength, c_j is cohesion, and ϕ_j is friction angle of discontinuity plane/set, respectively.

The method of solution for the above two conditions can generally be obtained through one of the iteration methods, such as Regula Falsi, Newton-Raphson, or explicitly for some simple particular cases. The Regula Falsi Method is employed in this subsection. As the expressions involve two variables, namely ξ and η, one of the variables must be kept constant during an iteration procedure for the other variable. For example, when the variable ξ is kept following its every increment, the iteration for the other variable η is expressed as:

$$\eta = \eta_i - (\eta_{i+1} - \eta_i)\frac{F_i}{F_{i+1} - F_i} \qquad (7.93)$$

The iteration terminates when the sign of $F_i \cdot F_{i+1}$ becomes negative.

As a particular case, circular openings are only considered herein; other cases can be dealt with in a similar way. The stresses at any point about a circular opening under the action

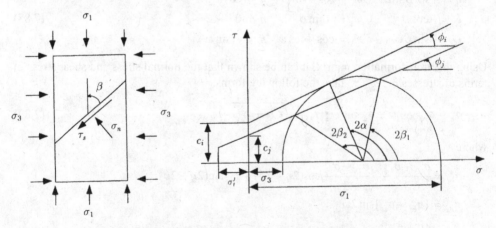

Figure 7.25 Mohr-Coulomb yield criterion with a tension cut.

of far-field stresses σ_{10}, σ_{30} and the internal pressure p_i are first obtained by Kirsch (1898). Adopting a polar coordinate system, the normalized stress components by the largest far-field stress $\sigma_{10} = \sigma_0$ are given by the following expressions (Fig. 7.26):

$$\frac{\sigma_r}{\sigma_0} = \frac{1+k}{2}\left(1-\left(\frac{a}{r}\right)^2\right) - \frac{1-k}{2}\left(1-4\left(\frac{a}{r}\right)^2 + 3\left(\frac{a}{r}\right)^4\right)\cos 2\theta + n_i\left(\frac{a}{r}\right)^2 \tag{7.94a}$$

$$\frac{\sigma_\theta}{\sigma_0} = \frac{1+k}{2}\left(1+\left(\frac{a}{r}\right)^2\right) - \frac{1-k}{2}\left(1+3\left(\frac{a}{r}\right)^4\right)\cos 2\theta - n_i\left(\frac{a}{r}\right)^2 \tag{7.94b}$$

$$\frac{\tau_{r\theta}}{\sigma_0} = \frac{1-k}{2}\left[1-4\left(\frac{a}{r}\right)^2 + 3\left(\frac{a}{r}\right)^4\right]\sin 2\theta \tag{7.94c}$$

where $k = \sigma_{30}/\sigma_0$ ·, σ_0 is maximum *in situ* stress, σ_{30} is minimum *in situ* stress, p_i is internal pressure, and $n_i = p_i/\sigma_0$, respectively.

The normal and shear stresses on a discontinuity plane are obtained after some manipulations as follows:

$$\frac{\sigma_n}{\sigma_0} = \frac{\sigma_{eq}}{\sigma_0}\cos^2\psi + \frac{\sigma_{zz}}{\sigma_0}\sin^2\psi \tag{7.95a}$$

$$\frac{\tau_s}{\sigma_0} = \sqrt{\left(\frac{\tau_{z'x'}}{\sigma_0}\right)^2 + \left(\frac{\tau_{z'y'}}{\sigma_0}\right)^2} \tag{7.95b}$$

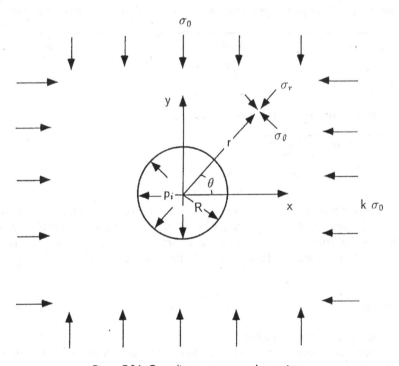

Figure 7.26 Coordinate system and notations.

where

$$\frac{\sigma_{eq}}{\sigma_0} = \frac{1+k}{2}\left(1+\frac{a}{r}\right)^2 \cos(2\theta-2\beta)) - \frac{1-k}{2}(\cos(4\theta-2\beta)$$

$$-4\left(\frac{a}{r}\right)^2 \sin(\theta+\beta)\sin(\theta-\beta) - 3\left(\frac{a}{r}\right)^4)\cos 2\beta) + n_i\left(\frac{a}{r}\right)^2)\cos(2\theta-2\beta)$$

$$\frac{\tau_{x'z'}}{\sigma_0} = \left[\frac{\sigma_{eq}}{\sigma_0} - \frac{\sigma_{zz}}{\sigma_0}\right]\sin^2\psi$$

$$\frac{\tau_{x'y'}}{\sigma_0} = \left[-\frac{1-k}{2}\left(\frac{a}{r}\right)^2 \sin(2\theta-2\beta) - \frac{1-k}{2}\left|\sin(4\theta-2\beta)\right.\right.$$

$$\left.\left.\left(\left[4\left(\frac{a}{r}\right)^2 - 3\left(\frac{a}{r}\right)^4\right]\right|\sin 2\beta\right| + n_i\left(\frac{a}{r}\right)^2\right|\sin(2\theta-2\beta)\right|\cos\psi$$

In a purely hydrostatic state of stress, the normal and shear stresses on a discontinuity plane can be shown to be:

$$\frac{\sigma_n}{\sigma_0} = 1 - (1-n_i)\left(\frac{a}{r}\right)^2 \cos(2\theta-2\beta)\cos^2\psi \tag{7.96a}$$

$$\frac{\tau_s}{\sigma_0} = (1-n_i)\left(\frac{a}{r}\right)^2 \cos\psi\sqrt{(1-cos^2\psi cos^2(2\theta-2\beta)}) \tag{7.96b}$$

Inserting these expressions in the slip condition yields the following expression for slip radius:

$$\frac{r}{a} = \sqrt{\frac{(1-n_i)[\sin\phi_j \cos(2\theta-2\beta)\cos^2\psi - \cos\psi\sqrt{(1-cos^2\psi \cos^2(2\theta-2\beta)}\cos\phi_j]}{\left(\frac{c_j}{\sigma_0}+1\right)\cos\phi_j}} \tag{7.97}$$

When $\psi = 0$, this expression is further simplified to:

$$\frac{r}{a} = \sqrt{\frac{(1-n_i)\sin(2\theta-2\beta-\phi_j)}{\left(\frac{c_j}{\sigma_0}+1\right)\cos\phi_j}} \tag{7.98}$$

Rockbolts generally provide an internal pressure on cavity walls, as well as shear reinforcement along the discontinuities as presented in Chapter 4. The internal pressure provided at the walls of the opening can be given in terms of the axial stress σ_{ax} at the bolt head and spacing e_t, e_l. If the bolts are installed radially, the internal pressure is:

$$p_i = \frac{\sigma_{ax}A_b}{e_t \times e_l} \tag{7.99}$$

where A_b is the cross-section of the bolt.

The total shear reinforcement offered by rockbolts to a discontinuity plane will depend upon the character of the axial stress in bolts, the installation angle, their type, and their dimensions. The grouted rockbolts offer a shear resistance not only from their axial response but also their shear response, and it will take the following form for a rockbolt (see Chapter 4 for details):

$$T_t = nA_b\sigma_b\left(1+\frac{1}{2}\tan\phi_j\sin 2\theta\right)$$

(7.100)

where n is bolting density parameter and σ_b is the strength of the bolt. The strength σ_b of the bar at yielding can be calculated from the von Mises criterion, written in the following form:

$$\sigma_b = \frac{\sigma_t}{\sqrt{\sin^2\theta + 3\cos^2\theta}}$$

(7.101)

where σ_t is tensile yield strength of bar. θ should be taken as the angle between the bolt axis and the normal of plane α for translational type movements and as the angle between the bolt axis and the plane α for separation type movements.

Figure 7.27 shows the application of the above technique when the *in situ* stress state is isotropic. Figure 7.27a shows the effect of rockbolts if they act as an internal pressure. The effect of rockbolts for this particular case is not very much pronounced. Figure 7.27b shows the effect of rockbolts as they provide an additional shear resistance to discontinuities. In this particular case, the effect of rockbolts are more pronounced. Figure 7.27c shows the effect of rockbolts when they provide both an internal pressure and shear resistance. As grouted rockbolts will act in this manner, the effect of bolting is well pronounced.

By superposing the separation zones and slip zones, it is also possible to study the effect of multiple discontinuity sets. Figure 7.28a shows the separation zones around underground openings when discontinuity sets are orthogonal to each other. The large separation zones are due to the vertical discontinuity set and when the contribution of the horizontal set is small. Nevertheless, it should be noted that the horizontal set will kinematically enable the separation zone to move into the opening.

Figure 7.28b shows the separation zones around underground openings when discontinuity sets are not orthogonal to each other (the apex angle is 60°) and inclined at an angle of 60° with respect to the horizontal. The separation zones are smaller in this case as compared with the previous case. The contributions of the both sets are same.

Figure 7.29 shows the slip zones around underground openings when discontinuity sets are orthogonal to each other for isotropic initial stress state. The large slip zones appear at corners when the friction angle is 15°. However, the slip zones decrease as the friction angles of the sets increase. Since the friction angle of sets is usually 30° or more, the regions prone to falling into the opening under gravitational forces are restricted to corners only. These zones will probably appear in underground excavations as overbreaks following the blasting operations.

Figure 7.30 shows the slip zones around underground openings when discontinuity sets are non-orthogonal (the apex angle is 60°) and inclined at an angle of 60° with respect to the horizontal. The slip zones in this case are more extensive. Wedges over the crown and at sidewalls appear, and these wedges will probably move into underground excavations as overbreaks following the blasting operations.

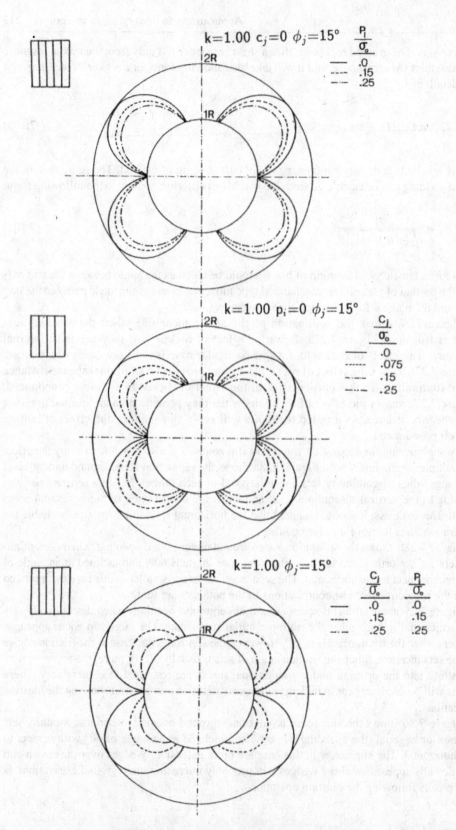

Figure 7.27 Reinforcement effect of rockbolts (k = 1.0).

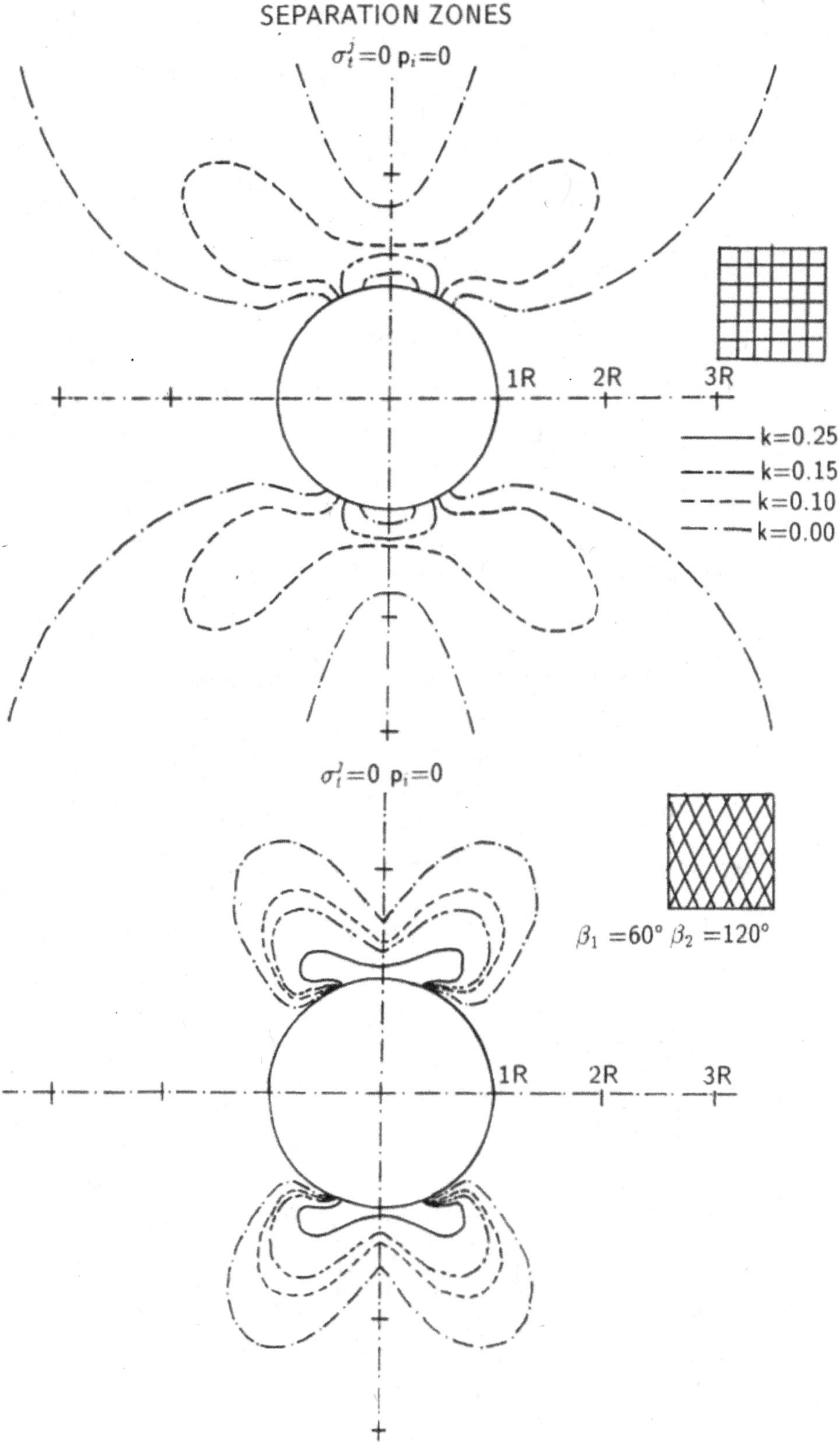

Figure 7.28 Separation zones around circular openings.

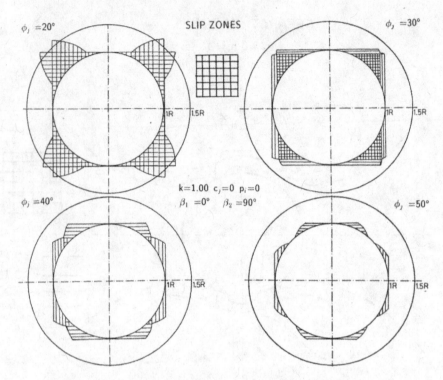

Figure 7.29 Slip zones around a circular opening (orthogonal sets).

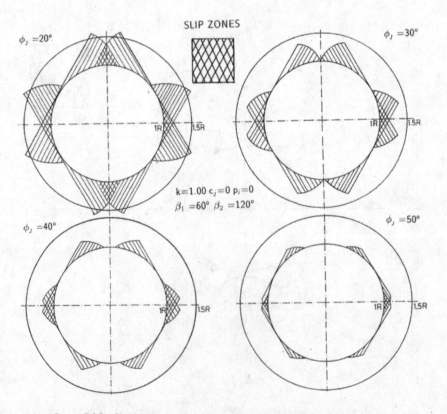

Figure 7.30a Slip zones around a circular opening (non-orthogonal sets).

7.3   NUMERICAL ANALYSES ON THE REINFORCEMENT AND SUPPORT EFFECTS IN CONTINUUM

In this subsection, several examples of applications analyzed by using the finite element model of rockbolts are given. The examples are again related to underground openings (Aydan *et al.*, 1986). Analyses were carried out on the following five particular cases that are thought to be relevant to field situations:

- Effect of bolt spacing
- Effect of the magnitude of the allowed displacement before the installation of the bolts
- Effect of elastic modulus of the surrounding rock
- Effect of equipping rockbolts with bearing plates
- Effect of bolting pattern

Material properties used in these analyses are shown in Table 7.5, and varied parameters are shown in the respective figures. The bolts were equipped with plates of 1 *cm* thick, having dimensions of 10 *cm* by 10 *cm*. The material behavior of the rock was assumed to be elastic-perfectly plastic. The shape of the tunnel is circular and is subjected to an initial hydrostatic state of stress.

7.3.1   Effect of bolt spacing

The bolt spacing was varied from $0.8 \times 0.8 \; m^2$ to $1.2 \times 1.2 \; m^2$. The calculated ground response-support reaction curves, axial and shear stresses in/along rockbolts, and the variation of the extent of elasto-plastic radius are shown in Figure 7.30. Numbers (2), (3), and (4) in Figures 7.30b and 7.30c indicate the distributions at the states of the 80%, 90%, and 100% release of the excavation force. The bolts were assumed to be installed after the 60% release of the excavation force in the analysis. As expected, increasing the density of bolting reduces not only the inward displacement and the extent of the plastic zone radius but also the magnitude of the axial and shear stresses in/along the bolts.

7.3.2   Effect of the magnitude of the allowed displacement before the installation of the bolts

To investigate the effect of the magnitude of the allowed displacement before the installation of bolts, two specific cases were considered:

- Bolts are installed when 60% of the total excavation load was released
- Bolts are installed when 80% of the total excavation load was released

Table 7.5 Material properties and dimensions.

E_b GPa	v_b	E_g GPa	v_g	E_r GPa	v_r	σ_c MPa	σ_0 MPa	ϕ (°)	D_b mm	D_h mm	L m	e_t m	E_l m
210	0.3	3.4	0.3	12.5	0.25	2.5	5	30	36	50	3	1	1

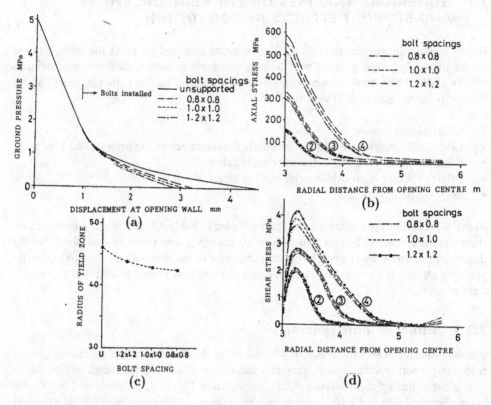

Figure 7.30b Ground response-support reaction curves and stress distributions in/along the bolts: bolt spacing variable.

The spacing of rockbolts were 0.8×0.8 m^2. Symbols u_1 and u_2 refer to the displacement of the tunnel wall at the time of the installation of bolts. The calculated ground response-support reaction curves and axial and shear stresses in/along rockbolts are shown in Figure 7.31. Numbers (2) and (3) in Figure 7.31 indicate the distributions at the states of 90% and 100% release of the excavation force. As expected, the magnitude of the axial stress in bolts is closely interrelated to the remaining deformation potential of the surrounding rock following the installation of the bolts and their stiffness. Nevertheless, it is possible to conclude from the calculated results that the larger the allowed displacement of surrounding rock is, the less the magnitude of the axial stress in bolts is.

7.3.3   Effect of elastic modulus of the surrounding rock

To investigate the effect of the magnitude of the elastic modulus of surrounding rock on the performance of rockbolts, two specific cases were considered:

- Elastic modulus of rock is 10.0 *GPa*
- Elastic modulus of rock 12.5 *GPa*

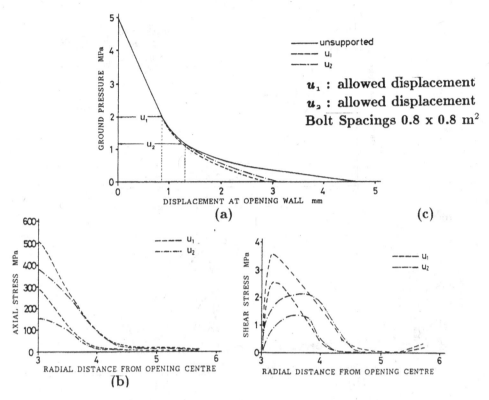

Figure 7.31 Ground response-support reaction curves and stress distributions in/along the bolts: bolt installation timing variable.

The calculated ground response-support reaction curves and axial and shear stresses in/ along rockbolts are shown in Figure 7.32. Numbers (2), (3), and (4) in Figure 7.32b and Figure 7.32c indicate the distributions at the states of 80%, 90%, and 100% releases of the excavation force. As expected, the magnitude of the axial stress in bolts is also closely interrelated to the elastic modulus of the surrounding rock. It is possible to conclude from the calculated results that the greater the elastic modulus of surrounding rock is, the less the magnitude of the axial stress in bolts is.

7.3.4 Effect of equipping rockbolts with bearing plates

To investigate the effect of equipping rockbolts with bearing plates on their performance, two specific cases were considered:

- Bolts are equipped with bearing plates
- Bolts are not equipped with bearing plates

The calculated axial and shear stresses in/along rockbolts and stresses in the surrounding rock are shown in Figure 7.33. Equipping the bolts with bearing plates results in the

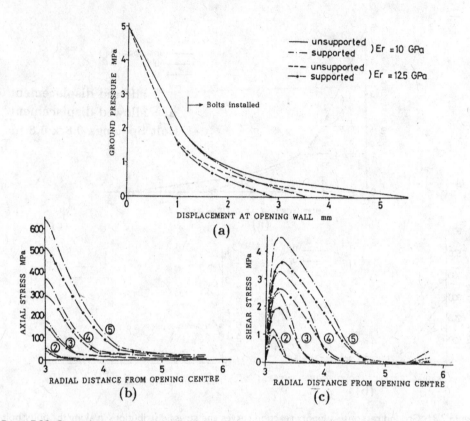

Figure 7.32 Ground response-support reaction curves and stress distributions in/along the bolts: elastic modulus of rock variable.

reduction of opening wall displacement and the extent of the plastic zone radius, compared with those for the case that bolts are not equipped with bearing plates, though the differences in the calculated results for both cases are not so pronounced. Nevertheless, it should be noted that there is a high shear stress concentration in the grout annulus in the close vicinity of the opening wall. Although no yielding was allowed in the annulus in this particular analysis, such high shear stresses will cause shearing along one of the interfaces that will result, in turn, in decreasing the effectiveness of the bolts.

This particular analysis is also interesting, as it invalidates the concept of neutral point proposed by Freeman (1978) to explain the reinforcement effect of grouted rockbolts. Freeman suggested that the load resulting in the region bounded by the opening wall and the neutral point was the load to be picked up and transferred to the region bounded by the neutral point and the far end of the bolt in rock. However, as noted from the axial stress distribution of the bolts and radial and tangential stress distributions in ground, the so-called neutral point has nothing to do with the resistance offered by rockbolts. It just indicates the location where the highest resistance of bolts is mobilized.

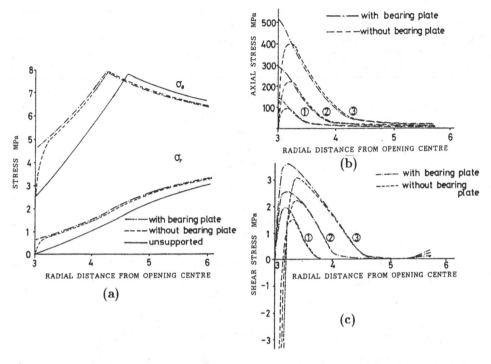

Figure 7.33 Ground-support reaction response curves and stress distributions in/along the bolts – effect of equipment of bearing plates.

7.3.5   Effect of bolting pattern

In addition to the effect of bolt spacing, the effect of floor bolting on the states of stress and deformation of the surrounding rock was analyzed by considering two specific cases (Aydan and Kawamoto, 1991):

- With floor bolting
- Without floor bolting

The shape of the opening was the horseshoe type, the opening was situated 44 *m* below the ground surface, and the lateral stress-coefficient was assumed to be 1.0. Material properties used in the analyses are given in Table 7.6, and the behavior of the surrounding medium was assumed to be elastic-perfectly plastic. Figure 7.34 shows the bolting pattern (a), the displacement of opening perimeter (b), the distribution of contours of safety factor about the opening (c), and axial stress distributions in selected rockbolts (d). As expected, the displacement of the lower half of the opening perimeter in the case of floor bolting is drastically decreased compared with that in the case of no floor bolting. It is interesting to note that the plastic region about the opening with floor bolting is much less than that without floor

Table 7.6 Material properties of steel bar and grout annulus.

E_b GPa	μ_b	σ_t^b MPa	σ_s^b MPa	K_t	K_s	E_{ga} GPa	μ_{ga}	c^p MPa	c^r MPa	ϕ^p (°)	ϕ^r (°)
210	0.3	500	288	0.1	0.1	5	0.2	15	1.5	35	35

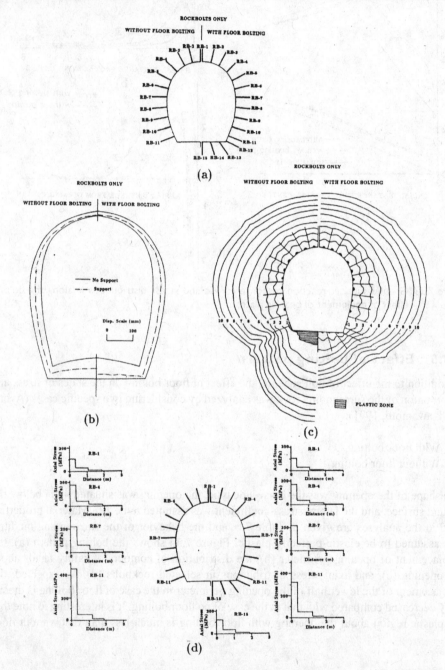

Figure 7.34 Deformation of perimeter of opening, safety factor distribution contours, and axial stress distributions in bolts.

bolting. The distribution of safety factor contours about the opening has a wavy form near the opening perimeter. The safety factor contours tend to go away from the opening perimeter at the mid-distance between two bolts while they approach the opening perimeter near the rockbolts. This is actually the arching action between rockbolts due to bolting, as noted in the model tests by Rabcewicz (1957a). It is also interesting to note that the all-around bolting results in more uniform axial stress distributions compared with those in the case of no floor bolting.

7.3.6   Applications to actual tunnel excavations

Following the above parametric studies, the next two examples are the simulation of excavations of two actual tunnels supported with a combination of rockbolts and other members. The first tunnel was situated 400 m below the ground surface. The primary support pattern (Support System 1) was a combination of rockbolts of 6 m long and shotcrete of 50 mm thick with slits to allow deformation of the tunnel. However, this support pattern was not able to stop the inward movement of the surrounding rock and rockbolts, and the shotcrete ruptured. Following some trial and error regarding the support pattern, it was found out that the support pattern (Support System 2) consisting of rockbolts of 9 m long or more with spacings of 0.6×0.6 m^2 and a shotcrete layer of 250 mm thick was the most effective pattern by which the deformation of surrounding rock could be stabilized.

A parametric study was made in order to investigate the effectiveness of the various support patterns employed in actual trial and error studies *in situ*. The mechanical properties of the surrounding rock are given in Table 7.7 and the details of the analyses are shown in Table 7.8. Five cases studies, numbered C-1, C-2, C-3, C-4, and C-5, were made with the consideration of the various support patterns used in the stabilization of the tunnel. The case study numbered C-2 corresponds to the initial support design, as the slits makes the shotcrete ineffective for a large amount of straining of the surrounding medium. Case studies numbered C-3 and C-4 correspond the trial support patterns. Finally, the case study C-5 corresponds to the support pattern finally decided and proved to be successful.

Table 7.7 Material properties of steel bar and shotcrete and rock.

E_b GPa	μ_b	σ_t^b MPa	E_s GPa	μ_s	σ_c MPa	E_r MPa	μ_r	σ_c MPa	ϕ^p (°)	σ_0 MPa
210	0.3	450	10	0.2	35	250	0.4	2	35	10

Table 7.8 Details of case studies.

Case (No)	Bolt length (m)	Bolt number	Shotcrete thickness (mm)	Comments
C-1	–	–	–	Unsupported
C-2	6	48	–	Rockbolts only
C-3	6	48	50	Rockbolts & Shotcrete
C-4	9	60	50	Rockbolts & Shotcrete
C-4	9	60	250	Rockbolts & Shotcrete

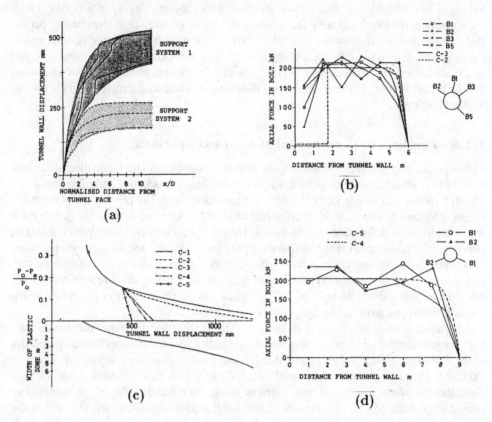

Figure 7.35 Ground-support reaction response curves and stress distributions in/along the bolts.

Figure 7.35 shows the deformation responses of the tunnel for two support systems (C-2 and C-5): (a) ground response-support reaction curves and (b) the comparison of the distributions of axial stress in rockbolts for different case studies together with those measured *in situ* (c), (d). Numbers C-1 to C-5 indicate the results obtained in the case studies listed in Table 7.8, and B1, B2, B3, B4, and B5 denote the rockbolts positioned about the opening in Figure 7.35. The calculated results clearly confirmed that the Support System 1 (C-1) was not suitable and the bolts were bound to rupture. As for the Support System 2 (C-5), though the yield strength of bolts could be exceeded, the support system would be able to stabilize the inward movement of the surrounding rock.

In the next example, a three-dimensional analysis was carried out to simulate the excavation of a shallow railway tunnel (Aydan *et al.*, 1988) As the rock was highly weathered, forepoles and rockbolts together with shotcrete and steel ribs were used as a support system. Material properties used in the analysis are given in Tables 7.9 and 7.10.

The forepoles in this particular case were also simulated by the bolt element. Figures 7.36 shows the geology (a) and the block diagram (b) of the analyzed domain. Figure 7.37 shows the comparison of the development of calculated axial stress in forepoles with those

Table 7.9 Material properties of ground.

Formations	E (MPa)	v	c (kPa)	ϕ (°)
Alluvial Deposit	50	0.3	50	35
V. Weathered Shale	200	0.3	50	30
Weathered Shale	800	0.25	950	35

Table 7.10 Dimensions and material properties of support members.

Support Members	Dimensions	E GPa	v	σ_c or σ_t MPa
Shotcrete	thickness t = 10 mm	5	0.2	10
Steel sets	Cross-section A = 21 cm²	210	0.3	300
Rockbolts and forepoles	Diameter of bolt D_b = 25 mm	210	0.3	300
Grout	Diameter of hole D_h = 37 mm	5	0.25	8

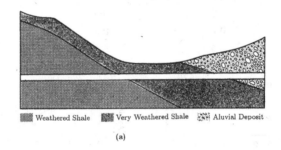

Weathered Shale   Very Weathered Shale   Aluvial Deposit

(a)

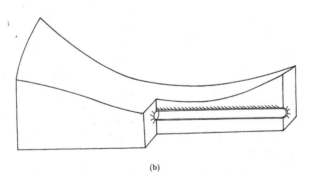

(b)

Figure 7.36 Geologic cross-section and block diagram of analyzed domain.

measured on site in relation to the distance from the tunnel face (a) and axial stress distributions in bolts (b), respectively. FP-1, FP-2, and FP-3 in Figure 7.37 denote the forepoles at crown and positioned at 0 m, 3 m, and 6 m from the tunnel face. B1 to B8 denote the rockbolts, installed about the tunnel in positions shown in Figure 7.37b. As noted from the figure,

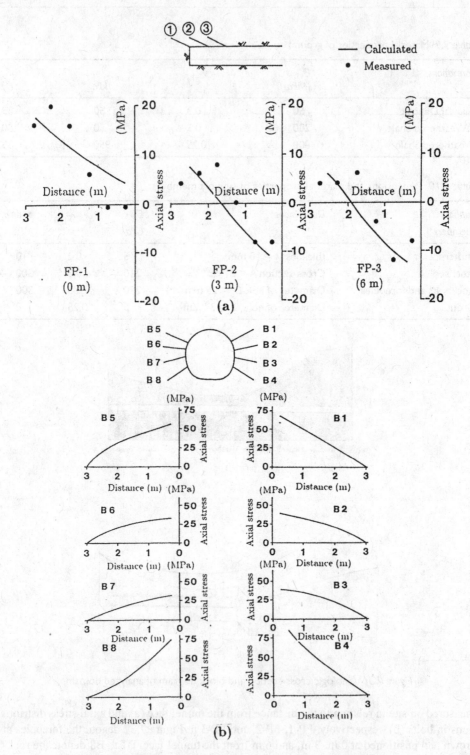

Figure 7.37 Axial stress distributions in the forepoles and rockbolts.

Table 7.11 Material properties used in analyses.

Analysis type	Uniaxial strength (MPa)	Elastic modulus (MPa)	Poisson's ratio	Depth (m)	Unit weight (kN/m³)	Friction angle (°)	Radius of tunnel (m)	Strength reduction ratio
Elastic	20	500	0.3	100	25	30	5	–
Elastic-plastic	2	500	0.3	100	25	30	5	0.8

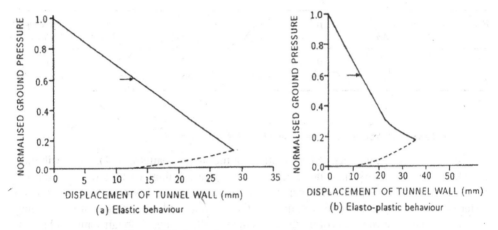

Figure 7.38 Ground-support system reaction curves.

the predicted response of forepoles are very well predicted. It is interesting to note that the forepoles are effective when the net axial stress in the forepoles is of tensile character, and their efficiency decreases as the character of the axial stress tends to become compressive.

The first example of application will be on the time-dependent effect of properties of shotcrete on the deformational response of a circular tunnel in a hydrostatic state of stress. The physical and mechanical properties of surrounding medium are given in Table 7.11. The far-field stress is assumed to be 2.5 MPa. Figure 7.38 shows the calculated ground response-support reaction curves for elastic behavior only and elasto-plastic behavior of medium. It is particularly interesting to note that the response of shotcrete is nonlinear. As the shotcrete hardens, the gradient of the support interaction curve tends to become stiffer.

7.3.7   Comparison of reinforcement effects of rockbolts and shotcrete

In this subsection, we compare the reinforcement effects of rockbolts with those of shotcrete and their combined performance for some typical cases relevant to the practical cases. The material properties of shotcrete and rock mass are given in Tables 7.12 and 7.13. The material properties and dimensions of rockbolts are the same as those given in Table 7.10. For the calculation of initial stress state, the initial *in situ* stress was assumed to be resulting from the overburden (44 *m* from the surface) and the lateral stress coefficient was assumed to be 1.

Table 7.12 Material properties of shotcrete.

E_s (GPa)	μ_s	σ_t^s (MPa)	σ_c^s (MPa)	c (MPa)	h_a
200	0.2	3	15	3	0.1

Table 7.13 Material properties of rock.

Rock Class	E_r (MPa)	μ_r	σ_t^r (kPa)	c (kPa)	ϕ (°)	Unit Weight (kN/m³)
1	75	0.3	0	0	35	19.0
2	180	0.3	120	40	37	20.5

1)   Unclosed shotcrete ring – without floor bolting

Figure 7.39 shows the geometrical illustrations of the support patterns (a-1), the distribution of safety factor contours and plastic regions about the openings (b-1), and the deformed configuration of the opening perimeter (c-1). Between two cases it seems there is not much difference in regard with the distributions of the contours and deformation of the opening perimeter except the waviness of the contours in the bolted zone, although some slight plastifications between bolts are apparent.

2)   Closed shotcrete ring – with floor bolting

Figure 7.40 shows the geometrical illustrations of the support patterns (a-2), the distribution of safety factor contours and plastified regions about the openings (b-2), and the deformed configuration of the opening perimeter (c-2). In contrast to the previous case, there is a remarkable difference. While the contours in the rockbolted case become symmetric about the center of the opening, the contours below the floor in the shotcrete-only case go away from the perimeter of the opening. This implies that rockbolts create an equivalent reinforced ring about the opening perimeter, while the action of shotcrete is only a resistance against the inward movement of the surrounding rock offered at the perimeter.

In this subsection, we compare a number of rockbolt and shotcrete combination patterns in order to investigate the optimum support system from the mechanical point of view.

3)   Comparison of the closed shotcrete ring and without floor bolting system with the unclosed shotcrete ring and with floor bolting system

Here, the main purpose is to investigate whether rockbolt or shotcrete is effective when the floor section is supported by only one type of support member. The comparison is made in terms of the distributions of the safety factor contours and the deformed configuration of the

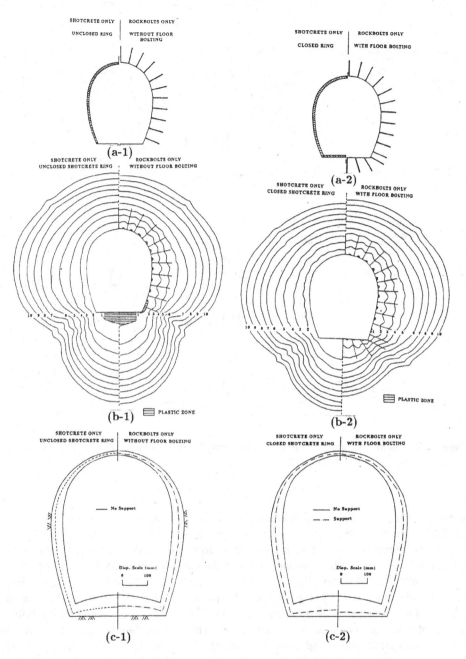

Figure 7.39 Deformation of perimeter of opening and safety factor distribution contours.

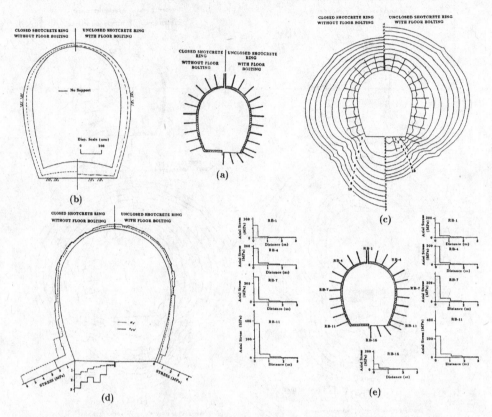

Figure 7.40 Deformation of perimeter of opening, safety factor distribution contours, and axial stress distributions in bolts.

opening perimeter and the magnitude of stresses developed in the support members. Figure 7.41 shows the distribution of safety factor contours (c), the deformed configuration of the opening perimeter (b), and the magnitude of stresses developed in the support members (d), (e). While the all-around bolting results in the contours distributed symmetrically, the contours in the case of without floor bolting are unsymmetrical. Particularly, the safety factor contours below the floor level are more widely spaced, and the area of surrounding medium influenced by the redistribution of stresses is much larger in the case without floor bolting, compared with that in the case with floor bolting. This implies that the support pattern consisted of all-around bolting and unclosed shotcrete ring is preferable to the support pattern consisting of closed shotcrete ring and without floor bolting. The deformed configurations of the opening perimeter for both cases shown in the same figure also confirm the above conclusion. Even if a closed shotcrete ring is used, stresses in shotcrete and rockbolts are larger near the corner of the opening for the case without floor bolting. On the other hand, when the all-around bolting is employed, the stresses in shotcrete and rockbolts become more uniformly distributed, even if the shotcrete ring is unclosed.

4) Comparison of the unclosed shotcrete ring and without floor bolting support system with the closed shotcrete ring and with floor bolting support system

Figure 7.41 shows the distributions of safety factor contours (c), the deformed configuration of the opening perimeter (b), and the magnitude of stresses developed in the support members (d), (e). As expected, when all-around bolting and closed shotcrete ring support system is used, optimum conditions are obtained. It should be also noted that this type of support pattern results in a favorable state of stress, even if the opening shape has a corner, which is highly undesirable from the point of stress concentration. When the combined bolt and shotcrete support system has no floor bolts and without invert shotcreting, a plastic zone appears and the safety factor contours are widely spaced, which indicates the large influence of the stress state (Fig. 7.41c). This implies that, whenever possible, the all-around bolting and closed shotcrete ring is desirable.

When the stresses in support members are compared, a more uniform stress state can be achieved in the complete support system compared with the incomplete support system (Fig. 7.41d,e). In addition, there are large stress concentrations in support members in the case of the incomplete support system near the corners, besides the stress concentrations in the ground.

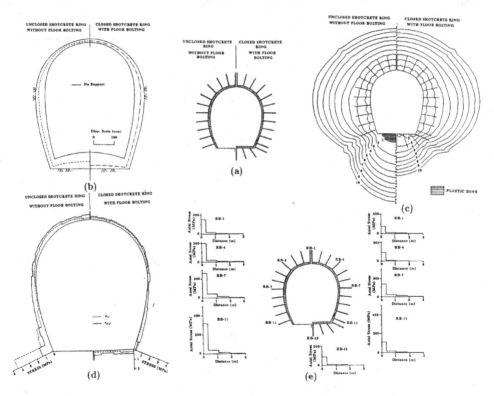

Figure 7.41 Deformation of perimeter of opening, safety factor distribution contours, and axial stress distributions in bolts.

7.3.8   Application to Tawarazaka Tunnel

As the deformation of the Tawarazaka tunnel, located in Nagasaki Prefecture in Japan, had been continuing for more than 1500 days after the completion of excavation along this section, it was understood that some investigations were necessary for the causes of time-dependent deformations. Additional laboratory tests and *in situ* pressure-meter tests and borings were conducted at certain locations along this section. Some swelling tests were also performed. The swelling tests indicated that rocks in this section had no swelling potential. This was unexpected.

The section of the Tawarazaka tunnel exhibiting time-dependent characteristics was analyzed using the elasto-visco-plastic finite element method (Aydan *et al.*, 1995). In the finite element analyses, the effect of support/reinforcement system was considered. First, the unsupported case is considered. The overburden was assumed to be 280 m at the respective location. The physical and mechanical characteristics of rock are given in Table 7.14. Figure 7.42 shows the deformed configuration of surrounding rock around the tunnel at 2500 days after excavation.

Table 7.14 Physical and mechanical properties used in finite element analyses.

ρ (kN/m³)	λ_0 MPa	μ_0 MPa	λ^* MPa day	μ^* MPa day	α_{st}	α_{lt}	K_{st} MPa	K_{lt} MPa
22	275	254	3.96	3.65	1.51	0.9	0.19	0.16

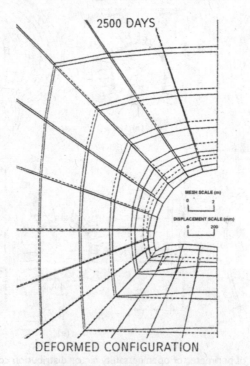

Figure 7.42 Computed deformed configuration.

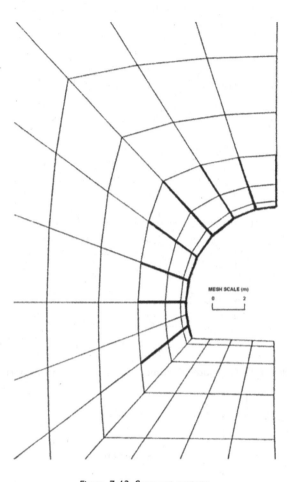

Figure 7.43 Support system.

Figure 7.43 shows the finite element mesh for supported case, in which bold lines correspond to bolts, shotcrete, and steel ribs. This support system was initially adopted for supporting the tunnel through this section. Rockbolts, steel ribs, and shotcrete were represented by rockbolt element (Aydan, 1989) and shotcrete element (Aydan *et al.*, 1990), respectively. Figure 7.44 shows the deformed configuration of surrounding rock around the tunnel at 2500 days after excavation.

Figure 7.45 compares the time history of displacements in the center of the floor after excavation (note that displacements due to excavation were subtracted from the figures) for supported and unsupported cases, together with measurements at the center of the floor. While the time-dependent deformation of the crown decreased from 48.9 mm to 33.7 mm at 2500 days due to the effect of the support pattern, no remarkable difference was observed for the floor as shown in Figure 7.45. This was thought to be due to the pattern of the support system, since it was a non-closed support ring.

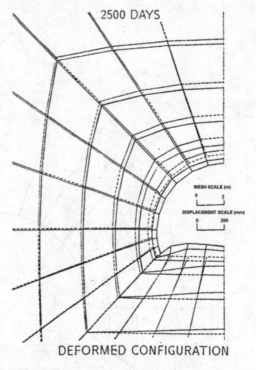

2500 DAYS

MESH SCALE (m)

0   2

DISPLACEMENT SCALE (mm)

0   200

DEFORMED CONFIGURATION

Figure 7.44 Computed deformed configuration (with support).

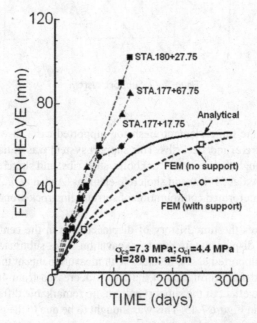

Figure 7.45 Comparison of computed responses with measurements.

7.4   MESH BOLTING IN COMPRESSED AIR ENERGY STORAGE SCHEMES

There is an increasing demand for underground openings for energy storage projects (i.e. CAES and SMES projects) and it is lately an active field of research. The shape of cavities is circular in cross-section with a diameter of 5–6 m. Cavities are subjected to high internal pressures varying between 4 to 8 MPa. As such high internal pressures will cause high tensile stresses in rock masses, which are generally weak against tensile stresses, the reinforcement of the opening by rockbolts is one of the options for resisting such high tensile stresses.

In this subsection, analytical solutions are presented to evaluate the stress state around the internally pressurized circular cavities reinforced by rockbolts having a non-radial pattern, which is different from the conventional rockbolting pattern. Solutions are used for studying the effect of rockbolting, and implications of the parametric studies are discussed. Furthermore, a series of finite element studies was carried out and results are compared with analytical solutions.

The contribution of mesh bolting to the properties of reinforced rock mass has been already presented in Subsection 4.10. A parametric study is carried out together with FEM analyses by Aydan *et al.* (1995a) and it is shown that the contributions of the bolt in its axial direction is the largest and its contribution is negligible in the direction perpendicular to the bolt axis. Besides these, the contribution of rockbolts to the Poisson's ratio of rock mass is negligible.

7.4.1   Analytical solution

In this subsection, a circular cavity is considered and the problem is assumed to be axisymmetric (Fig. 7.46):

- It is assumed that there exist three zones around the cavity:
  - Ruptured reinforced zone $a \le r \le R_p$
  - Elastic reinforced zone $R_p \le r \le b$
  - Unreinforced zone $b \le r \le \infty$

- The constitutive law between stresses and strains for reinforced zones and unreinforced zone are of the following forms:

$$\begin{Bmatrix} \sigma_r \\ \sigma_\theta \end{Bmatrix} = \begin{bmatrix} K_r & G \\ G & K_\theta \end{bmatrix} \begin{Bmatrix} \varepsilon_r \\ \varepsilon_\theta \end{Bmatrix} \tag{7.102}$$

It should be noted that the above constitutive law holds for a medium that is concentrically anisotropic about the cavity. The constants in Equation (7.102) may be written in the following forms for each zone respectively:

Ruptured reinforced zone:

$$K_r = \frac{E(1-\nu)}{(1+\nu)(1-2\nu)}, \quad G = \frac{E\nu}{2(1+\nu)(1-2\nu)}, \quad K_\theta = \frac{E\nu}{(1+\nu)(1-2\nu)} \tag{7.103}$$

Elastic reinforced zone:

$$K_r = \frac{E_r^*(1-\nu)}{(1+\nu)(1-2\nu)}, \quad G = \frac{(E_r^* + E_\theta^*)\nu}{2(1+\nu)(1-2\nu)}, \quad K_\theta = \frac{E_\theta^*(1-\nu)}{(1+\nu)(1-2\nu)}, \quad (7.104)$$

where

$$E_r^* = E + \Delta E_r, \ E_\theta^* = E + \Delta E_\theta$$

Elastic unreinforced zone:

$$K_r = \frac{E(1-\nu)}{(1+\nu)(1-2\nu)}, \quad G = \frac{E\nu}{(1+\nu)(1-2\nu)}, \quad K_\theta = \frac{E(1-\nu)}{(1+\nu)(1-2\nu)} \quad (7.105)$$

where
E = elastic modulus of rock
ν = Poisson's ratio of rock
ΔE_r = contribution in r direction
ΔE_θ = contribution in θ direction

Furthermore, the behavior of ruptured reinforced zone is elastic-brittle plastic, as illustrated in Figure 7.43.

- Ruptured reinforced zone:

Utilizing the residual tensile strength of mesh-bolted rock mass in Equation (7.4) with $n = 1$ and solving the resulting differential equation yields:

$$\sigma_r = \frac{C}{r} - \sigma_{tr}^* \quad (7.106)$$

By introducing the following boundary condition for the ruptured reinforced zone:

$$\sigma_r = p_i \text{ at } r = a$$

Integration constant C is obtained as:

$$C = (p_i + \sigma_{tr}^*)a$$

Accordingly, stresses are obtained as:

$$\sigma_r = (p_i + \sigma_{tr}^*)\frac{a}{r} - \sigma_{tr}^* \quad (7.107a)$$

$$\sigma_\theta = -\sigma_{tr}^* \quad (7.107b)$$

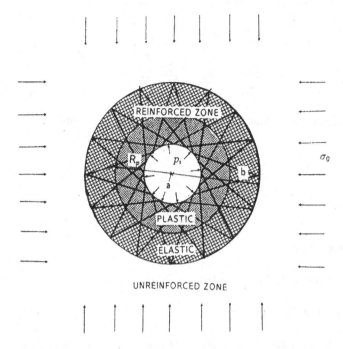

Figure 7.46 Notation of mesh-bolted axisymmetric underground opening.

- Elastic reinforced zone

   The stresses and displacements can be obtained by solving the governing equation (7.4) together with the constitutive law (7.102) and the boundary conditions. Substituting the constitutive law (7.102) in the governing equation (7.4; $n = 1$) together with the relationship (7.104), results in the following differential equation:

$$r^2 \frac{d^2u}{dr^2} + r \frac{du}{dr} - \lambda^2 u = 0 \quad \lambda = \sqrt{\frac{K_2}{K_1}} \tag{7.108}$$

The above differential equation is known as a Cauchy or Euler type differential equation. The general solutions of the above differential equation is of the following form:

$$u = A_1 r^{\lambda} + A_2 r^{-\lambda} \tag{7.109}$$

Note that when λ is equal to 1, it corresponds to isotropic case. By introducing the following boundary conditions for reinforced zone

$$\sigma_r = p_{rp} \text{ at } r = R_p \text{ and } \sigma_r = p_b \text{ at } r = b$$

Integration constants A_1, A_2 are obtained as:

$$A_1 = \frac{1}{K_r\lambda + G}\beta_1 \quad \beta_1 = \frac{p_b b^{\lambda+1} - p_a a^{\lambda+1}}{[b^{2\lambda} - a^{2\lambda}]}$$

$$A_2 = \frac{1}{K_r\lambda - G}\beta_2 \quad \beta_2 = \frac{p_b a^{\lambda-1} - p_a b^{\lambda-1}}{[b^{2\lambda} - a^{2\lambda}]}b^{\lambda+1}a^{\lambda+1}$$

Accordingly, stresses are obtained by inserting the above constants in the respective equations as:

$$\sigma_r = \beta_1 r^{\lambda-1} - \frac{\beta_2}{r^{\lambda+1}} \tag{7.110a}$$

$$\sigma_\theta = \beta_1 r^{\lambda-1} + \frac{\beta_2}{r^{\lambda+1}} \tag{7.110b}$$

• Unreinforced zone

By introducing the following boundary conditions

$$\sigma_r = p_b \text{ at } r = b \text{ and } u = 0 \text{ at } r = \infty$$

and assuming that outer zone is isotropic, one easily obtains the integration constants A_1, A_2 as:

$$A_1 = 0, \quad A_2 = -\frac{(1+\nu)}{E}b^2 p_b.$$

Then, stresses are obtained as:

$$\sigma_r = p_b\frac{b^2}{r^2} \tag{7.111a}$$

$$\sigma_\theta = -p_b\frac{b^2}{r^2} \tag{7.111b}$$

Interface pressures appearing in the above integration constants can be easily obtained from the conditions, such as the continuity of radial displacement at the respective interfaces and yielding conditions.

7.4.2 Applications

In this section, the applications of the presented close-form solutions are given. In the applications, the ratio of E_θ/E is varied with respect to the elastic modulus of medium E while the ratio of radius of the reinforced zone to the cavity radius is varied from 1.2 to 2.4 times (Figs. 7.47–7.49). The variation of E_θ/E corresponds to various rockbolting patterns. When $E_\theta = E$ and $E_r > E$, it corresponds to the radial pattern (Fig. 7.47). On the other hand, when $E_r = E$ and $E_\theta > E$, it corresponds to the tangential pattern (this case can be also considered as reinforced concrete lining) (Fig. 7.48). When $E_\theta = E_r$ and E_r, $E_\theta > E$, it corresponds to the

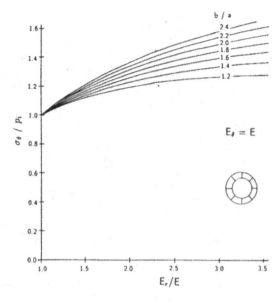

Figure 7.47 Effect of radial installation of rockbolts.

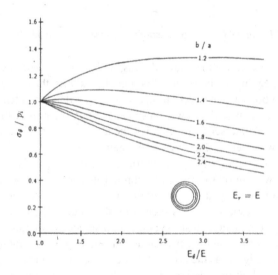

Figure 7.48 Effect of tangential installation of rockbolts.

inclined rockbolting pattern (Fig. 7.49). As seen from Figure 7.47, the intensity of tangential stress becomes higher than that of the isotropic case when the elastic modulus of the reinforced zone is increased radially and isotropically. On the other hand, increasing the tangential elastic modulus results in the decrease of the intensity of tangential stresses as the ratio of E_θ/E increases. It seems that it would be desirable to reinforce the rock tangentially if the intensity of tensile stress is to be decreased. The inclined rockbolting pattern does not result in any

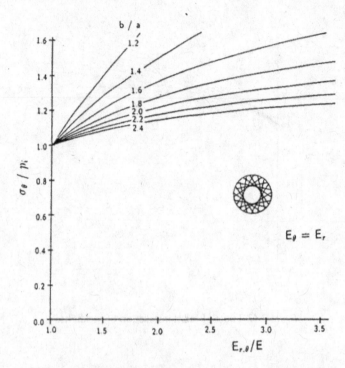

Figure 7.49 Effect of mesh-type installation of rockbolts.

reduction in tensile stress reduction. Nevertheless, it should be noted that the above stresses are total stresses. Depending on the volume fraction of rockbolts, the tensile and radial stresses will be shared by rockbolts which will, in turn, result in the reduction of tensile stresses in rock.

A series of finite element analyses were also carried out to compare the result of the analytical solution, as well as to see how effective the averaging techniques could be. In FEM analyses, rockbolts were represented by bar elements having axial stiffness only, which is not inappropriate for modeling fully grouted rockbolts. Bearing this in mind, the tangential stress at the tunnel wall normalized by the applied internal pressure and the tunnel wall displacement normalized by that of the unreinforced tunnel were obtained and shown in Figures 7.50 and 7.51, respectively. As seen from the figure, the finite element results are clearly different from the tangential stress and wall displacement obtained from the averaging model described in this volume, particularly in the case of the radial pattern, and this difference becomes negligible as the bolt installation tends to become tangential. Figure 7.52 compares the effect of bolt installation angle on the stress components and displacements at the tunnel wall calculated by the proposed method and the FEM. As is clearly seen from the figure, the results calculated by the proposed method are seemingly different from those calculated by the FEM, in terms of magnitude, while the tendencies are almost the same.

A series of analyses was carried out to see the effect of the ratio of residual tensile strength of rock on the extent of the plastic zone while changing the magnitude of peak tensile strength. The calculated results are also shown in Figure 7.52. As expected, as residual strength of the rock mass increases, the radius of plastic zone decreases. This implies that the dense bolting, particularly near the opening wall, may be quite effective in controlling the extent of the plastic zone.

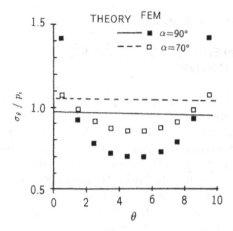

Figure 7.50 Distribution of tangential stress between two rockbolts at the cavity surface.

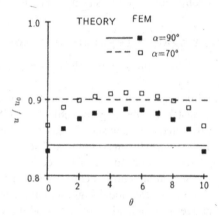

Figure 7.51 Distribution of radial deformation normalized by that of unbolted cavity.

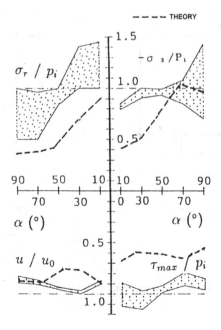

Figure 7.52 Effect of bolt installation angle on various parameters.

7.5   REINFORCEMENT EFFECTS OF ROCKBOLTS IN DISCONTINUUM

In light of presenting the fundamental aspects of rockbolt responses to various forms of movements of discontinuities, the reinforcement offered by rockbolts for some specific cases is evaluated both qualitatively and quantitatively in the next examples. Instabilities about underground openings in a layered or blocky rock mass generally involve block falls, block slides, flexural or block toppling, and combined block toppling and slides (Aydan, 2016). More complicated cases can be found in Aydan (1989) and Kawamoto *et al*. (1991). Though discontinuity patterns in rock mass are closely associated with rock type, the patterns in a blocky rock mass can be generally classified (Aydan *et al*., 1989) as cross-continuous pattern and intermittent pattern. The intermittent pattern actually represents the most likely pattern in all kinds of rock. If an intermittency parameter introduced by Aydan *et al*. (1989) is used, it can be seen that the cross-continuous pattern is a special case of the intermittent pattern.

7.5.1   Reinforcement against separation: suspension effect

Separation-type movements of discontinuities may be observed particularly in the case of underground openings. As a very specific case, the tendency of the blocks or layers in the roof under the gravitational loading to move into an opening will cause separation-type movements. To analyze the resistance offered by rockbolts, a specific case as shown in Figure 7.53a is analyzed. The calculated axial and shear stresses in/along the bolts are shown in Figure 7.53b. As noted from the figure, the maximum axial stress is observed at the discontinuity plane, and the stress state is similar to that observed in pull-out tests. Therefore, the design of rockbolts must be based on the pull-out capacity of the bolts, which can be easily obtained from the analytical or numerical model suggested in the previous sections.

As long as the bolt anchorage is sufficiently strong to lead the yielding of steel bar, it is instructive to have the right image of a column of rock that can be suspended by a rockbolt

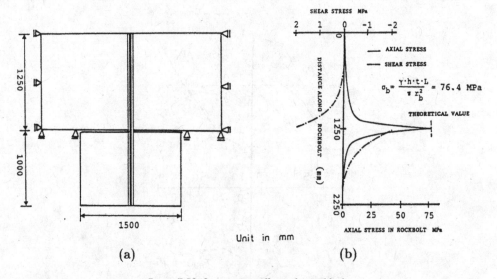

Figure 7.53 Suspension effect of a rockbolt.

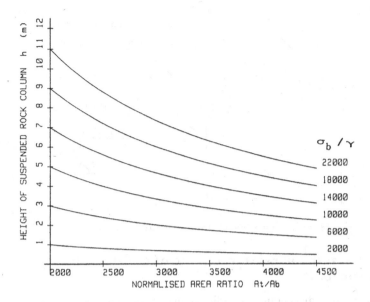

Figure 7.54 Suspension capacity of a rockbolt.

to a stable region further up. For this purpose, a parametric study is done, and the calculated results are shown in Figure 7.54, where σ_b and γ denote the tensile strength of the bolt and the unit weight of the suspended column, respectively. The symbols A_t and A_b have the same meaning as those given in Subsection 4.5.1. Assuming that unit weight of rock is 25 kN/m^3 and the steel bar of 25 mm in diameter with a tensile strength of 550 MPa, the height of a column of rock with a base area of 1 m^2 can be as much as 11 m ($A/A_b = 2043$). This calculation shows that the suspension effect of rockbolts can be significant when considering the size of the current underground openings.

Suspension loads arise when any frictional resistance on the critical bounding planes cannot be mobilized during movements of unstable region towards the opening. Practically, when the apex angle of the unstable region is greater than 90° and there is no possibility of this angle becoming less than 90° during movements due to asperities that may exist on the critical bounding planes, the load due to the dead weight of the unstable region is called the suspension load.

Let us consider that the unstable region is defined by two planes α_1 and α_2 as shown in Figure 7.55. The area of the potentially unstable region can be obtained using the geometry as:

$$A^r = \frac{L_a}{2}\left[L_a \frac{\tan\alpha_1 \tan\alpha_2^*}{\tan\alpha_1 + \tan\alpha_2^*} - R\left\{2\theta \frac{R}{L_a} - \cos\theta\right\}\right] \qquad (7.112)$$

where L_a is width, R, θ are radius and angle of arch, and $\alpha_2^* = \alpha_2 - \xi$.

For a given thickness t, the suspension load can be written in the following form:

$$F_{sus} = \gamma A^r \cdot t \qquad (7.113)$$

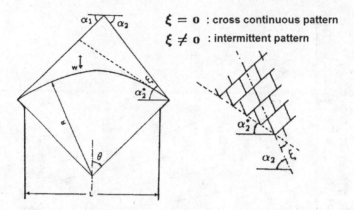

Figure 7.55 Notation for estimating suspension load in roof.

7.5.2 Pillars: shear reinforcement of a discontinuity by a rockbolt

Pillars between two adjacent tunnels, adits, or underground mines utilizing room and pillar technique is a common situation as illustrated in Figure 7.56. When rock mass is layered with a certain inclination, it is very likely that the pillars may fail by the sliding.

The model analyzed is a simulation of tests reported by Egger and Fernandes (1983). They tested a concrete pillar, reinforced by two bolts of 6 *mm* in diameter with a total tensile resistance of 14 *kN* and had a thoroughgoing discontinuity plane inclined at an angle of 45° with respect to pillar axis. The inclination of the bolts with respect to the normal of the discontinuity was varied between 0° and +65° during tests. The geometry and material properties adopted in the analysis are shown in Figure 7.57a and Table 7.15. In the analyses, the inclination of the installation angle of rockbolts with respect to the normal of the discontinuity plane is varied between −45° and +45°, and two bolts were replaced by an equivalent rockbolt of 8.5 *mm* in diameter. The axial and shear stress development in the bolts during shearing is shown in Figures 7.57c and 7.55b. As it is apparent from the figure, the axial and

Table 7.15 Material properties of steel bar and grout and pillar.

E_b (GPa)	μ_b	σ_t^b (MPa)	σ_s^b MPa	E_j	G_j GPa	ϕ (°)	E_r GPa	μ_r	E_{ga}	μ_{ga}
180	0.3	247.5	148	12	.5	33	25	.2	15	0.2

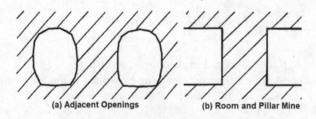

(a) Adjacent Openings   (b) Room and Pillar Mine

Figure 7.56 Pillars susceptible to fail by sliding.

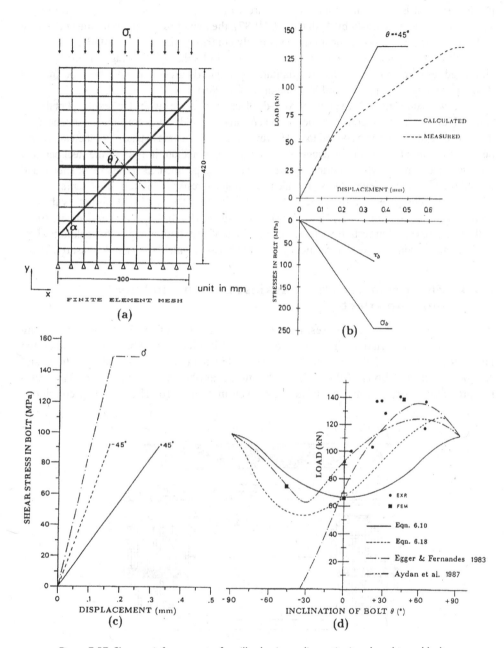

Figure 7.57 Shear reinforcement of a pillar having a discontinuity plane by rockbolts.

shear stress development in the bolts at the vicinity of the discontinuity greatly depend upon the inclination of the bolts.

When the inclination angle is negative, the axial stress development is purely of compressive character. On the other hand, when the inclination is $0°$ the axial stress in the bolt is almost nil. However, the shear stress developing in the bolt during shearing is the greatest in this case compared with that in other installation angles. When the installation angle is positive (here it is +45°), the axial stress is purely of tensile character. As discussed in Sub-section 4.5.2 and previously by Aydan *et al.* (1987), the axial force in the bolt must be tensile if the axial load in the bolt is desired to positively contribute the shear resistance of the dis-continuities. As no measurement of axial stresses in bolts during shearing was made in the simulated tests, the above results are qualitatively in good agreement with the experimental data reported by Haas (1976) and Yoshinaka *et al.* (1986).

Next, we compare the calculated load-displacement curve with the experimental one (Fig. 7.57c). The initial responses of the calculated and measured curves are almost the same up to the half value of the ultimate applied load. However, a deviation starts to take place thereafter. This deviation is attributable to the breakage of the surrounding medium in front of the steel bar near the discontinuity plane, which is often reported by various experimenters (Haas, 1976; Hibino and Motojima, 1981, etc.). Nevertheless, the calculated ultimate resistance is almost the same.

Finally, we compare the ultimate resistances calculated by the FEM analysis and by the the-oretical models (Equations (4.41) and (4.44); Aydan *et al.*, 1987b; Egger and Fernandes, 1983) with those measured in tests for various inclinations of installation angles of rockbolts (Fig. 7.57d). Once again, the good agreement is notable between calculations and measurements.

7.5.3 Shear reinforcement against bending and beam building effect

Particularly in layered rock masses, the beam building effect of rockbolts against bending is of paramount importance (Fig. 7.58). To analyze this effect of rockbolts, we consider two layers of the same thickness put on one another, subjected to a gravitational force field, and reinforced with rockbolts of three different patterns, as shown in Figure 7.59a. The mate-rial properties employed in the analyses are given in Table 7.16. The boundary conditions

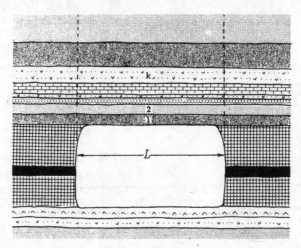

Figure 7.58 Layered rock mass above an underground opening.

Table 7.16 Material properties of rock and discontinuity plane.

E_r (GPa)	μ_r	E_d (GPa)	G_d (MPa)	c (MPa)	ϕ (°)	Unit Weight (kN/m³)	t (mm)
3.15	0.2	3.15	300	0	0	22.0	2.5

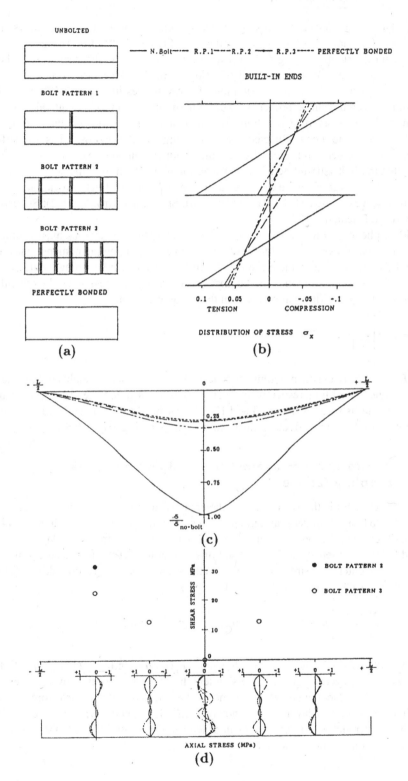

Figure 7.59 Shear reinforcement against bending and beam building effect.

employed in the reported case correspond to built-in conditions. Figure 7.59b shows the fiber stresses parallel to layering at the center of the span. Figures 7.59c, d, and e show the vertical displacement distribution along the layering at the neutral axis, shear stresses in the bolts at the bedding plane, and axial stress distributions in the bolts for various rockbolt patterns, respectively. As noted from the distribution of fiber stresses in Figure 7.59b, the behavior of the multi-layered beam approaches that of a monolithic beam of an equivalent thickness through stitching the layers by rockbolts. It is also interesting to note that the bolt in the center of the span has not been subjected to any stressing. In other words, it has no reinforcing function. On the other hand, the bolts near the abutments undergo very high stresses. This implies that the bolts should be more dense at abutments than near the center of the span. The reported results confirm the existence of the beam building effect of rockbolts as experimentally observed in the tests by Panek (1962a,b); Snyder and Krohn (1982); Snyder(1983) and Roko and Daemen (1983).

Buckling phenomenon is also observed in rock excavations when rock mass is thinly layered. In schistose rocks, a special form of buckling called kinking phenomenon may also be observed. The stress state in the close vicinity of the underground opening would be similar to the uniaxial stress condition (Fig. 7.60a). For such a simple stress state, the critical stress (σ_{cr}) for buckling condition can be given in the following form:

$$\sigma_{cr} = \frac{E}{\alpha}\left(\frac{\pi t}{L}\right)^2 \tag{7.114}$$

where E, t, and L are elastic modulus, thickness of layer, and span (height) of the side of the opening prone to buckling, respectively. Factor (α) depends upon the end constrain condition of the layer and its value would be 3 and 12 for built-in ends and hinged ends respectively. Rockbolts would be most effective means to prevent buckling (Fig. 7.60b).

7.5.4   Reinforcement against flexural and columnar toppling failure

When the rock mass is thinly layered, the layers may not be strong enough to resist tensile stresses due to bending under gravitational forces. Seismic forces may also induce additional loads, easing flexural toppling failure. In such cases, tensile stresses in layers should be reduced below their tensile strength, if stability is required. If the roof consists of n layers (Fig. 7.61), the outer fiber stresses for layers in the sidewall and in the roof take the following forms (Aydan, 1989; Aydan and Kawamoto, 1992):

$$\sigma_t^i = \pm\frac{N_i}{A_i} + \frac{6t_i}{I_i}\left[P_{i+1}\eta h_i - T_{i+1}\frac{t_i}{2} - P_{i-1}\eta h_{i-1} - T_{i-1}\frac{t_i}{2} + S_i\right] \tag{7.115}$$

where $N_i = W_i\cos\alpha - E_i\sin(\alpha + \beta)$; $S_i = W_i\sin\alpha + E_i\cos(\alpha + \beta)$; $W_i = \gamma_i t_i b(h_{i-1} + h_i)/2$; $A_i = t_i b$; γ_i: unit weight of layer; h_i and h_{i+1} are side lengths of the interfaces between layers $i - 1,i$ and layers $i, i + 1$, respectively; P_{i-1}, P_{i+1} and T_{i-1}, T_{i+1} are normal and shear forces acting on of the interfaces between layers $i - 1,i$ and layers $i, i + 1$, respectively; t_i: thickness of layer I; b: width; α: layer inclination; η: coefficient of load action location. Sign (+) stands for layers in roof and (−)for layers in sidewalls.

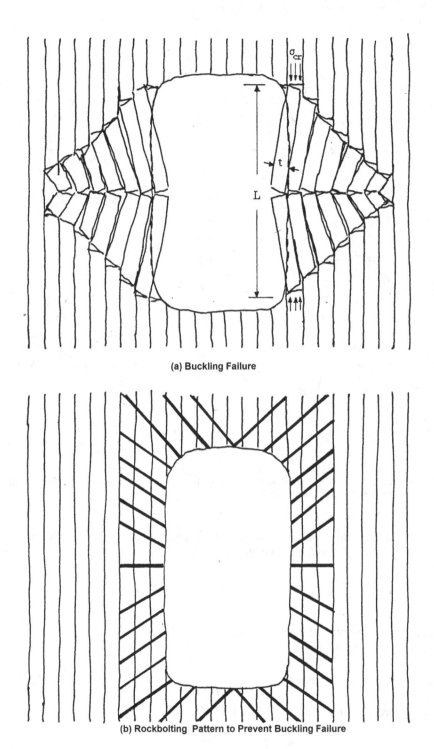

(a) Buckling Failure

(b) Rockbolting Pattern to Prevent Buckling Failure

Figure 7.60 Modeling of buckling failure and prevention by rockbolts.

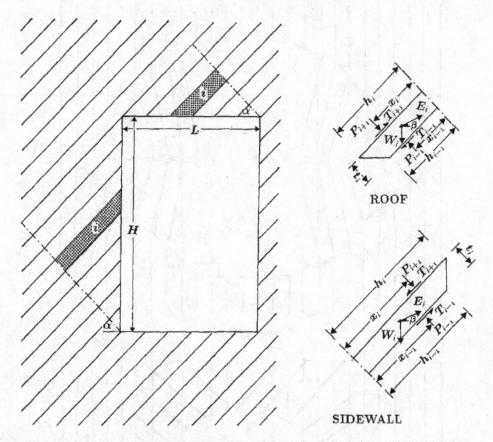

Figure 7.61 Model for limiting equilibrium analysis of flexural toppling of an underground opening (from Aydan and Kawamoto, 1992).

Introducing the yield condition such that the outer fiber stress of the layer is equal to the tensile strength σ_T of the rock with a factor of safety *SF* as:

$$\sigma_t^i \leq \frac{\sigma_T}{SF} \tag{7.116}$$

and assuming the normal and shear forces acting interfaces of layers through frictional yielding condition (ϕ: is friction angle):

$$T_{i+1} = P_{i+1}\mu; \ T_{i-1} = P_{i-1}\mu; \ \mu = \tan\phi \tag{7.117}$$

the normal forces acting on layer $i - 1$ can be easily obtained as:

$$P_{i-1} = \frac{P_{i+1}\left(\eta h_i - \mu\dfrac{t_i}{2}\right) + S_i\dfrac{h_i}{2} - \dfrac{2I_i}{t_i}\left(\dfrac{\sigma_T}{SF} \pm \dfrac{N_i}{A_i}\right)}{\left(\eta h_{i-1} + \mu\dfrac{t_i}{2}\right)} \tag{7.118}$$

The equation above is solved by a "step-by-step method," and rock load P_s is obtained from the following criterion:

$$P_0 > 0 \text{ and } P_0 = P_s \tag{7.119}$$

If $P_0 > 0$, it is interpreted that some support measures are necessary.

The stabilization of underground openings against flexural toppling failure requires reduction in the magnitude of the moment and an increase in the compressive normal forces acting on interfaces. There are a number of ways to provide such an effect through the use of artificial support. Pre-stressed cables and/or fully grouted rockbolts are effective solutions. When the pre-stressed cables are used, they should be anchored beyond the basal plane, otherwise pre-stress forces may cause much higher bending stresses in layers. The alternative is to use fully grouted rockbolts or "dowels." The fully grouted rockbolts would be more economical than rockanchors, as they are shorter and do not need to be pre-stressed. In this subsection, a procedure proposed by Aydan (Aydan, 1989; Aydan and Kawamoto, 1992) is introduced on how to consider the reinforcement effect of the fully grouted rockbolts against flexural toppling failure.

Fully grouted rockbolts contribute to the shear resistance of discontinuities directly through the shear resistance of steel bar itself and indirectly through the shear and frictional components of the axial force in the bar. The reinforcing effect of n rockbolts on the shear resistance of a discontinuity plane can be expressed in the following form (Aydan, 1989) (Fig. 7.62):

$$T_T = nA_b\sigma_b\left(1 + \frac{1}{2}\sin 2\theta \tan \phi\right) \tag{7.120}$$

where σ_b: axial stress; A_b: cross-section of bar; θ: angle between the bar axis and discontinuity (installation angle).

The results of laboratory and field studies and numerical modeling show that the stress in the bar quickly develops and becomes almost equal to its yielding strength at very small displacements in the order of 0.1 mm to 3 mm. It is reasonable, therefore, to assume that the stresses in the bars are the same and equal to their yield strength. The yielding strength

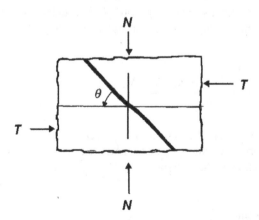

Figure 7.62 A model for rockbolt crossing a discontinuity (modified from Aydan and Kawamoto, 1992).

of a bar subjected to tension or compression and shearing in Equation (7.120) may be given as:

$$\sigma_b = \frac{\sigma_{tb}}{\sqrt{\cos^2 + 3\sin^2\theta}} \tag{7.121}$$

where σ_{tb}: tensile strength of rockbolt.

Considering the contribution of bolts as an addition to the shear resistance across interfaces of layers, one obtains the following:

$$P_{i-1} = \frac{P_{i+1}\left(\eta h_i - \mu\dfrac{t_i}{2}\right) + \left(T_T^{i+1} + T_T^{i-1}\right)\dfrac{t_i}{2} + S_i\dfrac{h_i}{2} - \dfrac{2I_i}{t_i}\left(\dfrac{\sigma_T}{SF} \pm \dfrac{N_i}{A_i}\right)}{\left(\eta h_{i-1} + \mu\dfrac{t_i}{2}\right)} \tag{7.122}$$

where T_T^{i-1} and T_T^{i+1} are resistance provided by rockbolts at the interfaces between layers $i-1$, i and layers i, $i+1$, respectively.

A parametric study on the reinforcement effect of rockbolts on roof and sidewall stabilization of underground openings is carried out and the results are shown in Figure 7.63. Material properties used in the analyses are given in Table 7.17. The opening shape was assumed to be a square of 10 m wide and 10 m high. In the calculations, the ratio of the height and width of the opening to the thickness of layers was varied and the necessary required rockbolt forces per unit width (1 m) for the stability in roof and sidewall was calculated by varying the inclination of the normal of layers from horizontal. Figure 7.63a and 7.63b clearly illustrate that the required reinforcement effect of rockbolts in the roof is much larger than that in the sidewalls. The figure also shows that the mobilized shear resistance of bars will be larger in the case of thinly layered media as compared with the thickly layered media, provided that rock in the vicinity of the bars is not failing.

In the case of columnar toppling, the columns can freely rotate about their base. Therefore, this type of failure is only possible when the shear resistance of the bases is larger than the disturbing shear force and can only occur in the sidewalls of the openings and slopes. The reinforcement effect of rockbolts can be similarly taken as an increase in the shear resistance of interfacing sides of the columns. In this particular case, the following expression will be obtained for sidewalls:

$$P_{i-1} = \frac{P_{i+1}\left(\eta h_i - \mu t_i\right) - T_T^{i+1}t_i + \left(S_i^w\eta\bar{h}_i - N_i^w t_i\right)}{\eta h_{i-1}} \tag{7.123}$$

The meanings of parameters are the same as those given above. The above expressions must be solved for given slope or opening geometry and material properties together with a bolting pattern and be checked by the criteria:

$P_0 < 0$ stable
$P_0 = 0$ at limiting state
$P_0 > 0$ unstable

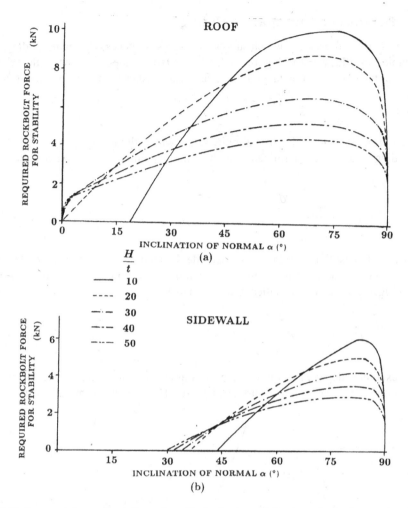

Figure 7.63 Shear reinforcement effect of rockbolts in the stabilization of openings against flexural toppling failure.

Table 7.17 Mechanical properties and dimensions.

Materials	σ_t (MPa)	γ (kN/m³)	ϕ (°)	H m	t m	d mm	Ω	θ (°)
Rock	0.25	25	–	10	variable	–	–	–
Discontinuity	–	–	15	–	–	–	–	–
Rockbolts	–	–	–	–	–	25	0.56	30

7.5.5   Reinforcement against sliding

There are a number of cases in which support measures are necessary against sliding. Typical examples are illustrated in Figure 7.64. Sliding failure is possible when the frictional resistance of the critically orientated discontinuity set satisfies the following condition:

$$\tan \alpha_i < \tan \phi_i \qquad (7.124)$$

where subscript i represents the discontinuity set on which sliding is likely.

The sliding load can be determined from the limiting equilibrium approach in the following form:

$$F_{slid} = \gamma S_{slid} \cdot t \frac{\sin(\alpha_i - \phi_i)}{\cos \phi_i} \qquad (7.125)$$

The area S_{slid} of the sliding region can be easily determined from the geometry of the region prone to sliding in relation to the geometry of opening. For example, the area of sliding body shown in Figure 7.65 can be specifically obtained as follows:

$$S_{slid} = \frac{H_s}{2} \left[H_s \frac{\tan \alpha_1 \tan \alpha_2^*}{\tan \alpha_1 + \tan \alpha_2^*} \right] \qquad (7.126)$$

In the examples depicted in Figure 7.65, it is assumed that a sliding takes place along plane α_1 only and a separation occurs at plane α_2. The specific forms of the resistances

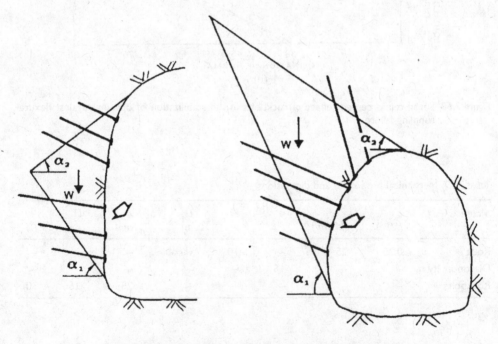

Figure 7.64 Possible sliding modes in underground excavations.

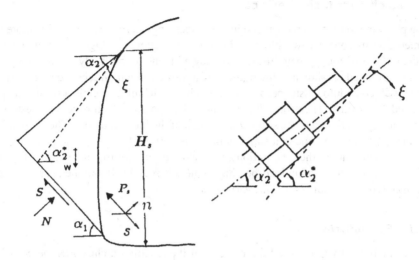

Figure 7.65 Computational models for support pressures for sliding-induced loads.

$\Sigma T_b^{\alpha 1}, \Sigma T_b^{\alpha 2}$ offered by rockbolts on planes α_1 and α_2 can be given in the following forms on the basis of discussions in previous sections, provided that the angles $\theta_{\alpha 1}, \theta_{\alpha 2}$ between bolt axis and the normal of critical planes remain to be the same at the respective planes.

On plane α_1:

$$\Sigma T_b^{\alpha_1} = n_{\alpha 1} A_b^{\alpha_1} \sigma_b^{\alpha_1} \left(1 + \frac{1}{2} \sin 2\theta_{\alpha_1} \tan\phi\right) \tag{7.127}$$

where n_{α_1} and $A_b^{\alpha_1}, \sigma_b^{\alpha_1}$ are the total number of bolts and their cross-section and strength, respectively.

On plane α_2:

$$\Sigma T_b^{\alpha_2} = n_{\alpha_2} A_b^{\alpha_2} \sigma_b^{\alpha_2} \tag{7.128}$$

where n_{α_2} and $A_b^{\alpha_2}, \sigma_b^{\alpha_2}$ are the total number of bolts and their cross-section and strength, respectively.

Then, the factor of safety against sliding takes the following form after some algebraic manipulations:

$$SF = \frac{W \cos\alpha_1 \tan\phi + \Sigma T_b^{\alpha_1} + \Sigma T_b^{\alpha_2}}{W \sin\alpha_1} \tag{7.129}$$

where W is the weight of the sliding body, which can be easily calculated from its geometry together with the unit weight of rock.

The above formulations are based on the static loads. If dynamic loads resulting from earthquakes or turbines present, for example, the method suggested by Aydan et al. (2012), which is also described in Chapter 10, can be utilized.

7.5.6 Arch formation effect

Arching phenomenon is well recognized in many rock engineering works. To illustrate these phenomena, let us consider two blocks with a gap δ as shown in Figure 7.66 in a gravitational field. If these blocks are not restrained, they will tend to fall freely. On the other hand, if they are restrained, as shown in the figure, and allowed to rotate freely, the so-called arching action will come into existence. The arch will be stable unless the blocks are cracked by the induced stress state. Therefore, the arching is only possible with necessary restraining, the allowance of rotation, and the sufficient strength of the block material. The most suitable supporting material is the one that satisfies the necessary conditions of the arch action. As rockbolts superbly satisfy these conditions, they will be superior to other support members. It should be noted that the main load-bearing element is rock itself, not the support member that helps the rock to mobilize its intrinsic resistance.

7.5.6.1 Formulation

To design the bolts, two possible failure forms of the resulting arches must be considered (Fig. 7.67):

- Compressive failure at crest or abutments
- Shear failure at abutments or on a discontinuity plane within the arch

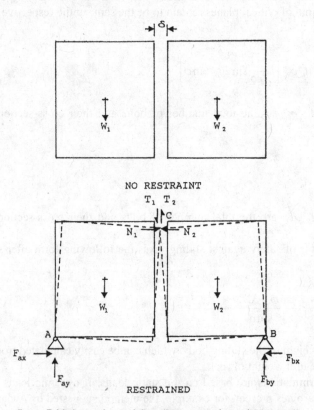

Figure 7.66 A simple model to illustrate arching phenomena.

COMPRESSION FAILURE AT CROWN AND
ABUTMENTS

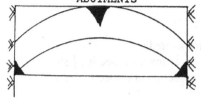

VERTICAL SHEARING AT ABUTMENTS

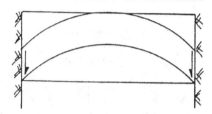

HORIZONTAL SHEARING AT ABUTMENTS

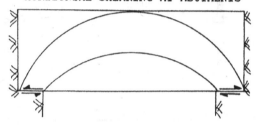

SLIDING ALONG A DISCONTINUITY WITHIN THE ROCK ARCH

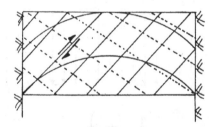

Figure 7.67 Failure modes of rock arches.

In the following presentations, the arch is assumed to be equivalent to a three-hinged arch from the structural engineering point of view (Fig. 7.68). For a given symmetrical loading function $F_2(x) - F_1(x) = f(x)$ with respect to the center line of the opening per unit width, one can write the following equilibrium equations for the adopted coordinate system:

$$\sum F_x = F_{ax} - F_{bx} = 0 \rightarrow F_{ax} = F_{bx} = H \tag{7.130a}$$

$$\sum F_y = W_t \frac{L}{2} - F_{by} L = 0 \rightarrow F_{by} = \frac{W_t}{2} \tag{7.130b}$$

$$\sum M_B = W_t \frac{L}{2} - F_{ay}L = 0 \rightarrow F_{ay} = \frac{W_t}{2} \qquad (7.130c)$$

where
$W_t = \int_0^L \gamma f(x)dx$
γ = unit weight of loose medium
F_x = reactions in x-direction at respective points
F_y = reactions in y-direction at respective points
H = horizontal thrust
L = span

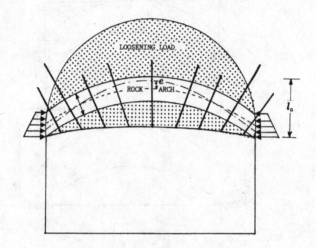

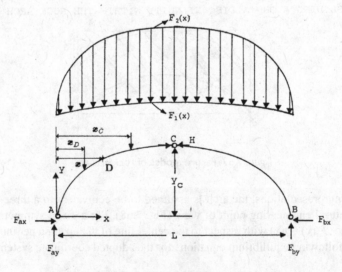

Figure 7.68 Structural model for rock arches.

Taking the moment with respect to point C (left or right), one can write the following:

$$\sum M_C = F_{ax} y_c - F_{ay} \frac{L}{2} + \frac{W_t}{2} \left(\frac{L}{2} - \overline{x}_C \right) = 0 \tag{7.131}$$

where $\overline{x}_C = \int_0^{L/2} f(x)x dx / \int_0^{L/2} f(x) dx$ is average distance of the load vector from the origin, and y_c is the rise of arch. Accordingly, the horizontal thrust H is obtained from Equations (7.130) and (7.131) as:

$$H = \frac{W}{2y_c} \overline{x}_C \tag{7.132}$$

Taking the moment with respect to point D (to the left) results in:

$$\sum M_D = F_{ax} y - F_{ay} x + W_x (x - \overline{x}_D) = 0 \tag{7.133}$$

where $W_x = \int_0^x \gamma f(x) dx$; $\overline{x}_D = \int_0^x f(x)x dx / \int_0^x f(x) dx$; x is distance of point D from the origin. Rearranging the above equation yields the position y of the thrust line as:

$$y = \frac{F_{ax}}{F_{ay}} x - \frac{W_x}{F_{ay}} (x - \overline{x}_D) \tag{7.134}$$

Bending in the arch results from the deviation between the center line of the arch and the influence line of thrust T. Thus, the distribution of the axial stress in the arch per unit width can be calculated from the following expression in terms of an average stress, defined as $\sigma_0 = T/t$:

$$\sigma_a = \frac{T}{t} + \frac{Tey}{I} = \sigma_0 \left(1 + 12 \frac{e}{t} \frac{y}{t} \right) \tag{7.135}$$

where
t = thickness of arch
e = distance between center line and influence line
y = distance of a point from the center line
T = thrust

Let us denote the maximum compressive stress by σ_{cr}, which occurs at the upper side of the arch ($y = t/2$) and e/t by n_e. Then, the relationship between σ_{cr} and σ_0 becomes:

$$\sigma_0 = \frac{\sigma_{cr}}{1 + n_e} \tag{7.136}$$

At the center C of the span, the following relationship must hold:

$$T = H \tag{7.137}$$

As $T = \sigma_0 t = \sigma_{cr} t/(1 + n_e)$, and H is given explicitly in Equation (7.132), the following relationship can be written as:

$$\frac{\sigma_{cr} t}{1+6n_e} = \frac{W}{2y_c}\bar{x}_C \tag{7.138}$$

Introducing a normalizing parameter $\xi = t/l_a$ between the thickness t of arch and the maximum height l_a, the rise y_c of the arch is expressed from the geometry of the arch as:

$$y_c = l_a - 2\left(\frac{t}{2} - e\right) = \left(1 - \xi(1 - 2n_e)\right)l_a \tag{7.139}$$

Inserting Equation (7.139) into Equation (7.138) and rearranging the resulting expression yields the maximum height of the arch as:

$$l_a = \sqrt{\frac{W\bar{x}(1+6n_e)}{2\sigma_{cr}\xi(1-\xi(1-2n_e))}} \tag{7.140}$$

Deriving this expression with respect to ξ yields:

$$\xi = \frac{1}{2(1-2n_e)} \tag{7.141}$$

Substituting the above result in Equation (6.138) gives:

$$y_c = \frac{l_a}{2} \tag{7.142}$$

Then, thickness t of the arch is obtained by inserting the above expression in Equation (7.139) and rearranging the resulting expression as:

$$t = \sqrt{\frac{W\bar{x}}{2\sigma_{cr}} \cdot \frac{1+6n_e}{1-2n_e}} \tag{7.143}$$

The length of rockbolts may then be assumed to be equal to the maximum height of the arch. The thickness of the arch can be specifically determined by introducing some criteria for the stability of the arch against various forms of failure. These criteria are as follows:

1) *Compressive failure at crest*: This condition is introduced by equating the compressive strength of the rock to the maximum compressive stress at the crest of the arch:

$$\sigma_c = \sigma_{cr} \tag{7.144}$$

In this particular case, the action of rockbolts is just to provide a restraining effect and to allow block rotations. The strength of bolts cannot contribute to the resistance of the arch.

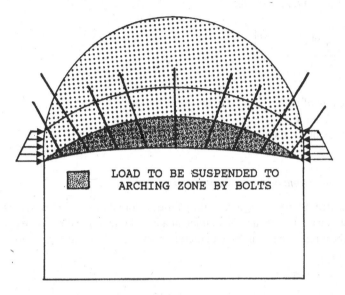

Figure 7.69 Illustration of the load to be suspended by bolts to rock arch.

Therefore, the determination of the necessary bolt number is indeterminate. Nevertheless, the following formula is suggested to calculate the number of bolts (it is assumed that the bolts should suspend the rock mass bound by the opening geometry and the lower side of the arch) (Fig. 7.69).

$$n = \frac{\gamma \int_0^L (F_2 - F_1) dx}{T_b} \qquad (7.145)$$

where
$F_1(x)$ = equation of roof line
$F_2(x)$ = equation of the lower side of the arch
T_b = pull-out capacity of a rockbolt
γ = unit weight of rock

2) *Shear failure at abutments or along a discontinuity plane*: The stability of the arch should be checked against vertical and horizontal shearing at abutments. For each case, the necessary rockbolt number is calculated from the following expressions by assuming that rock obeys to the Mohr-Coulomb yield criterion and bolts are subjected to the same magnitude of bolt stress:

i) *Shearing at abutments*:

a *Vertical shearing*:

$$n \geq \frac{\frac{W}{2} - (c \cdot kt + H \tan \phi)}{T_b} \qquad (7.146)$$

b   *Horizontal shearing*:

$$n \geq \frac{H - \left(\dfrac{W}{2}\tan\phi + c \cdot kt\right)}{T_b} \qquad (7.147)$$

where
c = cohesion of rock
ϕ = friction angle
k = coefficient for the assumed stress distribution
$k = 1/(1+n_e)$

ii)   *Shearing at one of discontinuity planes*:

The most critical plane in the arch with well-developed discontinuity sets is the one that emanates from the abutments and is inclined at a low angle to the thrust line (Fig. 7.70). The required rockbolt resistance can be calculated from the following expression, as shown in previous sections:

$$\frac{H}{\cos\psi}\left(\cos(\alpha-\psi) - \sin(\alpha-\psi)\tan\phi\right) \leq \sum_{i=1}^{n_{pl}}\left(T_b^i\left(1 + \frac{1}{2}\tan\phi\sin 2\theta_i\right)\right) \qquad (7.148)$$

where
ψ = inclination of the thrust line from horizontal
ϕ = friction angle of discontinuity plane
α = inclination of critical discontinuity plane
H = horizontal thrust
n_{pl} = number of bolts crossing the critical plane
T_b^i = force acting in the bolt
θ_i = inclination of bolt with respect to the normal of the critical plane

To dimension the bolt length, the height of the arch should be determined. For this purpose, the following procedure can be followed. For the overall stability of the arch, the following condition must be satisfied:

$$\int_0^{\psi_0}\frac{H}{\cos\psi}\left(\cos(\alpha-\psi) - \sin(\alpha-\psi)\tan\phi\right)d\psi \leq \sum_{i=1}^{n}\left(T_b^i\left(1 + \frac{1}{2}\tan\phi\sin 2\theta_i\right)\right) \qquad (7.149)$$

where
ψ_0 = inclination of the thrust line from horizontal at abutments
n = total number of bolts per unit width

If bolts are assumed to be strained to the same stress level $T_b = T_b^i$ and their inclination with respect to the critical discontinuity set remains the same ($\theta_i = \theta$), the rise y_c of the arch can be found by inserting H given in Equation (7.132):

$$y_c = \frac{W\overline{x}}{2}\frac{\cos\alpha\left(\psi_0 - \ln(\cos\psi_0)\tan\phi\right) - \sin\alpha\left(\ln(\cos\psi_0) + \psi_0\tan\phi\right)}{nT_b\left(1 + \dfrac{1}{2}\tan\phi\sin\theta\right)} \qquad (7.150)$$

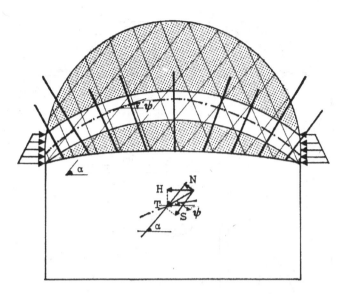

Figure 7.70 Rock arch having well-defined discontinuity sets and force system acting on a discontinuity plane.

Derivation of the expression for thrust line y with respect to x at $x = 0$ yields ψ_0 as:

$$\psi_0 = \tan^{-1}\left(\frac{dy}{dx}\right)_{x=0} \qquad (7.151)$$

The above expression will be a function of y_c and L only.

ψ_0 is first obtained by applying an iteration scheme, such as the Newton-Raphson technique, to the function defined as:

$$g = y_c(Eqn.(7.151)) - y_c(Eqn.(7.150)) = 0 \qquad (7.152)$$

for a given set of n, T_b, θ, α, $f(x)$, and L. The procedure for the iteration scheme is the same as that described in previous subsections. Once ψ_0 becomes known, then y_c, and subsequently l_a and t, can be easily calculated.

For example, if the load W_x is a uniformly distributed load (i.e. $Wx = \gamma x l_a$), then $y = 4y_c(x/L)$ $(1 - x/L)$. The derivation of the above equation at $x = 0$, $\tan \psi_0 = 4y_c/L$ or $y_c = L \tan \psi_0/4$.

7.5.6.2 Model tests and comparisons

In this subjection, a comparison between calculations by the analytical model described in previous subsections and measurements made on model underground openings in laboratory only are given. A model study was made to study the most likely modes of failure experimentally and the effect of rockbolts in arch formation. Discontinuous rock mass was simulated using acrylic blocks of $15 \times 25 \times 60$ *mm*³, and a thin slice of adhesive tape was used as a model grouted rockbolt. The base friction apparatus available in the laboratory of the

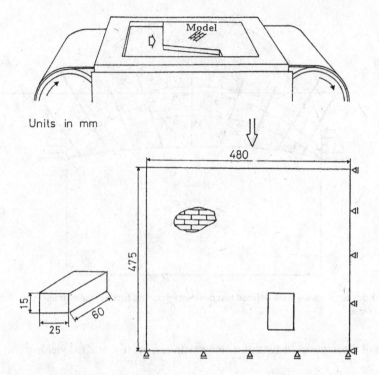

Figure 7.71 Schematic diagram of testing apparatus and model layout.

Department of Geotechnical Engineering, Nagoya University, was used (Kawamoto *et al.*, 1983). Discontinuity sets in rock mass were simulated by piling up the blocks in various patterns. A typical layout of the model and dimensions are shown in Figure 7.71.

Bonding strength between model bolts and blocks was 90 *kPa*. The width of the tape was 3 *mm* and the maximum bolted height of the arch 50 *mm*. Bolts crossed the abutments had a 30 *mm* anchorage length. All installation patterns were (Fig. 7.72):

- Bolts perpendicular to roof line and not crossing the abutments
- Radial pattern and crossing the abutments

The number of bolts was changed until stable opening conditions were established. Some of test configurations are shown in Figure 7.73.

Before presenting the results, it would be better to summarize some significant findings and to discuss these in relation to the arching phenomena from these experiments.

- When rockbolts were installed perpendicularly to the roof line and were not crossing the sliding planes at abutments, no arching could take place and all openings were unstable, irrespective of the length and number of bolts.
- On the other hand, when bolts crossed such potential sliding planes, the arching action could be initiated. However, the stability of the opening was then dependent of the number and length of the bolts.

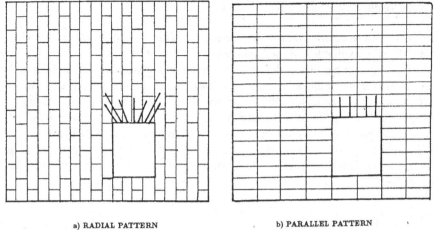

a) RADIAL PATTERN   b) PARALLEL PATTERN

Figure 7.72 Bolting patterns used in model tests.

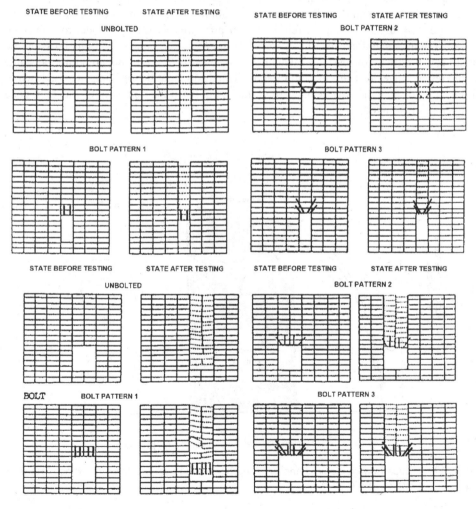

Figure 7.73 States of some model openings in cross-continuously jointed rock mass before and after testing.

These conclusions clearly confirm that the arching action could be activated when some form of restraint is available and some amount of rotation of blocks is made to occur. When bolts are orientated in parallel to the possible direction of movement, no arching action is possible. However, when the bolts cross the critical planes of potential failure to which the bolts are oblique, then an arching action will come into existence and the opening can be stable, provided that sufficient resistance by the bolts is available. In the particular tests, failure always took place along the vertical throughgoing discontinuity planes emanating from the upper corners of the opening. Therefore, the calculations are made by taking this type failure mode into consideration. The expression (7.138) was transformed to the following form:

$$\frac{\sum T_b}{W} = \frac{1}{2}\left[1 - \frac{L}{2L_a}\tan\phi\right] \tag{7.153}$$

The calculated normalized required rockbolt force by the weight of the sliding body for various normalized span over the bolted arch height by using the above expression is shown in Figure 7.74, together with the reinforcement pressure provided by model rockbolts for different numbers and constant anchorage length. When the number of bolts crossing the abutments was three or less, the calculations indicate that the opening could not be stable in the tested range. Test results confirmed the conclusion that the openings that had two or three bolts crossing the abutments were unstable, but when the bolt number was four, none of openings had failed.

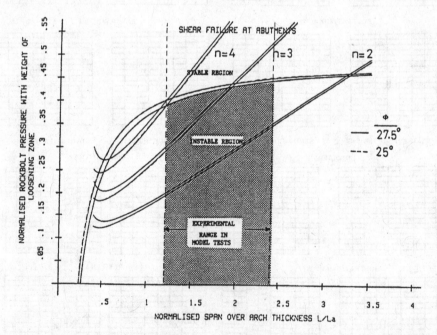

Figure 7.74 Comparison of required and provided normalized rockbolt force for various range of width to span ratio.

7.6 SUPPORT OF SUBSEA TUNNELS

The first subaqueous tunnel was built by the Sumerians of Central Asian origin who settled in Mesopotamia. Recently, there is a great deal of interest in building subsea tunnels all over the world. Most of these tunnels in rock is constructed using a support/reinforcement system consisting of rockbolts and shotcrete with wire mesh (Nilsen et al., 1999). The shield-type excavations are preferred over conventional drill-blast tunnels due to excavation velocity. Although the drill-blast technique utilizing rockbolts, shotcrete with wire mesh, steel ribs, and concrete liner (e.g. Seikan Tunnel) is preferred in some countries (Tsuchiya et al., 2009), the construction of tunnels utilizing shield and concrete liners resistant to water pressure is desirable when the long-term operation of the subsea tunnels are considered. The drainage of water becomes a heavy burden on operation costs when the tunnel is relatively long. Such a tunnel was recently built in the İstanbul Strait. The maximum depth of this tunnel was 106 m with a shield thickness of 400 mm and 800 mm thick concrete liner.

If the load on the lining of the subsea tunnel is due to water pressure and the tunnel is circular, the tunnel lining may be modeled as a thick-wall cylinder. If the liner behaves elastically, the maximum tangential stress at the tunnel perimeter can be given as (see for example Aydan, 1982, 1989):

$$\sigma_\theta = 2p_o \frac{r_o^2}{r_o^2 - r_i^2} \tag{7.154}$$

where p_o: water pressure; r_o: outer radius; r_i: inner radius.

Water pressure (p_o) can be related to depth (h) and unit weight (γ_w) of sea water as:

$$p_o = \gamma_w h \tag{7.155}$$

The outer radius (r_o) can be related to inner radius (r_i) and liner thickness (t) as:

$$r_o = r_i + t \tag{7.156}$$

If relationships given by Equations (7.155) and (7.156) are inserted in Equation (7.154) and tangential stress at the tunnel perimeter is equal to the compressive strength of concrete liner with a safety factor (SF), the lining thickness could be determined from the following relationship:

$$t = r_i \frac{\gamma_w h}{\sigma_c / SF - 2\gamma_w h} \tag{7.157}$$

Figure 7.75 shows the tunnel liner thickness for a tunnel with an internal diameter of 12 m for an SF = 1. As noted from the figure, the required tunnel liner thickness decreases as the compressive strength of concrete liner increases. If the compressive strength of concrete liner is 40 MPa, the required liner thickness is 300 mm for a depth of 150 m, while it is 1000 mm for a depth of 600 m.

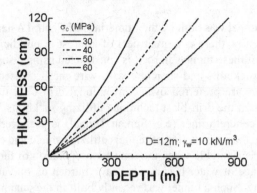

Figure 7.75 The required thickness of concrete liner as a function of depth.

7.7   REINFORCEMENT AND SUPPORT OF SHAFTS

Shafts are a part of an exploitation scheme in underground mining, access and ventilation in roadway and railway tunnels, underground powerhouses, and nuclear waste disposal projects. They are generally vertical underground openings built with the purpose of transportation, access, and/or ventilation, and they may be sometimes inclined. They are basically similar to other structures, and their reinforcement and support would be basically similar to those explained in previous section. A major difference would be the *in situ* stress state, and they may be subjected to high groundwater pressures. For vertical shafts, the major *in situ* stress would be horizontal. The horizontal stress may be given as a fraction of vertical stress as given below:

$$\sigma_h = k\sigma_v \tag{7.158}$$

The value of coefficient k may range from the well-known relationship derived by Terzaghi and Richart (1952) and the empirical relationship developed by Aydan and Pasamehmetoglu (1994) (see Figure 7.76):

$$k_{min} = \frac{v}{1-v} \text{ and } k_{max} = \frac{2000}{h} + 1.0 \tag{7.159}$$

where v and h are Poisson's ratio of rock mass and depth (in m) from the ground surface. As seen from Figure 7.76, the horizontal stress may be several times the vertical stress, and it decreases as the depth increases and become asymptotic to the value of 1.

The functional duties of the shaft and environmental conditions determine the requirements for their reinforcement and support systems. Figure 7.77 shows the general scheme of the concrete lining design of shafts. Figure 7.78 show an example of groundwater pressure build-up in North Selby Mine in Yorkshire, UK, during the thawing of frozen ground. Prediction of the pressure imposed on a shaft lining is the critical item in designing a concrete shaft. Surrounding rocks and water can exert horizontal forces that act against the lining

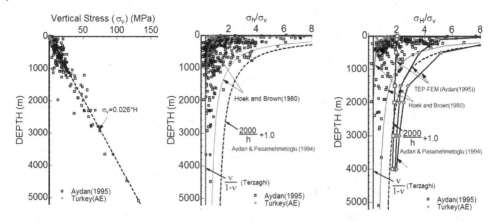

Figure 7.76 Variation of vertical and minimum and maximum horizontal stresses (from Aydan, 2016b).

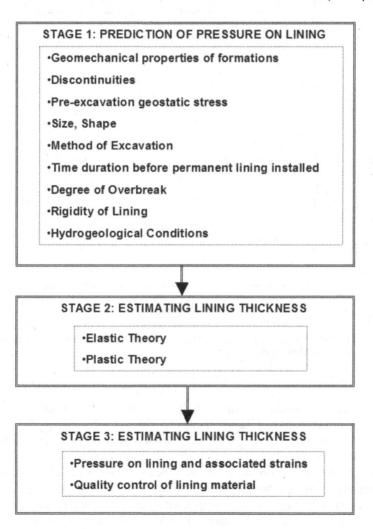

Figure 7.77 Flow chart of shaft lining design (after Aydan, 1982).

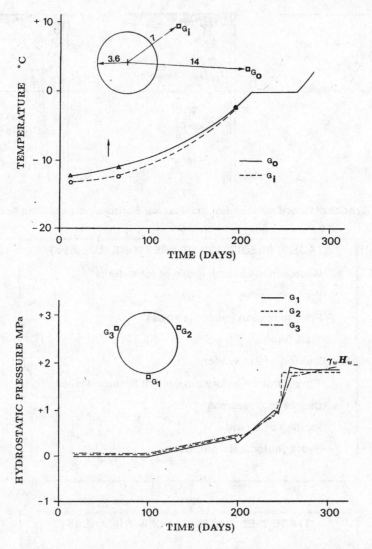

Figure 7.78 Groundwater pressure build-up at North Selby Mine during the thawing of frozen sandstone (from Aydan, 1982).

structure. One can find several approaches for the determination of ground pressures on the support system of the shafts (e.g. Brady and Brown, 2005; Coates, 1981).

Water-bearing strata, which consist of pervious sedimentary rocks, exert a hydrostatic pressure equal the piezometric head of groundwater. Therefore, the reinforcement and support systems should be designed to resist both rock and groundwater pressure, generally. If the rock pressure is taken by the reinforcement system consisting of rockbolts, shotcrete, and steel ribs, the concrete lining, which is constructed much later, would generally resist against groundwater pressure if the shaft is required to be dry. The response of surrounding ground during excavation can be evaluated using analytical procedures described in previous sections if horizontal stresses are isotropic. If the concrete lining is designed against groundwater pressures, Equation (157) can be used to determine the thickness of the concrete lining. In this approach, the dynamic loads are not considered. However, they may be dealt with through the increase of safety factor or dynamic numerical analyses.

7.8 SPECIAL FORM OF ROCK SUPPORT: BACKFILLING OF ABANDONED ROOM AND PILLAR MINES

The Great East Japan earthquake with a moment magnitude 9.0 caused gigantic tsunami waves, which destroyed many cities and towns along the shores of the Tohoku and Kanto Regions of Japan (Aydan and Tano, 2012a,b,c). Besides the structural damage on ground surface, this earthquake triggered the collapses of abandoned lignite and coal mines and underground stone quarries and caused associated damage to super structures at 316 localities (Figs. 7.79 and 7.80). Similar events occurred in the previous 1978 off-Miyagi earthquake, the 2003 Miyagi-Hokubu earthquake, and the 2008 Iwate-Miyagi intraplate earthquake (Aydan, 2004; Aydan and Kawamoto, 2004).

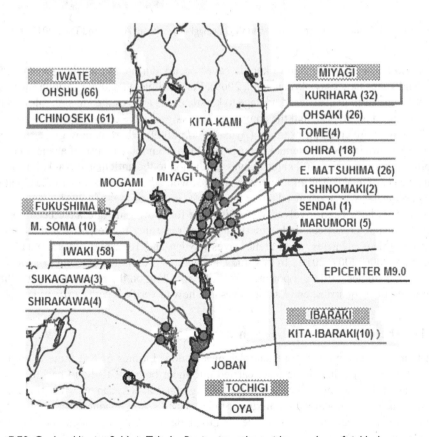

Figure 7.79 Coal and lignite fields in Tohoku Region, together with a number of sinkhole events caused by the 2011 Japanese earthquake (from Aydan and Tano, 2012a,b,c).

Figure 7.80 Some views of sinkholes in Wakayanagi town (from Aydan and Tano, 2012a,b,c).

It is also known that damage to karstic caves occurs during earthquakes (Aydan, 2008; Aydan and Tokashiki, 2007; Aydan *et al.*, 2009). For example, the 2009 L'Aquila earthquake and 2005 Nias earthquake caused damage to karstic caves and resulted in sinkholes. Similar events occurred in the past, and the damage to the Ishigaki cave on Ishigaki Island of Japan in 1771 may be given as an example (Aydan *et al.*, 2009). There is now great concern in Japan on how to deal with potential damage resulting from the collapse of abandoned room and pillar mines, quarries, and karstic caves in relation to the anticipated Nankai-Tonankai-Tokai mega earthquake. The anticipated magnitude would be similar to that of the Great East Japan earthquake, which occurred along the Tohoku region of Japan in 2011.

The authors have been involved with this issue for some time and have been conducting experimental, analytical, and numerical studies regarding the response and stability of abandoned mines, quarries, and karstic caves (Aydan and Tokashiki, 2011; Aydan *et al.*, 2003, 2005, 2006, 2007, 2010). In this study, the authors present the outcomes of a series of experimental studies on the supporting effect of backfilling on the response and stability of abandoned mines, quarries, and karstic caves and how to verify it with *in situ* monitoring.

7.8.1 Short-term experiments

Short-term experiments on lignite samples of abandoned mines of Nagakute town and of Oya tuff and Ryukyu limestone were carried out under both unfilled and backfilled states using granular backfilling and cohesive backfilling materials (Aydan and Tokashiki, 2013). Cohesive backfilling material is a mixture of clayey and sandy residue from ceramic factories and cement. Experiments were carried out to 28 days with the consideration of cement's hydration process. This backfill material is commonly used to fill abandoned lignite mines in the Tokai region in Japan; it is referred to as NSK backfilling material in this article.

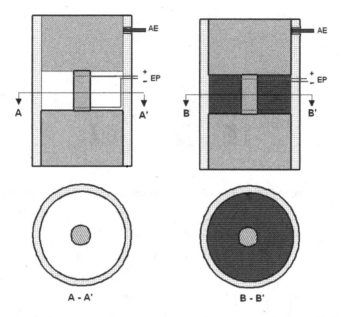

A - A'   B - B'

Figure 7.81 Experimental setup for short-term large-scale experiments for unfilled and backfilled states (from Aydan and Tokashiki, 2013).

7.8.1.1   *Experimental setup and instrumentation*

A 10-mm-thick cylindrical acrylic container with an internal diameter of 100 mm and a height of 250 mm was used in large-scale experiments for investigating the supporting effect of granular and cohesive backfilling on pillar samples of lignite from abandoned lignite mines of Nagakute town, tuff from Oya town (Tochigi Prefecture), and Ryukyu limestone from Okinawa Prefecture. Size of pillars were 46–50 mm in diameter and 96–100 mm in height. Pillars were sandwiched between two 100-mm-high cylindrical rock blocks with a diameter of 98 mm. Top and bottom blocks were Oya tuff for experiments on Nagakute lignite and Oya tuff samples while Ryukyu limestone was made of Ryukyu limestone blocks for Ryukyu limestone pillars. Load, displacement, and acoustic emission (AE) were monitored continuously using a setup as shown in Figure 7.81. In some experiments, electrical resistivity was also measured.

Two or three experiments were carried out for each pillar material under unfilled and backfilled states for both granular backfill material and cohesive backfill material (NSK material). Typical experimental results for each case are presented in the following subsections.

7.8.1.2   *Experiments on Nagakute lignite pillars*

Figures 7.82, 7.83, and 7.84 show stress, strain, and AE responses of unfilled lignite pillar and lignite pillars backfilled with granular and cohesive (NSK) filling materials during cyclic compression, respectively. As noted from the figures, the unfilled lignite sample fails in a brittle manner while the backfilled sample continues to sustain higher loads. Furthermore, permanent straining is noted after each loading-unloading cycle and cumulative AE

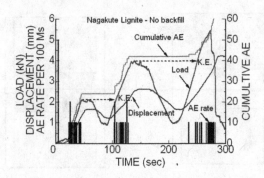

Figure 7.82 Stress, strain, and cumulative AE count responses of an unfilled lignite pillar during cyclic compression.

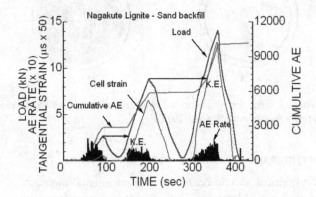

Figure 7.83 Stress, strain, and cumulative AE count responses of a Nagakute lignite sample backfilled with sand during cyclic compression.

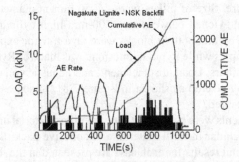

Figure 7.84 Stress, strain, and cumulative AE count responses of Nagakute lignite pillar backfilled with cohesive NSK backfill material during cyclic compression.

response is closely associated with straining and a somewhat Kaiser effect is noted from the measured responses, as noted in the small-scale, short-term experiments. The tangential strain of the acrylic container is also shown in Figures 7.82–7.83. The tangential strain starts after a certain strain level, which is closely associated with the deformation and initiation of yielding of pillars.

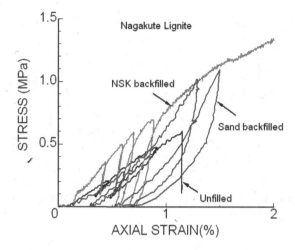

Figure 7.85 Comparison of strain-stress responses of unfilled pillar and backfilled pillars with granular and cohesive backfill materials during cyclic compression.

Figure 7.85 compares the average strain-stress relations for unfilled pillars and pillars backfilled with granular and cohesive backfill materials. It is interesting to note that the bearing capacity of backfilled pillars is increased to about 1.3–1.5 times that of the unfilled sample at the same strain level. Furthermore, the behavior of backfilled pillars is elasto-plastic without any softening, even after the failure of the pillars. Furthermore, the bearing capacity of the backfilled pillars was of great significance. The bearing capacity of the pillar backfilled with NSK backfill material is greater than that of the pillar backfilled with granular backfill material.

7.8.1.3 Experiments on Oya tuff pillars

There are many abandoned and active underground quarries at Oya town in Tochigi Prefecture. Quarrying has been active in this area for more than 100 years. As briefly mentioned in the introduction, very big sinkhole formations occurred in Oya town from time to time. There is growing attention paid to whether to fill the abandoned underground quarries.

Figures 7.86, 7.87, and 7.88 show stress, strain, and AE responses of unfilled tuff pillar and tuff pillars backfilled with granular and cohesive (NSK) filling materials during cyclic compression, respectively. As noted from the figures, the unfilled lignite sample fails in a brittle manner while the backfilled sample continues to take up higher loads. Furthermore, permanent straining is noted after each loading-unloading cycle, cumulative AE response is closely associated with straining, and the Kaiser effect is noted from the measured responses. Experiments on Oya tuff is carried out under both dry and saturated conditions. As the experiments on pillars with NSK backfilling material were carried out to 28 days of curing in water, it is expected that the strength of Oya tuff would be lower than that under dry conditions (Aydan, 2003; Aydan *et al.*, 2011b). Experiments showed that the uniaxial compressive strength of Oya tuff is reduced to 30–50% of the uniaxial compressive strength under dry conditions.

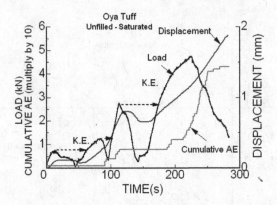

Figure 7.86 Stress, strain, and cumulative AE count responses of unfilled saturated Oya tuff pillar during cyclic compression.

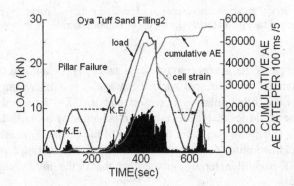

Figure 7.87 Stress, strain, and cumulative AE count responses of Oya tuff pillar backfilled with granular backfill material during cyclic compression.

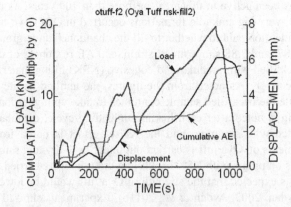

Figure 7.88 Stress, strain, and cumulative AE count responses of Oya tuff pillar backfilled with cohesive NSK backfill material during cyclic compression.

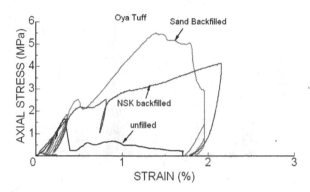

Figure 7.89 Comparison of strain-stress responses of unfilled pillar, backfilled pillars with granular and cohesive backfill materials during cyclic compression.

In Figure 7.83, the tangential strain of the acrylic cell is also shown. As it is clearly noted from this figure, the tangential strain of the cell drastically increased as soon as the pillar starts to fail and exhibit dilatant behavior. Nevertheless, the confinement effect of the back-filling material assisted by the constraint of the acrylic cell increases the bearing capacity of the pillar, in spite of slight softening behavior following the yielding of the pillar.

Figure 7.89 compares the average strain-stress relations for unfilled and pillars backfilled with granular and cohesive backfill materials. It is interesting to note that the bearing capacity of backfilled pillars is increased about 1.3–1.5 times that of the unfilled sample at the same strain level. Furthermore, the behavior of backfilled pillars is elasto-plastic with slight softening following the yielding of the pillars and strain hardening thereafter. Although the cohesive backfill material provides better resistance for pillars, the difference between the average strain-stress response of the pillar (saturated) backfilled with NSK backfilling material from the pillar (dry) with backfilled with granular sandy material is due to the strength difference of tuff pillars (Aydan *et al.*, 2011b). The tuff pillar is dry in the sand-backfilled case, while it is saturated in the NSK-backfilled case.

7.8.1.4 Experiments on Ryukyu limestone pillars

Ryukyu limestone is widely distributed in the Ryukyu Islands, and this formation contains various sizes of solution cavities. These solution (karstic) cavities cause some sinkhole problems from time to time in Ryukyu limestone regions. Therefore, some experiments were carried out to investigate the effect of backfilling on the bearing capacity of Ryukyu limestone formation.

Figures 7.90, 7.91, and 7.92 show stress, strain, and AE responses of unfilled limestone pillar and limestone pillars backfilled with granular and cohesive (NSK) filling materials during cyclic compression, respectively. Although Ryukyu limestone is much stronger rock compared with lignite and tuff, the overall responses and observations are quite similar to those of experiments on lignite and tuff samples. As noted from the figures, unfilled limestone sample fails in a brittle manner, while backfilled samples continue to take up higher loads. Furthermore, permanent straining is noted after each loading-unloading cycle and cumulative AE response is closely associated with straining. Kaiser effects are clearly observed in unfilled and backfilled samples.

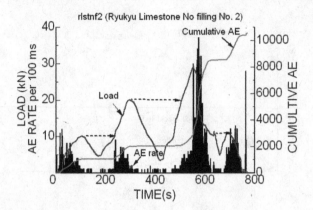

Figure 7.90 Stress, strain, and cumulative AE count responses of unfilled Ryukyu limestone pillar during cyclic compression.

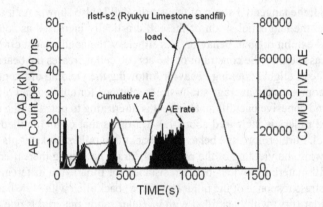

Figure 7.91 Stress, strain, and cumulative AE count responses of Ryukyu limestone backfilled with granular backfill material during cyclic compression.

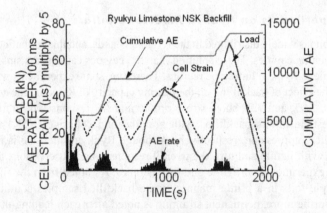

Figure 7.92 Stress, strain, and cumulative AE count responses of Ryukyu limestone pillar backfilled with cohesive NSK backfill material during cyclic compression.

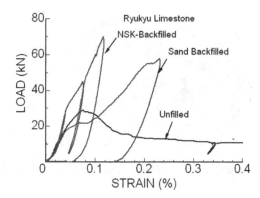

Figure 7.93 Comparison of strain-stress responses of unfilled pillar and backfilled pillars with granular and cohesive backfill materials during cyclic compression.

Figure 7.93 compares the average strain-stress relations for unfilled pillars and pillars backfilled with granular and cohesive backfill materials. It is interesting to note that the bearing capacity of backfilled pillars is increased about 1.2–1.3 times that of the unfilled sample at the same strain level. Furthermore, the behavior of backfilled pillars is elasto-plastic without any softening after the yielding of the pillar. When the limestone pillar is backfilled with NSK backfilling material, the overall strength of pillars is greatly increased, and the experiments were terminated to prevent the bursting of the acrylic cell, which might result in undesirable accidents.

7.8.2   Long-term experiments

Several experiments on lignite samples from Mitake town on the long-term behavior subjected to atmospheric degradation and saturation conditions were carried out. It is well known that lignite is vulnerable to water content variations – it starts cracking due to shrinkage and its strength decreases upon saturation (Aydan *et al.*, 2005; Aydan *et al.*, 2011b). While one of the samples (ATC3) was directly exposed to the environment, the other sample (ATC1) was backfilled after 36 hours of exposure to the atmospheric degradation. Both unfilled and backfilled samples were fully saturated at 1360 hours (about two months) after the commencement of the experiments. Figure 7.94 shows the views of the unfilled and backfilled samples. During long-term experiments, axial displacement, acoustic emissions (AE), and environmental conditions such as temperature, humidity, and air pressure were monitored continuously. The monitoring devices were entirely battery operated, and the creep devices were of the cantilever type so that they were not susceptible to power failures.

Figures 7.95 and 7.96 show the measured displacement and acoustic emission responses with time. Both experiments were terminated at the same time when the ATC3 sample failed. It is interesting to note that the displacement and cumulative AE count of the sample exposed to the atmosphere increased with elapsed time. Following the saturation of the sample at 1360 hours from the commencement of the experiment, the sample failed almost in 2 days after the saturation. On the other hand, the rate of AE occurrences of the ATC1 sample is almost none except for backfilling, loading, and saturation procedures. The AE occurrence

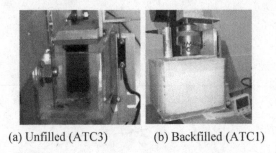

(a) Unfilled (ATC3)   (b) Backfilled (ATC1)

Figure 7.94 Views of unfilled and backfilled lignite samples.

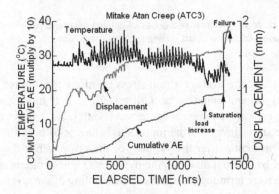

Figure 7.95 Displacement and cumulative AE count responses of ATC3 sample.

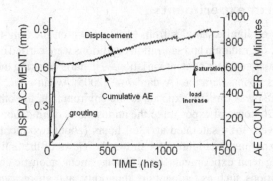

Figure 7.96 Displacement and cumulative AE count responses of ATC3 sample.

is drastically reduced following the backfilling of the sample. The AE occurrence is almost none following the saturation while unfilled samples failed after saturation. The main mechanisms for the reduction of AE occurrence in the backfilled sample can be considered to be due to preventing exposing the sample to the variations of environmental parameters, such as temperature and humidity, which govern the water content variation and subsequent degradation of lignite, and reducing deviatoric stress, which may be related to the well-known Kaiser effect. Therefore, these two mechanisms may be utilized to verify the effect of backfilling abandoned lignite mines, as well as other types of backfilled cavities through *in situ* monitoring of AE responses.

7.8.3   Verification of the effect of backfilling through in situ monitoring

7.8.3.1   Backfilling project of Takenoyama abandoned lignite mine area

Nagakute and Nisshin cities near Nagoya in Aichi Prefecture have been developing residential areas to accommodate newcomers. As there are many abandoned lignite mines in these two cities, the stability of these abandoned mines is a serious concern for the developers. Therefore, backfilling is now commonly adopted to ensure the safety of areas against sinkhole formations and subsidence. The authors were asked how to verify the effect of backfilling of abandoned mines *in situ*. On the bases of the outcomes of experiments under well-controlled laboratory conditions, it is expected that if the backfilling is effective, it should stop the further degradation of rocks surrounding the abandoned mines, reducing the velocity of seepage of groundwater and deviatoric stress due to the confinement effect in view of the Kaiser effect. Therefore, the AE occurrences before and after backfilling would be entirely different. If there are no technical problems of the devices used for monitoring acoustic emissions, the AE counts would drastically decrease following the backfilling, as illustrated in Figure 7.97.

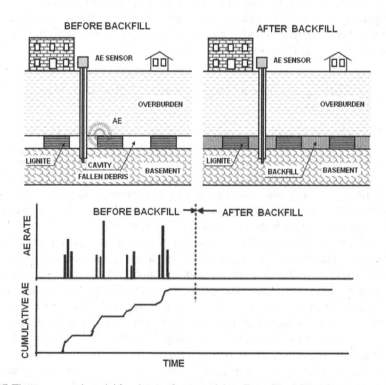

Figure 7.97 The conceptual model for the verification of the effect of backfilling from acoustic emission response before, during, and after backfilling.

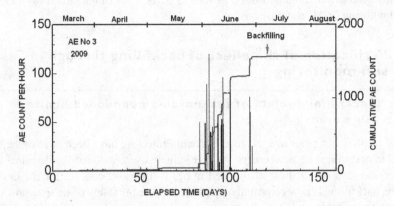

Figure 7.98 Acoustic emission responses of AE No 3 station during the period between March and August 2009.

Three boreholes were drilled in the area of abandoned lignite mines in Takenoyama region of Nisshin before the commencement of the backfilling operation, and wave guides reaching the cavities equipped with AE sensors were installed and the measurements carried out for a period of six months. Figure 7.98 shows the monitoring results for borehole No. 3. As noted from the figure, the monitoring results are in accordance with anticipated acoustic emission (AE) responses before, during, and after filling. AE occurrences drastically disappear following the completion of backfilling operation.

7.8.3.2 Backfilling project of karstic caves below Gushikawa Castle remains

Two karstic caves exist beneath the Gushikawa Castle remains in Itoman City of Okinawa Island of Japan (Fig. 7.99). An extensive monitoring program was undertaken to observe the performance of the caves before and after filling. The monitoring program involved measuring the displacement response of a continuous crack passing through Cave A by six gap gauges, tilting of pillars and ground surface by inclinometers, longitudinal and traverse strains and acoustic emissions of the roof of Caves A and B, temperature, humidity, and rainfall at the ground surface in the Gushikawa Castle remains.

The instruments were installed about one month before the construction of filling started. Particularly, the acoustic emission response is of great significance for seeing the effect of filling on the response of the caves before and after filling. As cavities subjected to daily variations of temperature, humidity, tidal loading, and rainfall, they experience gradual degradation. As a result of that, some acoustic emission events associated with this gradual degradation take place. Filling the cavities should terminate this degradation process so that the acoustic emission events must subside as time goes by. Figure 7.100 shows the acoustic emission responses of Caves A and B. As expected, the number of acoustic emission events drastically decreased following the completion of construction. This trend continued for a considerable period following the construction, which verifies that the backfilling was effective.

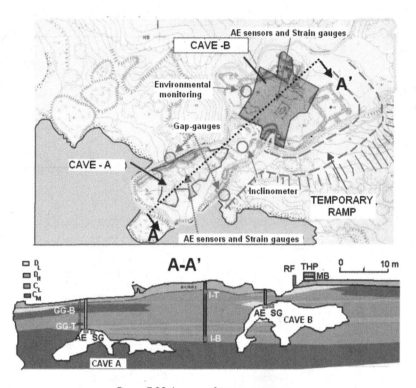

Figure 7.99 Layout of instrumentation.

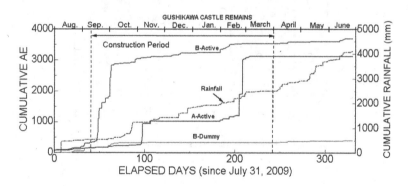

Figure 7.100 Acoustic emission responses of caves during Aug 2009–June 2010.

7.8.4   Analysis of backfilling of abandoned mines

7.8.4.1   Stability of pillars

A series of analyses using DFEM was carried out on the supporting effect of backfilling on a fractured pillar of an abandoned lignite mine. Overburden was 6 m, pillar height was 4 m, and width was 2 m. Table 7.18 gives material properties used in the analyses. Figure 7.101

Table 7.18 Material properties used in DFEM analyses.

Material	$\lambda(E_n)$ (MPa)	$\mu(E_s)$ (MPa)	γ (kN/m³)	c (kPa)	ϕ (°)	σ_t (kPa)
Top layer	200	200	22			
Lignite	200	200	14			
Lower layer	300	300	22			
Backfilling	100	100	14			
Fracture plane	200	10	–	10	10	5

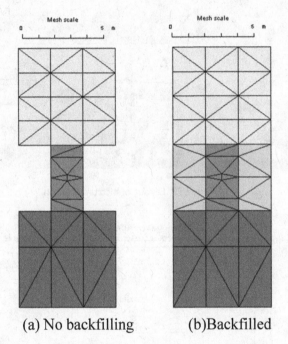

(a) No backfilling   (b)Backfilled

Figure 7.101 Finite element meshes used in DFEM analyses.

shows the finite element meshes used in analyses. The pseudo-dynamic version of DFEM analyses was used in the analyses, and computations were carried out up to 12 steps.

Figure 7.102 shows the deformed configuration of non-backfilled and backfilled pillars at the computation step of 12. Figure 7.103 shows the settlement of the pillar. As noted from the figures, when the pillar is not supported by backfilling, sliding occurs along the fracture plane and a vertical crack appears in the middle of the pillar. However, if the pillar is supported by backfilling, the sliding and separation of fracture planes is restrained and the pillar is stabilized by backfilling material. Although the backfilling material is soft, its supporting effect is remarkable.

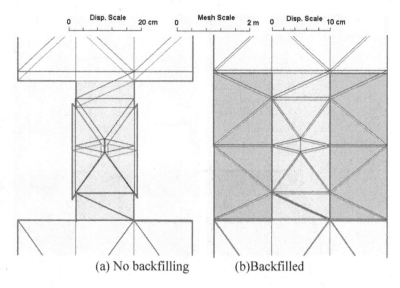

(a) No backfilling   (b)Backfilled

Figure 7.102 Deformed configurations of the pillar for non-backfilling and backfilling cases.

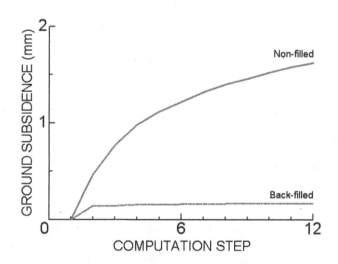

Figure 7.103 Settlement of the pillar for non-backfilling and backfilling cases.

7.8.4.2   The stability of an abandoned room and pillar mine next to steep cliff

An abandoned lignite mine with an overburden of 4 m next to a steep cliff was considered as shown in Figure 7.104. Pillar height was assumed to be 2 m with a width of 1 m. The computations were carried out up to 12 steps and deformed configuration of the abandoned mine for two cases are shown in Figure 7.105. The material properties used were the same as

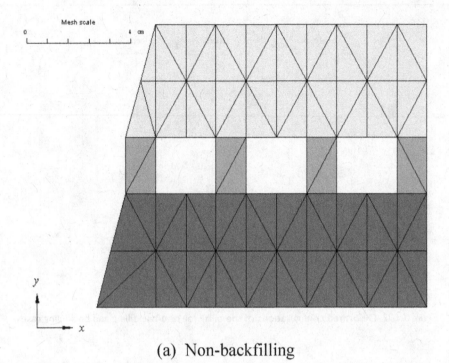

(a) Non-backfilling

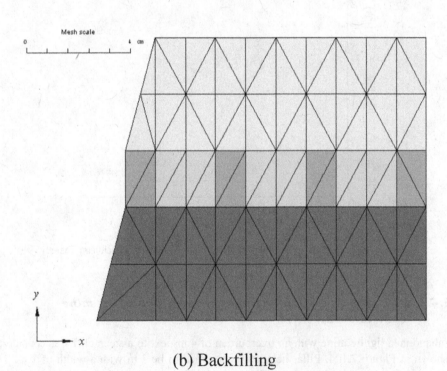

(b) Backfilling

Figure 7.104 Finite element meshed used in DFEM analyses.

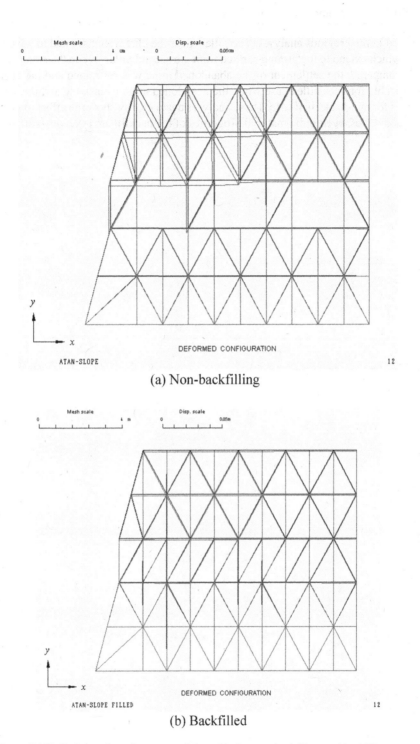

Mesh scale   0   4 cm   Disp. scale   0   0.05cm

DEFORMED CONFIGURATION

ATAN-SLOPE   12

(a) Non-backfilling

Mesh scale   0   4 m   Disp. scale   0   0.05m

DEFORMED CONFIGURATION

ATAN-SLOPE FILLED   12

(b) Backfilled

Figure 7.105 Deformed configuration of the pillar for non-backfilling and backfilling cases.

those used in the previous analyses. The pillars near the cliff were assumed to have fracture planes, which extend to the ground surface from the second pillar. When the results of analyses are compared, the settlement of the abandoned mine was restrained and the abandoned mine is stable for back-filled case while the abandoned mine is unstable and the settlement continues for non-backfilled case. These analyses again clearly show the effect of backfilling abandoned mines, even if the material properties of the backfill are relatively soft and weak.

Chapter 8

Reinforcement and support of rock slopes

8.1 INTRODUCTION

Fundamentally, rock support for rock slopes may only be in the form of retaining walls. Although shotcrete with/without wire mesh or fibers is often used, the structural effect of shotcrete is fundamentally almost none. However, shotcrete does prevent the deterioration of rock, the loosing of rock mass due to movements of small blocks, and internal erosion due to atmospheric agents and seepage in the long term, so its effect can be tremendous.

Piles of various types, rockbolts, and rockanchors that pass through the potential failure plane may be viewed as reinforcement. While retaining walls for rock slopes would have only a tiny effect on ordinary rock slopes, the emphasis in this chapter is given to reinforcement members, such as rockbolts and rockanchors. Most formulations are basically similar to those presented for underground structures. Therefore, detailed derivations are not presented and readers should consult related chapters and sections for them. Nevertheless, reinforcement and support effects are described with the consideration of common modes of failure of rock slopes. The methods are fundamentally based on limit equilibrium methods (LEM), and this chapter provides examples of analyses utilizing LEM, the finite element method, and the discrete finite element method (DFEM) incorporating Aydan's rockbolt element. Furthermore, some model experiments of jointed rock slopes with different reinforcement patterns are also presented.

8.2 REINFORCEMENT AGAINST PLANAR SLIDING

Planar sliding is one of the common forms of slope failure associated with thoroughgoing discontinuities, such as bedding planes, schistosity, or sheeting joint, as illustrated in Figure 8.1. Depending upon the position of potential failure planes, rockbolts/rockanchors may be required to provide shear and tensile resistance against planar failures. In any case, the reinforcement actions to be induced on the potential failure planes would involve the increase of frictional forces and apparent cohesion and the decrease of shear forces. Such actions can be evaluated using the procedures described in Chapters 4 and 7.

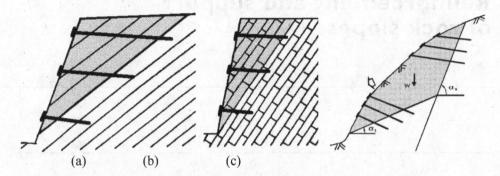

(a)   (b)   (c)

Figure 8.1 Possible modes of planar sliding and likely patterns of reinforcement.

Table 8.1 Material properties of rock and discontinuity plane.

E_r GPa	μ_r	E_d GPa	G_d MPa	c MPa	ϕ (°)	Unit Weight kN/m³	t mm
3.15	0.2	3.15	300	0	0	30.0	2.5

8.2.1   Finite element analysis

The stability of a rock slab as a simple slope stability problem against planar sliding is considered herein. Let us assume one rock slab is placed upon another, as shown in Figure 8.2a, inclined at various angles with respect to the horizontal, and reinforced with rockbolts of various patterns, as shown in Figure 8.2b with the material properties given in Table 8.1. Figure 8.2c shows the calculated shear stress in rockbolts at discontinuity plane for the inclination angle of 45° for three different rockbolt patterns. As is noticed from the figure, the shear stresses in the bolts differ depending upon the location and number of bolts. Nevertheless, the sum of the shear stresses is almost the same for the all patterns (Fig. 2e).

Figure 8.2e shows the sum of shear stresses in bolts at the discontinuity plane for bolt pattern 2 and for various inclinations of discontinuity plane. As the inclination increases, the calculated shear stresses increase in magnitude, as expected. The required reinforcements calculated from the numerical analyses for the stability of this slab for various inclination angles of the discontinuity plane are compared with those calculated from the theoretical model given below in Figure 8.2e:

$$SF = \frac{W \cos\alpha_1 \tan\phi + \sum T_b^{\alpha_1}}{W \sin\alpha_1} \tag{8.1}$$

where W is the dead weight of the slab, α_1, ϕ are inclination of failure plane and friction angle. The resistance given by rockbolts is given by the following equation (see Subsection 4.5 for details):

$$\sum T_b^{\alpha_1} = n_{\alpha_1} A_b^{\alpha_1} \sigma_b^{\alpha_1} (1 + \frac{1}{2}\sin 2\theta_{\alpha_1} \tan\phi) \tag{8.2}$$

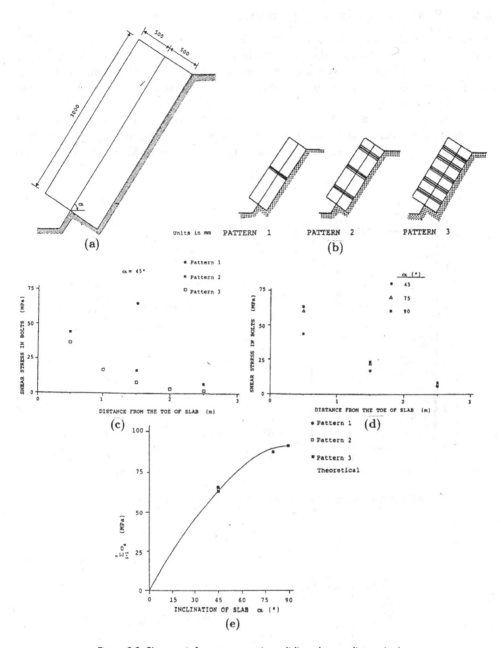

Figure 8.2 Shear reinforcement against sliding along a discontinuity.

Although at first glance it seems simple to calculate the total required reinforcement using the theoretical model, the stress to which the bolts are subjected at various locations always remains unknown. This might be a critical factor for the design of rockbolts in stabilization calculations, and numerical analyses may be necessary for the evaluation of shear resistance of the rockbolts.

8.2.2   Physical model experiments

Another important issue is the length and number of rockbolts/rockanchors. It is essential that rockbolts/rockanchors should have sufficient anchorage length in the stable region of slopes. Figure 8.3 shows the shear strength characteristics of discontinuities utilized in physical model experiments. The bond strength of model rockbolts ranged between 70 kPa and 100 kPa, as shown in Figure 8.4.

Figures 8.5, 8.6, 8.7, and 8.8 show several examples of model tests on rock slopes using different model materials and discontinuity patterns prone to planar sliding. As noted from the figures, the anchorage length, inclination, and number of rockbolts/rockanchors are

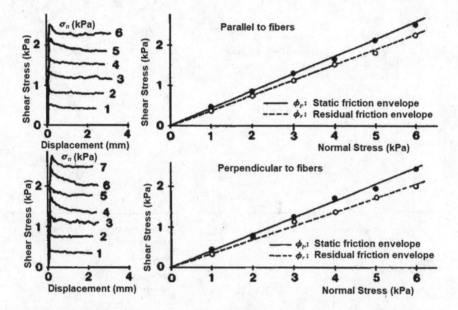

Figure 8.3 Shear strength characteristics of discontinuities in physical model tests.

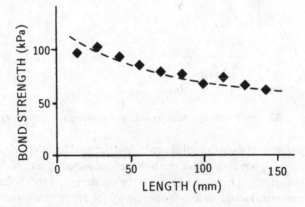

Figure 8.4 Variation of bond strength of model rockbolts.

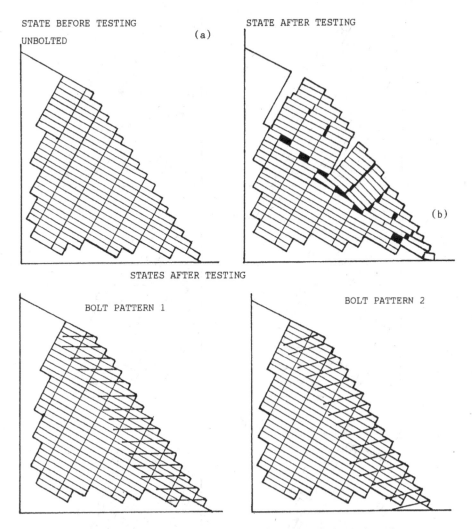

STATE BEFORE TESTING

UNBOLTED

(a)

STATE AFTER TESTING

(b)

STATES AFTER TESTING

BOLT PATTERN 1

BOLT PATTERN 2

Figure 8.5 Effect of bolting orientation on the stability of model rock slopes (base friction apparatus; $\alpha_1 = 30°$; $\alpha_2 = 120°$; $L_1/L_2 = 4$; cross-continuous pattern (CCP)).

essential parameters for the stabilization of rock slopes against planar sliding. Inclination of rockbolts should be such that maximum shear resistance should be achieved. The next parameters are the number and anchorage length beyond the potential failure plane. The same conclusions are also valid for slopes prone to wedge sliding failure.

8.2.3   Discrete finite element analyses

In the next examples, a series of discrete finite element analyses are carried out. A 5-m-high rock slope with a potential failure plane consisting of a plane with an inclination of 25° and tension crack is considered. Material properties are given in Table 8.2. Figure 8.9 shows

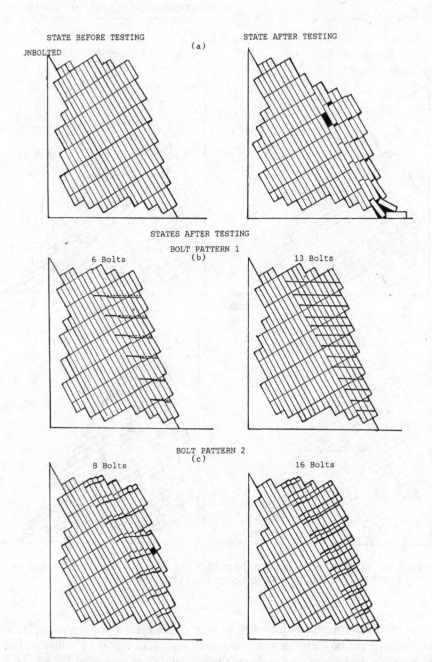

STATE BEFORE TESTING　　　(a)　　STATE AFTER TESTING

JNBOLTED

STATES AFTER TESTING

BOLT PATTERN 1
(b)

6 Bolts　　　　　　　13 Bolts

BOLT PATTERN 2
(c)

8 Bolts　　　　　　　16 Bolts

Figure 8.6 Effect of number and orientation rockbolts on the stability of model rock slopes (base friction apparatus; $\alpha_1 = 60o$; $\alpha_2 = 150o$; $L_1/L_2 = 4$; cross-continuous pattern (CCP)).

$$\alpha_1 = 30°; \alpha_2 = 120°; L_1/L_1 = 1; \xi = 26.6°.$$

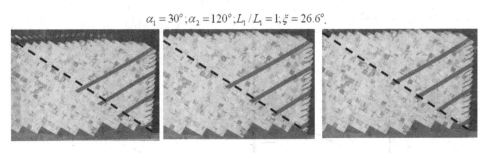

Figure 8.7 Effect of length of rockbolts on the stability of model rock slopes (gravitational model apparatus; α_1 = 30o; α_2 = 120o; L_1/L_2 = 1; intermittent pattern (IP)).

$$\alpha_1 = 45°; \alpha_2 = 135°; L_1/L_1 = 1; \xi = 26.6°$$

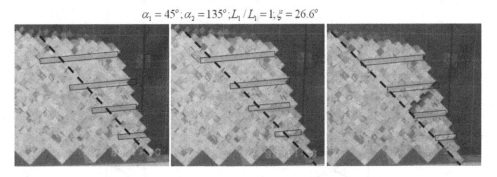

Figure 8.8 Effect of bolt length on the stability of model rock slopes (gravitational model apparatus; α_1 = 45°; α_2 = 135°; L_1/L_2 = 1; intermittent pattern (IP)).

Table 8.2 Material properties of rock and discontinuity plane.

E_r MPa	μ_r	E_d MPa	G_d MPa	c MPa	ϕ (°)	γ kN/m³	t mm	E_b GPa	D_b mm	D_b mm	G_g GPa
250	0.25	10	1	0	26	25.0	1.0	210	25	36	2

the DFEM mesh with/without bolt and deformed configurations of the slope at computation step of 10. The inclination of the bolt with respect to the discontinuity plane is +35 °. As noted from the figure, the rockbolt stabilizes unstable block and deformation of the block is restrained.

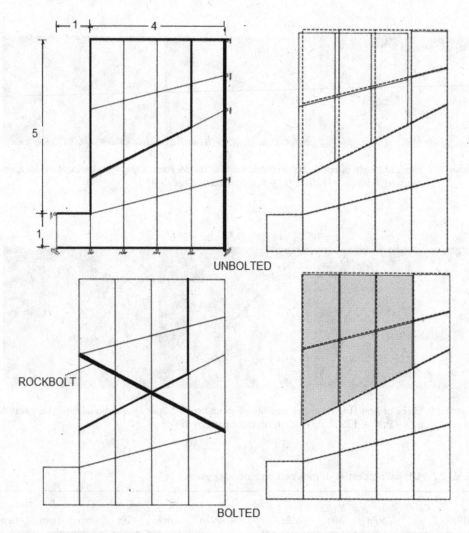

Figure 8.9 DFEM meshed deformed configurations of unbolted and bolted rock slope models.

8.3   REINFORCEMENT AGAINST FLEXURAL TOPPLING FAILURE

8.3.1   Limit equilibrium method

Aydan and Kawamoto (1987, 1992) studied the flexural toppling of rock slopes and developed a method to analyze the stability of rock slopes against flexural toppling failure. This method was developed for rock slopes in the 1989 publication and it was extended to underground openings in the 1992 publication. The fundamental procedure is basically the same as that given in Subsection 7.5.4 for sidewalls, except for the geometry of the slope. The consideration of incorporating the effect of rockbolts/rockanchors is the same as that in Subsection 7.5.4 and is illustrated in Figure 8.10.

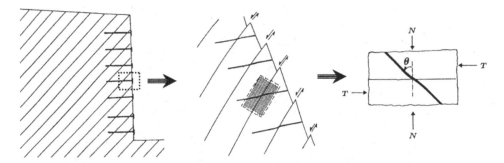

Figure 8.10 The fundamental concept for the consideration of rockbolts for reinforcing rock slopes against flexural failure.

Table 8.3 Mechanical properties and dimensions.

Materials	σ_t MPa	γ kN/m³	ϕ (°)	H m	t m	d mm	Ω (°)	θ (°)	n
Rock	0.5	25	–	20	0.5	–	–	–	–
Discontinuity	–	–	30	–	–	–	–	–	–
Rockbolts	300	–	–	–	–	25	0.56	30	2

An example of an application of the expressions, presented in Subsection 7.5.4 for the reinforcement effect of rockbolts on the required slope angle for the stability, was given by Aydan (1989) and Aydan and Kawamoto (1992). Material properties used in the analysis are given in Table 8.3, and the calculated results are shown in Figure 8.11. In the calculations, the inclination of the normal of layers from horizontal was varied from 0° to 90°. To see the effect of bolting, two case studies were made: (i) bolted slope and (ii) unbolted slope. As is seen from Figure 8.11c, the slope angle should be greatly reduced for a great range of layer inclination when no bolting is employed. On the other hand, when rockbolts are installed, it is possible to increase the slope angle remarkably. Figure 8.11c clearly illustrates this remarkable effect of rockbolts, and it implies that rockbolts can have a great stabilizing effect against flexural toppling failure on slopes.

8.3.2   Finite element method

To investigate the effect of rockbolting, it is assumed that two layers inclined at various angles with respect to the horizontal are subjected to a gravitational force field (Fig. 8.12a), and material properties given in Table 8.1 were utilized in computations. Four different case studies were made to see the differences among the various bolting patterns (Fig. 8.12b). Figures 8.12c-1 and c-2 show the fiber stress distributions at the base of the layers for four different cases for two inclinations (45° and 75°). As noted from the figures, the behavior of multi-layered column approaches that of a monolithic column of equivalent thickness through rockbolting. Figure 8.12d-1, d-2, e-1, and e-2 show the shear stresses in bolts at the discontinuity plane and the axial stress distributions in the bolts, respectively. The

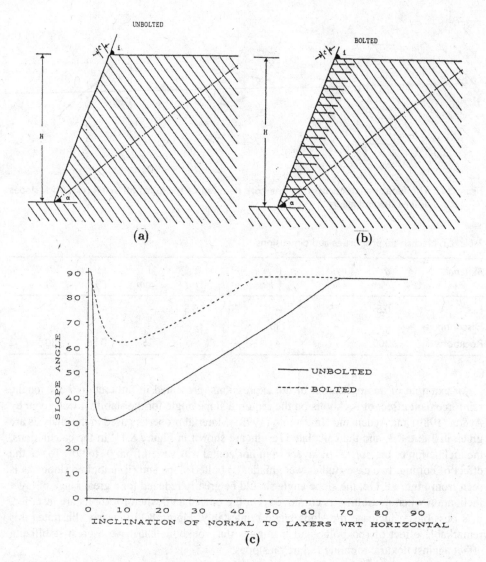

Figure 8.11 Reinforcement effect of rockbolts on the angle of slopes against flexural toppling failure.

distribution of stresses in bolts differs depending upon the location. It is expected that the reinforcement offered by rockbolts by thickening the layers should also be effective against the buckling failure of rock slopes.

8.3.3   Discrete finite element analyses

In the next example, a 4-m-high rock slope with a potential flexural toppling failure was considered. The layer was dipping mountain side with an inclination of 80°. It was assumed that a horizontally installed rockbolt stitching the first three layers together was installed.

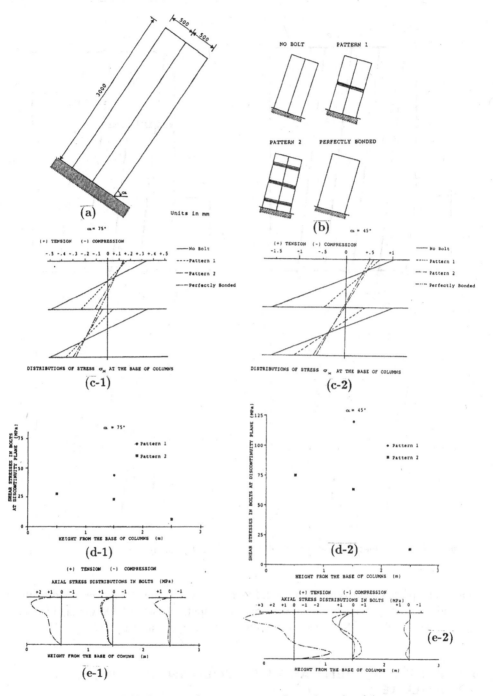

Figure 8.12 Shear reinforcement against flexural toppling failure.

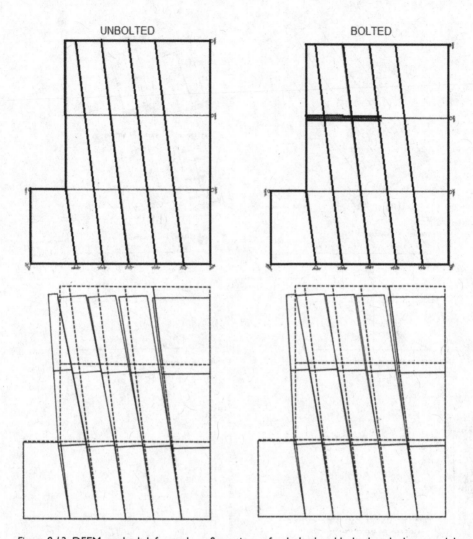

Figure 8.13 DFEM meshed deformed configurations of unbolted and bolted rock slope models.

Material properties are given in Table 8.2. Figure 8.13 shows the DFEM mesh with/without bolt and deformed configurations of the slope at computation step of 10. As noted from the figure, the rockbolts greatly restrain the deformation of layers, as expected.

8.4   REINFORCEMENT AGAINST COLUMNAR TOPPLING FAILURE

Columnar toppling and block-flexural toppling failure are also possible modes of failure when rock mass has cross-cutting discontinuities besides the thoroughgoing discontinuity set. Possible situations are illustrated in Figure 8.14. Even if the rock mass is stabilized

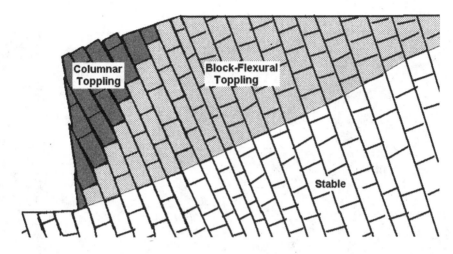

Figure 8.14 Illustration of potential zones of columnar toppling and block-flexural toppling failure.

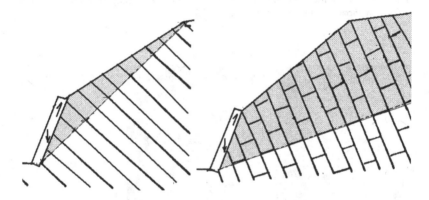

Figure 8.15 Consideration of supporting effects of retaining walls against toppling failure.

against columnar failure, there may be a possibility of block-flexural toppling failure, and rockbolting should also be implemented for such situations.

The effect of retaining walls at the toe of the slope or the lower part of rock slopes may also be taken into account by considering the weight of retaining walls and the friction between the wall and rock layers, as illustrated in Figure 8.15. The support action of the retaining walls can be implemented through the equations presented in Subsection 7.5.4 by considering the geometry of retaining walls and the position of rock layers, which may be somewhat cumbersome.

8.4.1   Physical model experiments

Figures 8.16, 8.17, and 8.18 show several examples of model tests on rock slopes using different model materials subjected to columnar toppling failure. In these model experiments, note that the anchorage length, inclination, and number of rockbolts/rockanchors are

$$\alpha_1 = 120°; \alpha_2 = 30°; L_1 / L_2 = 4; \xi = 45°$$

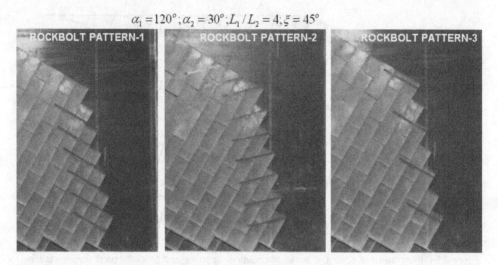

Figure 8.16 Effect of orientation and number of rockbolts on the stability of model rock slopes (base friction apparatus; intermittent pattern (IP)).

$$\alpha_1 = 135°; \alpha_2 = 45°; L_1 / L_1 = 1; \xi = 26.6°$$

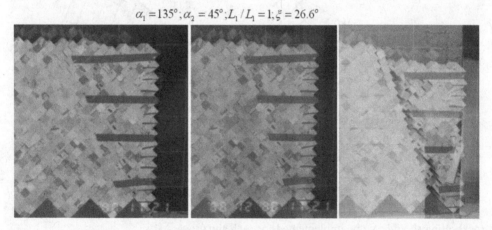

Figure 8.17 Effect of length of rockbolts on the stability of model rock slopes (gravitational model apparatus; intermittent pattern (IP)).

essential parameters to stabilize rock slopes against columnar toppling failure. Provided that sliding failure is not possible, the inclination of rockbolts/rockanchors should be such that maximum shear resistance is achieved. Furthermore, the number and anchorage length of rockbolts beyond the potential failure planes should be sufficient.

8.4.2 Discrete finite element analyses

In the next example, a 6-m-high rock slope with a potential combined columnar toppling and sliding failure was considered (Fig. 8.14). The layer was dipping mountainside with an

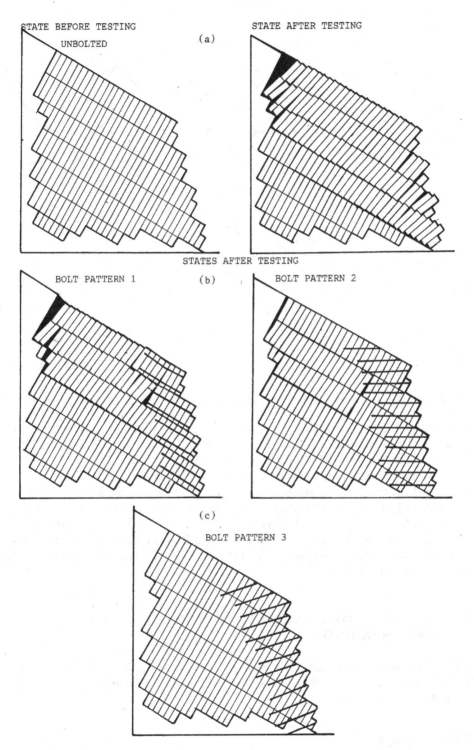

Figure 8.18 Effect of length and orientation of bolting on the stability of model rock slopes (base friction apparatus: $\alpha_1 = 120°$; $\alpha_2 = 60°$; $L_2/L_1 = 4$; cross-continuous pattern (CCP)).

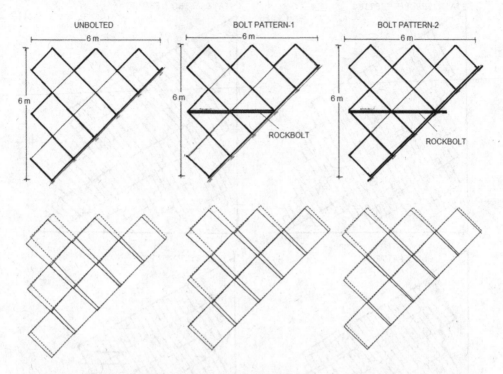

Figure 8.19 DFEM meshed deformed configurations of unbolted and bolted rock slope models.

inclination of 135°. A horizontally installed rockbolt stitching the first two column layers was assumed to be installed. In bolt pattern 1, the rockbolt does not cross the potential failure plane. However, bolt pattern 2 is fixed to beyond the potential failure plane.

Material properties are given in Table 8.2. Figure 8.19 shows the DFEM mesh with/without bolt and deformed configurations of the slope at computation step of 10. As noted from the figure, the blocks topple while they slide on the plane inclined at a 45° angle. The movement of blocks is not restrained. However, the inter-block sliding is restrained by the rockbolt. As for bolt pattern 2, the movement of columns is greatly restrained by the rockbolt anchored to the stable part. As the bolt is of the passive type, there is some relative slide along the base and columns.

8.5 REINFORCEMENT AGAINST COMBINED SLIDING AND SHEARING

In this subsection, stability analysis methods developed by Aydan *et al.* (1992) for the stabilization of rock slopes against the combined shearing and sliding failure modes illustrated in Figure 8.20 are presented.

8.5.1 Formulation

The theory presented in the following is based on the principles of limiting equilibrium theory.

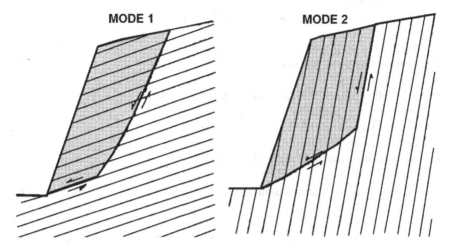

Figure 8.20 Modes of combined shearing and sliding failure.

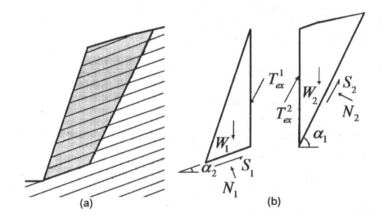

Figure 8.21 Mechanical model for mode I of combined shearing and sliding failure.

i)   Mode I

The slope illustrated in Figure 8.21a is assumed to fail by involving sliding on a thoroughgoing set and shearing on a failure plane within the intact mass. By assuming that the sheared surface is planar, one may contemplate force systems acting on blocks as illustrated in Figure 8.21b. For this particular case, the concept of excess force and excess resistance utilized in conventional theory of plasticity is introduced. The excess force is due to the active block numbered (1) and the excess resistance is due to the passive block numbered (2). In addition, it is assumed that the excess force acts parallel to the shear plane α_2. This is acceptable, provided that there is no fracturing at the interface between the blocks.

The force equilibrium conditions for the active block numbered (2) can be written as:

$$\sum S^2 = S_2 + T_{ex}^2 - W_2 \sin \alpha_2 = 0 \tag{8.3}$$

$$\Sigma N^2 = N_2 - W_2 \cos \alpha_2 \tag{8.4}$$

where
S_1 = shear force on plane (1)
N_1 = normal force on plane (1)
W_1 = weight of block (1)
α_1 = inclination of the basal plane of block (1)
S_2 = shear force on plane (2)
N_2 = normal force on plane (2)
W_2 = weight of block (2)
α_2 = inclination of the basal plane of block (2)

Let us assume that the Mohr-Coulomb criterion is satisfied on the shear plane, which is given as:

$$S_2 = c_2 L_2 + N_2 \tan \phi_2 \tag{8.5}$$

where
c_2 = cohesion of the basal plane of block (2)
L_2 = length of the basal plane of block (2)
ϕ_2 = friction angle of the basal plane of block (2)

The excess force T_{ex}^2 for block (2) takes the following form:

$$T_{ex}^2 = -c_2 L_2 + W_2 (\sin \alpha_2 - \cos \alpha_2 \tan \phi_2) \tag{8.6}$$

In a similar way, the force equilibrium conditions for the passive block numbered (1) can be written as:

$$\Sigma S^1 = S_1 - T_{ex}^1 \cos(\alpha_2 - \alpha_1) - W_1 \sin \alpha_1 = 0 \tag{8.7}$$

$$\Sigma N^1 = N_1 - T_{ex}^1 \sin(\alpha_2 - \alpha_1) - W_1 \cos \alpha_1 = 0 \tag{8.8}$$

Let us also assume that the Mohr-Coulomb criterion is satisfied on the sliding plane (i.e. bedding plane), which is given as:

$$S_1 = c_1 L_1 + N_1 \tan \phi_1 \tag{8.9}$$

where
c_1 = cohesion of the basal plane of block (1)
L_1 = length of the basal plane of block (1)
ϕ_1 = friction angle of the basal plane of block (1)

Thus, the excess resistance T_{ex}^1 for block (1) takes the following form:

$$T_{ex}^1 = \frac{c_1 L_1 - W_1 (\sin \alpha_1 - \cos \alpha_1 \tan \phi_1)}{\cos(\alpha_2 - \alpha_1) - \sin(\alpha_2 - \alpha_1) \tan \phi_1} \tag{8.9}$$

The stability of the slope can be evaluated using the following criteria:

- $T_{ex}^1 > T_{ex}^2$: stable
- $T_{ex}^1 = T_{ex}^2$: limiting state
- $T_{ex}^1 > T_{ex}^2$: unstable

ii) Mode 2

The slope illustrated in Figure 8.22a is assumed to fail by involving shearing on a failure plane within the intact mass and interlayer sliding on a thoroughgoing set. By assuming that the shear surface is planar, one may contemplate a force system acting on a typical block numbered (i) as illustrated in Figure 8.22b. The force equilibrium equations for the block are as follows:

$$\sum S^i = S_b^i - (N_f^i - N_h^i)\sin(\alpha_l - \alpha_b) + (S_f^i - S_h^i)\cos(\alpha_l - \alpha_b) - W_i \sin\alpha_f = 0 \qquad (8.10)$$

$$\sum N^i = N_b^i + (N_f^i - N_h^i)\cos(\alpha_l - \alpha_b) - (S_f^i - S_h^i)\sin(\alpha_l - \alpha_b) - W_i \cos\alpha_b = 0 \qquad (8.11)$$

where
S_b^i = shear force on the basal plane
N_b^i = normal force on the basal plane
S_h^i = shear force on the hanging-wall side plane
N_h^i = normal force on the hanging-wall side plane
S_f^i = shear force on the footwall side plane
N_f^i = normal force on the footwall side plane
W_i = weight of block
α_b = inclination of the basal plane
α_l = inclination of the thoroughgoing set

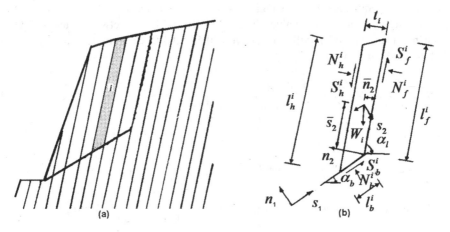

Figure 8.22 Mechanical model for mode 2 of combined shearing and sliding failure.

And the moment equilibrium with respect to the upper point of the base is:

$$\sum M^i = N_b^i \eta_b l_b^i - N_f^i \eta_l l_f^i - S_h^i t_i + W_i(\cos\alpha_l \bar{s}_2^i - \sin\alpha_l \bar{n}_2^i)$$ (8.12)
$$+ N_h^i(\eta l_h^i - t_i \cos(\alpha_l - \alpha_f)) = 0$$

where
l_b^i = length of the basal plane
l_h^i = length of hanging-wall side plane
l_f^i = length of footwall side plane
t^i = thickness of block
η_b = force position coefficient on the basal plane
η_l = force position coefficient on the side planes

Let us also assume that the Mohr-Coulomb criterion is satisfied on the sliding planes as:

$$S_f^i = c_l l_f^i + N_f^i \tan\phi_l$$ (8.13)

$$S_h^i = c_l l_f^i + N_h^i \tan\phi_l$$ (8.14)

where
c_l = cohesion of the thoroughgoing set
ϕ_l = friction angle of the thoroughgoing set

The stability condition for the block is given as:

$$S_b^i \le c_b l_b^i + N_b^i \tan\phi_b$$ (8.15)

where
c_b = cohesion of the basal plane
ϕ_b = friction angle of the basal plane

In the above six equations there are eight unknowns. The unknowns η_b^i and η_l may be eliminated by assuming some given values. The value of η_b^i will vary between 0 and 0.5. On the other hand, The value of η_l will vary between 0.5 and 1.0. Combining Equations (8.11)–(8.15) yields the following:

$$N_f^i = \frac{W^i \cdot A_1 + N_h^i \cdot A_2}{[\cos(\alpha_l - \alpha_b) + \sin(\alpha_l - \alpha_b)]\eta_b l_b^i + \eta_l l_f^i}$$ (8.16)

where

$$A_1 = \cos\alpha_b \eta_b l_b^i + \cos\alpha_b \bar{s}_2 - \sin\alpha_l \bar{n}_2$$

$$A_2 = [\cos(\alpha_l - \alpha_b)\sin(\alpha_l - \alpha_b)\tan\phi_l]\eta_b l_b^i + \eta_l l_h^i - \frac{t^i}{\tan(\alpha_l - \alpha_b)} - t^i \tan\phi_l$$

The stability condition of the column is obtained from Equations (8.11), (8.12), and (8.16) as:

$$N_f^i \leq N_h^i + \frac{c_b l_b^i + W_i(\cos\alpha_b \tan\phi_b - \sin\alpha_b)}{\sin(\alpha_l - \alpha_b) \cdot P_1 - \cos(\alpha_l - \alpha_b) \cdot P_2} \tag{8.17}$$

where

$$P_1 = 1 + \tan\phi_b \tan\phi_l$$
$$P_2 = \tan\phi_l - \tan\phi_b$$

The above simultaneous equation system will be solved by the step-by-step method, using the following boundary condition at the toe of the slope:

$$S_h^0 = 0 \quad and \quad N_h^0 = 0$$

The calculation starts from the column numbered (1) and proceeds upward. By comparing the value of N_f^i given by Equation (8.16) with that given by Equation (8.17), the stability analysis will proceed. If the value of N_f^i given by Equation (8.16) remains less than that given by Equation (8.17) for the region susceptible to failure, then the slope is assessed as stable, otherwise as unstable.

8.5.2   Stabilization

Chapter 4 and Chapter 7 already presented how to consider the stabilization effect of rock-bolts in limiting equilibrium analyses. The total shear reinforcement offered by rockbolts to a discontinuity plane will depend upon the character of the axial stress in bolts, their instal-lation angle, and their type, as well as their dimensions. The grouted rockbolts will offer a shear resistance not only from their axial response, but also their shear response will take the form given by Equation (4.52): The strength σ_b of the bar at yielding can be calculated from the von Mises criterion written in the following form (see Subsection 4.5.4):

$$\sigma_b = \frac{\sigma_t}{\sqrt{\sin^2\theta + 3\cos^2\theta}} \tag{8.18}$$

where σ_t is the tensile yield strength of the bar. The angle θ should be taken as the angle between the bolt axis and the normal of plane α for translational type movements and as the angle between the bolt axis and the plane α for separation type movements.

8.5.3   Applications

The first example of applications will be on mode 1. Aydan et al. (1992) have carried out a series of model tests using wooden blocks, the corners of which were rounded to prevent point contacts at the tips (Fig. 8.23). The failure planes were cohesionless and the friction angles were the same at 17°. The angle difference $\alpha_2 - \alpha_1$ was kept the same in all tests at 45° and the base of the model was tilted until the blocks slid. One purpose of the tests was

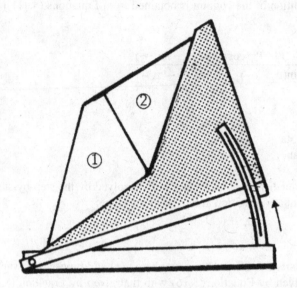

Figure 8.23 Setup model experiments.

to check the validity of the inclination of the excess force T_{ex}. If we introduce an arbitrary inclination θ, we will obtain the following expressions for the cohesionless case:

$$T_{ex}^2 = W_2 \frac{\sin(\alpha_2 - \phi_2)}{\cos(\alpha_2 - \theta - \phi_2)}, \tag{8.19}$$

$$T_{ex}^1 = W_1 \frac{\sin(\phi_1 - \alpha_1)}{\cos(\theta + \phi_1 - \alpha_1)}, \tag{8.20}$$

Limiting state yields the following:

$$\frac{W_1}{W_2} = \frac{\sin(\alpha_2 - \phi_2)}{\sin(\phi_1 - \alpha_1)} \cdot \frac{\cos(\theta + \phi_1 - \alpha_1)}{\cos(\alpha_2 - \theta - \phi_2)} \tag{8.21}$$

Figure 8.24 shows test results together with calculated limiting state curves of W_1/W_2 for various values of θ. As seen from the figure, the experimental results confirm the suggestion that θ should be taken as equal to α_2 and also the validity of the proposed method. This conclusion will certainly have important implications on the evaluation of interslice forces in the commonly used slice method in soil and rock mechanics.

The next example of applications is associated with the method presented for mode 2. For another validation of the proposed method, a series of model tests were carried out. In the tests, again the failure planes were cohesionless and model slopes were created by piling up rhomboid blocks, but the slope angle, slope height, and the inclination of discontinuity sets were varied. Direct shear tests were performed on the discontinuity planes. The inclination of the thoroughgoing set α_1 was varied between 60° and 75°, and the angle difference $\alpha_1 - \alpha_b$ was kept as 60°. The corresponding variation of the second set was 0° to 15°. The

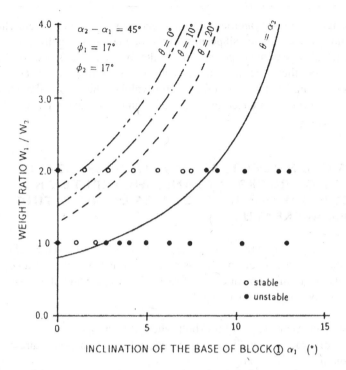

Figure 8.24 Comparison of experimental results with theoretical estimations.

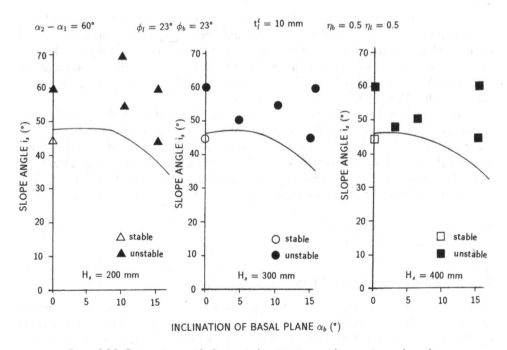

Figure 8.25 Comparison with theoretical estimations with experimental results.

initial slope angles were 45° and 60°. In the calculations, η_l and η_b were chosen as 0.5. Figure 8.20 shows the calculated slope angle – thoroughgoing set inclination relation at the limiting state together with the experimental results for the model slopes of three different heights. As seen from the figure, the limiting state curves for slopes of different heights well predict the experimental results and the test results validate the proposed method. However, further tests are necessary to check the value of η_l and η_b in relation to the discontinuous nature of rock masses.

8.6   PHYSICAL MODEL TESTS ON THE STABILIZATION EFFECT OF ROCKBOLTS AND SHOTCRETE ON DISCONTINUOUS ROCK SLOPES USING TILTING FRAME APPARATUS

Shimizu *et al.* (1992) reported results of some physical model tests on the stabilization effect of rockbolts and shotcrete on discontinuous rock slopes using a tilting frame apparatus. In the experiments, the reinforcement effect of rockbolting with and without shotcreting have been thoroughly investigated by varying the

- installation pattern and angle of rockbolts and their number
- inclination of throughgoing discontinuity sets and discontinuous nature of rock mass
- inclination of initial slope angle

8.6.1   Model materials and their properties

In all tests, wooden blocks were used. The reason for selecting this material was that the same blocks can be used several times without any apparent damage to them. We carried out direct shear tests to determine the frictional properties of interfaces between blocks. Figure 8.4 shows the results of direct shear tests on the interfaces. The static friction angle was about 23–25° and the residual friction angle was about 19–21°. The size of blocks used in tests were 10 × 10 × 100 mm and 10 × 20 × 100 mm.

Bolts were modeled by thin stripes of cello tape. The length and width of model bolts were 60 mm and 7 mm respectively. The inclination of rockbolts with respect to the horizontal was varied from −15° to −60° by −15°. The number of bolts were three and six. The bonding strength of rockbolts was measured and plotted in Figure 8.5 as a function of bolt length.

Shotcrete lining was also simulated by cello tape. This type of modeling corresponds to shotcreting with wire mesh. The discontinuity patterns were

- Cross-continuous pattern (CCP)
- Intermittent pattern (IP)

8.6.2   Apparatuses and testing procedure

An automatically tiltable frame model test apparatus was used. The apparatus allows one to investigate the failure of slopes, which are initially stable, and the base of the model slopes is tilted until slope failure occurs. Figure 8.26 shows a drawing of the tiltable frame device and the variation of the throughgoing set inclination and the slope angle in relation to the rotation angle of the model's base. When the rotation angle is 0°, the slope angle, the inclination of the throughgoing set, and the secondary set are 63°, 45°, and 135°, respectively.

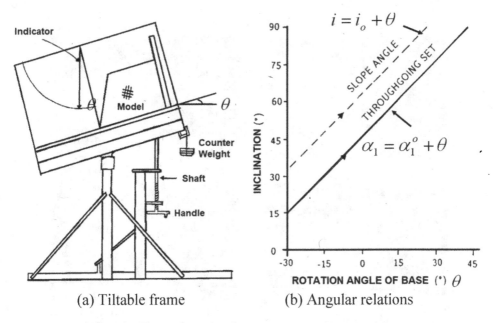

(a) Tiltable frame   (b) Angular relations

Figure 8.26 Illustration of tiltable frame device and related angular relations.

8.6.3 Test cases

Six different cases were investigated for each discontinuity pattern, and the inclination of bolts were varied in bolted models as follows:

- Case 1: No supporting
- Case 2: Shotcreting only
- Case 3: Rockbolting only (three bolts)
- Case 4: Rockbolting only (six bolts)
- Case 5: Rockbolting with shotcreting (three bolts)
- Case 6: Rockbolting with shotcreting (six bolts)

8.6.4 Results and discussions

The common failure form for IP slopes was sliding, while the CCP slopes failed by either toppling or toppling and sliding. Figure 8.27 shows pictures of some tests. The general features in each case are as follows:

Case 2: The shotcrete layer acts as a thin wall and topples by the force exerted by the unstable mass behind.

Case 3 and Case 4: When there is no shotcreting, the blocks near the slope surface among the bolts fail long before the global failure of the slopes. In addition, the integrity of the supported mass is disrupted. When the bolt density increases, the performance of supported slopes improves.

Case 5 and Case 6: When the support system consists of rockbolts and shotcrete, the performance of slopes is excellent. The supported zone acts like a retaining wall.

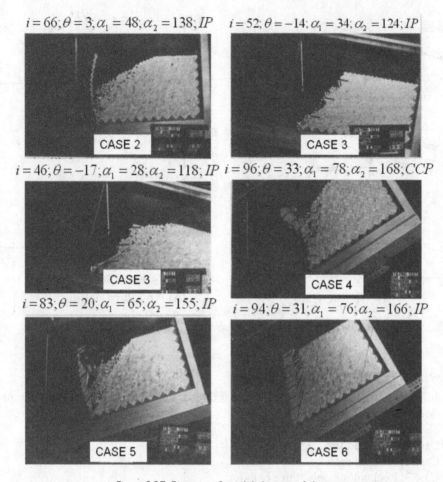

Figure 8.27 Pictures of model slopes at failure.

This conclusion has important engineering implications in the design of reinforce-ment systems of rockbolts, particularly in the event of toppling failure. Nevertheless, one must be very careful to consider the large forces near the toe of slopes, which would require installing heavy reinforcement at those locations.

Tests results are expressed in terms of measured tilting angle at the time of failure and the installation angle of rockbolts for CCP and IP slopes (Fig. 8.28). As noted from the figure, the reinforcement effect of support systems can be listed based on better performance as follows:

- Case 6: Six rockbolts with shotcreting
- Case 5: Three rockbolts with shotcreting
- Case 4: Six rockbolts without shotcreting

- Case 3: Three rockbolts without shotcreting
- Case 2: Shotcreting only

These experiments confirm the superiority of reinforcement systems consisting of rockbolts and shotcrete with a wire mesh over other support systems. In addition, the reinforcement effect is more pronounced in the CCP slopes than the IP slopes, as rockbolts allow rock mass to mobilize its inherent resistance by accommodating deformations while keeping the integrity of the mass.

The experimental results on the effect of the rockbolt installation angle indicate that the angle should lie between 0° and 45°. The variation between the installation angle and the normal of discontinuity sets is shown in Figure 8.28a. The reinforcement effect of rockbolts related to the sense of shearing depends upon the inclination of the bolt axis with respect to the normal of discontinuity and the direction of shearing. The expected normalized shear resistance of a single bolt is shown in Figure 8.28b. For IP slopes, the maximum resistance

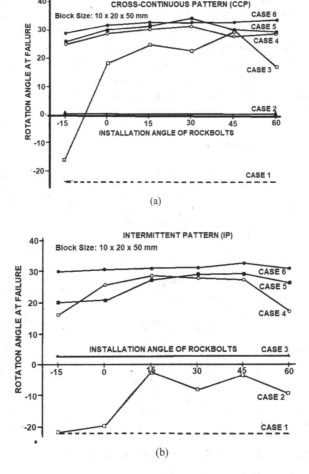

Figure 8.28 Variation of rotation angle at failure related to bolt installation angle.

should be about $-15°$. However, the test results disagree with this conclusion. The discrepancy is due to the decrease of bonding length in the anchorage zone. On the other hand, the optimum installation angle for CCP slopes should be 45°. In this case, the experimental results agree with those expected from the theoretical considerations.

8.7   STABILIZATION OF SLOPE AGAINST BUCKLING FAILURE

Although it is quite rare to observe the buckling failure of rock slopes, it has been reported (e.g. Cavers, 1981; Aydan *et al.*, 1996; Ulusay *et al.*, 1995). Cavers (1981) also studied a mode of buckling called "block buckling," which is fundamentally similar to the arching theory presented in Chapter 7. The formula (Eq. 7.114) presented in Chapter 7 is used fundamentally to check the possibility of slopes against buckling failure. Kutter (1974) suggested using the value of α as 5.88 by assuming that the lower end has a built-in condition, while the other end is hinged together with a distributed load. On the other hand, Goodman (1976) suggested using the formula based on the assumption of a buckling layer with built-in conditions at both ends, which implies that the value of α is 3. However, Cavers suggested that the value of α should be 12 due to cross joints, which implies that both ends are hinged. This probably would correspond to the most conservative situation.

Equation (7.114) is applied to the slope by assuming that the lower part (L_b) of the layer with a length $(L = L_a + L_u)$ next to the slope surface buckles while the upper part (L_u) induces critical stress on the lower part. The critical stress acting on the buckling part is assumed to be at the limiting state with a frictional resistance as given below:

$$\sigma_{cr} = \gamma_r L_u \cos \alpha (\tan \alpha - \tan \phi). \tag{8.22}$$

The stabilization effect of rockbolts on rock slopes prone to buckling failure would be achieved by increasing the shear resistance of interface and the thickness of the layers subjected to buckling through the stitching action of rockbolts. Figure 8.29 illustrates the possible rockbolting pattern of rock slopes against buckling failure.

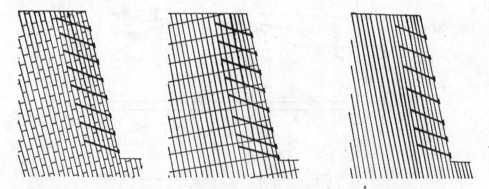

Figure 8.29 Possible rockbolting pattern of rock slopes against buckling failure.

Chapter 9

Foundations

9.1 INTRODUCTION

Foundations of bridges, pylons, and dams on rock mass may be subjected to tension or compressive forces. For example, foundations of pylons and tunnel-type anchorage for suspension bridges are examples of foundations subjected to tension, while sockets for stabilizing bridges and dam foundations would be under compression. This chapter covers foundations of rock engineering structures subjected to tension or compression, introduces current design practices and techniques, and provides specific examples.

9.2 FOUNDATIONS UNDER TENSION

Estimating the uplift (pull-out) capacity of anchors and anchorages embedded in rock is important in geotechnical engineering applications, such as the design of anchorages for suspension bridges, towers of any kind, gravity dams, and tsunami-protection walls for important civil engineering structures (Fig. 9.1). The uplift capacity of anchors in rock has been extensively investigated both theoretically and experimentally in the laboratory and *in situ*. Despite these investigations, no unified approach exists – standards and practices differ in many countries throughout the world based on their empirical experiences.

In this section, the various approaches and their fundamental concepts to the design of anchor rocks with the consideration of anchor geometry and loading conditions are presented.

9.2.1 Pylons

There is an increasing demand for the use of rockanchors as foundations in many geotechnical engineering structures, such as pylons, dams, suspension bridges, etc. Particularly, the design of rockanchors as pylon foundations is one of the intensive fields of research, as the present design concept is fundamentally based on the design of anchors in soils, which is overconservative and costly. Figure 9.2 shows a typical pylon design utilizing rockanchors as foundations against uplift due to wind or seismic loadings. Using rockanchors as foundations is cost effective, as the volume of excavation is small compared with pile-type foundations. This is an important element in foundation design, as many pylons are constructed in mountainous areas where construction is difficult in view of material transport, excavation, etc. Nevertheless, experience in using rockanchors as foundations is scarce, and their long-term performance is in doubt, as little experimental data is available.

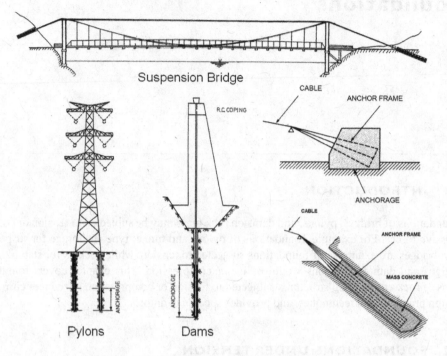

Suspension Bridge

Pylons   Dams

Figure 9.1 Examples of foundations under tension.

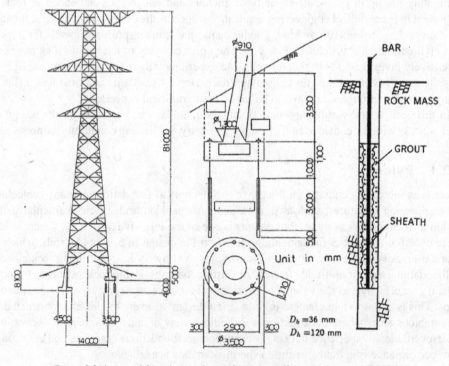

Figure 9.2 A typical foundation design for pylons (from Aydan *et al.*, 1995).

Aydan *et al.* (1994, 1995) reported a research program concerning laboratory and *in situ* experimental tests and theoretical and numerical studies on the mechanics and load-bearing capacity of rockanchors. Based on the outcomes of this research program, Aydan et al. drew some guidelines and proposed a unified design methodology for rockanchor foundations. This approach is described in the following subsections.

9.2.1.1   Classifications of anchors and anchorages

In this section, anchors are classified from several aspects, such as

* Anchor geometry
* Loading conditions

and associated modes of failure are described by considering the above items.

Anchors can be generally classified from the point of their geometry, as follows (see Fig. 9.3):

* *Cylindrical anchor*: This type of anchor is commonly used in rocks and its surface may be smooth or rough (i.e. ribs).
* *Reamed or bulbous anchor*: Though this type of anchor is circular, the anchor shaft is enlarged at the bottom or at several locations to increase the anchorage capacity.
* *Tunnel anchorage*: This type of anchor is generally used to anchor suspension bridge cables. For this purpose, an inclined tunnel is excavated and concrete is cast with a cable frame. Anchors may be loaded at the top, the bottom, or multiple points.

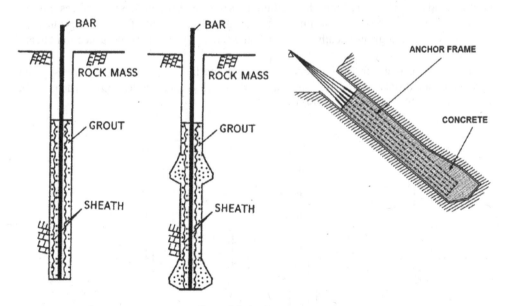

Figure 9.3 Anchor types.

9.2.1.2 Failure modes of rockanchors

This section briefly describes the failure modes observed in *in situ* pull-out tests, their possible mechanisms, and their associated conditions. Failure modes of anchors are closely interdependent with the anchor type, kind of medium, loading conditions, and introduced surfaces of weakness due to constructional and structural factors. In general, the failure of anchors may involve one or a combined form of the following failure modes (Fig. 9.4):

1) Failure of intact medium along

    i)   a shear failure plane (band) or
    ii) tensile failure plane (band)

2) Failure along a surface of weakness
3) Failure of anchor material in tension

(1) Failure of intact medium (truncated conical form)

Although this is a commonly quoted form of failure of anchors installed in a soil-like medium, it is an unlikely form of failure for cylindrical anchors unless the bottom of the anchors is enlarged.

(2) Failure along a surface of weakness

As anchors are a human-made structures, some surfaces of weakness are inevitably introduced. As a result, these surfaces have lower adhesive strengths than the medium. Since applied loads are transferred into the medium by shearing through these interfaces, failure occurs by shearing at one of the interfaces (i.e. grout-tendon interface (GT), grout-rock interface (GR), and grout-sheath-grout interface (GSG) (if a corrosion protection sheath is used).

The mechanism and fracturing state along these interfaces were previously investigated by the authors, discussed in detail, and classified in their earlier publications (Aydan, 1989; Aydan *et al.*, 1990; Ebisu *et al.*, 1993).

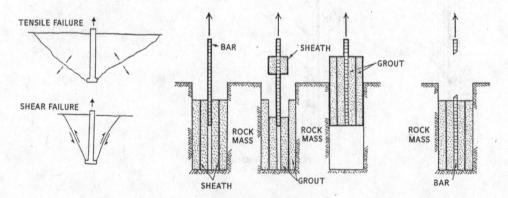

Figure 9.4 Failure modes of rockanchors.

(3) Failure of tendon

The main element transferring the load applied by the super structure to the ground is steel bar. The common failure of steel bar is tensile. It should be noted that the partial failure of interfaces or the medium may take place before the failure of the steel bar.

9.2.1.3 Constitutive laws of rockanchor systems

The mechanical behavior of steel has been well known for many centuries and is explained in Chapter 4. Its mechanical behavior is generally elasto-plastic, as illustrated in Figure 9.5a.

As the interfaces are found as cylindrical surfaces in rockanchor systems, it is diffi-cult to study their mechanical behavior. Therefore, Aydan (1989) has rolled these surfaces onto two-dimensional planes and carried out both short-term and long-term tests. Typical short-term shear responses are illustrated in Figure 9.5b and the shear strength of various interfaces is shown in Figure 9.6.

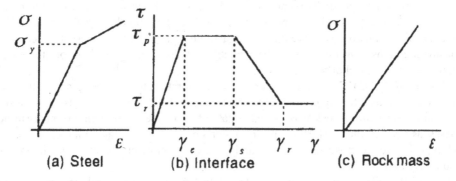

Figure 9.5 Illustration of constitutive models of materials in rockanchor systems.

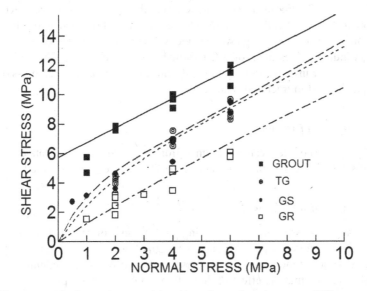

Figure 9.6 Shear strength of grouting material and interfaces of tendon-grout (TG), grout-sheath (GS), and grout-rock (GR).

Considering the extreme conditions, winds such as typhoons have a period of 10 seconds with a velocity of 30–50 m/s. As the shear strength of interfaces decreases with cyclic loads of longer period, a sinusoidal cyclic loading with a period of 20 s was chosen.

Although the mechanical behavior of intact rocks has been well known for about a century, the mechanical behavior of rock masses is less understood, since it is very expensive to carry out tests on them *in situ*. The rock mass around boreholes in which rockanchors are installed may behave elastically or in an elasto-plastic manner.

9.2.1.4   Design method for rock-anchor foundations

The design procedure for rock-anchor foundations is illustrated in Figure 9.7 and described in the following subsections.

(1) Characterization of rock mass

Outcrop surveying is generally recommended, but detailed information is difficult to obtain since the rock mass is usually covered with topsoil. Therefore, borehole drilling is necessary to get more information. We have used RQD and other parameters to assess the rock mass. Our investigations show that most of the available techniques are not satisfactory for assessing rock masses as anchor foundations.

Ebisu *et al.* (1992) used various geophysical exploration techniques to assess the rock mass and found that the exploration technique based on the Rayleigh waves is the most sensitive to the structure of rock masses and suitable for the site investigation of near-surface masses, as it yields the most detailed information on the physical state of a rock mass. The technique is also the most cost-effective and is easy to operate on the site, since it does not require any boreholes.

Aydan *et al.* (1995) and Ebisu *et al.* (1992) reported a good correlation between the deformability of rock masses by the borehole jack test and the Rayleigh wave velocity Vr as shown in Figure 9.8a. For measuring the deformability of rock mass, they recommended using the borehole jack test or the pressuremeter test in principle at each pylon site. However, the correlation between the Rayleigh wave velocity V_r and deformability of rock masses may be used as an indirect assessment of the deformability of rock mass, which is expected to reduce the cost of site investigation. Figure 9.8b shows a response of rock mass obtained in a borehole-jacking test in fractured sandstone.

(2) Characterization of tendon

Mechanical properties of steel can be obtained from standard tests. The data on mechanical properties of steel will be generally the same unless the type of steel is varied.

(3) Shear strength of interfaces in boreholes

The normal (radial) stress on interfaces during pull-out tests of rockanchors or rockbolts is an important element for estimating the shear strength of interfaces in boreholes. The normal stress is likely to be a function of the stiffness of the surrounding medium, since experimental results indicate that the effect of the stiffness on *in situ* tests plays a big role on the

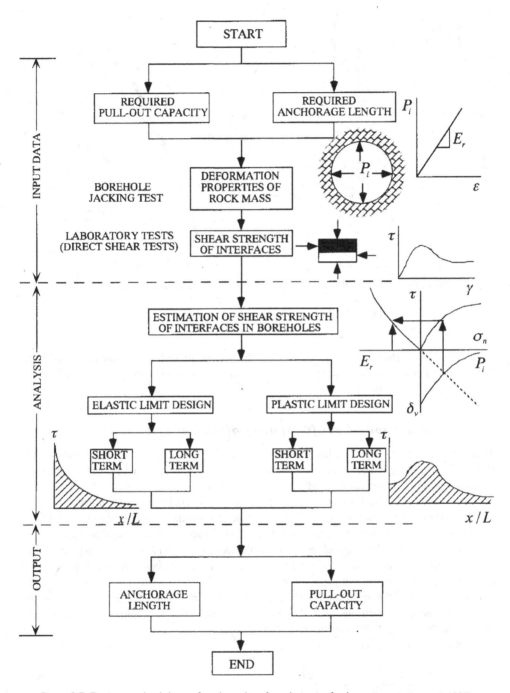

Figure 9.7 Design methodology of rock-anchor foundations of pylons. (from Aydan *et al.*, 1995).

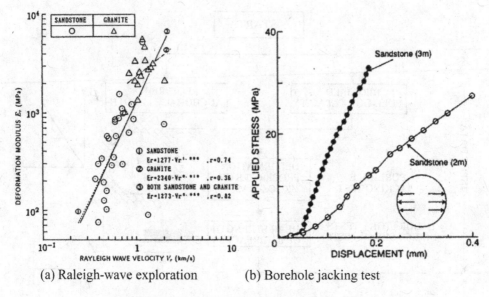

(a) Raleigh-wave exploration   (b) Borehole jacking test

Figure 9.8 Characterization of rock mass properties.

interface strength. However, the assessment of the normal stress is still an unsolved problem. For the specification of the shear strength of interfaces in boreholes, the procedure described in Subsection 4.7 can be utilized.

9.2.1.5   *Estimation of uplift capacity of anchor systems*

There are, in general, four fundamental approaches for the design of anchors, namely, (1) standards and empirical approaches, (2) limiting equilibrium approaches, (3) closed-form solutions, and (4) numerical methods.

State-of-the-art methods for estimating uplift capacity of rockanchors have been reviewed by the authors. From this review, we found that the standards and limiting equilibrium approaches are unsatisfactory. For the design of rockanchors and rockbolts, Aydan *et al.* (1985a, 1985b, 1989) have developed several theoretical solutions for various kind of failure forms of interfaces and yielding of steel bar, and they are presented in Chapter 4. The final solutions for the pull-out capacity of rockanchors by considering the elastic and elasto-plastic behaviors of the grout-tendon interface illustrated in Figure 9.9 only are given herein. For more complex conditions, it is suggested to use numerical techniques such as the one developed by Aydan (1989) and described in Chapter 4.

Elastic limit (Fig. 9.9a):

$$P_0 = \frac{\tau_p^{gt} \alpha E_a}{K_g} \tanh \alpha L, \quad \alpha = \sqrt{\frac{2K_g}{E_a r_a}} \tag{9.1}$$

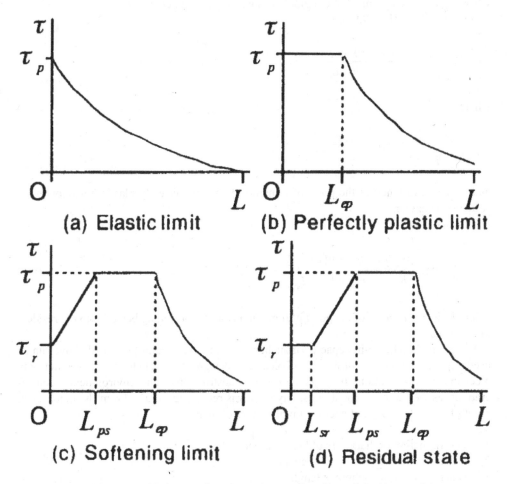

Figure 9.9 Assumed behavior of tendon-grout interface for each limiting condition.

where τ_p^{gt} is the peak shear strength of the grout-tendon interface; r_a: radius of anchor; r_h: radius of borehole; r_0: radius of the fictitious rigid boundary ($r_0 = 2L$); E_a: elastic modulus of tendon; G_g: shear modulus of grout; G_m: shear modulus of medium.

$$K_g = \frac{G_g}{r_a \ln\left(\frac{r_h}{r_a}\right)} \frac{K-1}{K}, \quad K_m = \frac{G_m}{r_h \ln\left(\frac{r_0}{r_h}\right)} \quad K = \frac{G_m}{G_g} \frac{\ln\left(\frac{r_h}{r_a}\right)}{\ln\left(\frac{r_0}{r_h}\right)} + 1.$$

Perfectly plastic limit (Fig. 9.9b):

$$P_0 = \tau_p^{gt}\left[(\xi_s - 1)\frac{E_a}{K_g L_{ep}} + \frac{L_{ep}}{r_a}\right] \qquad (9.2)$$

where L_{ep} is the length of plastic region measured from the anchor head.

Softening limit (Fig. 9.9c):

$$P_0 = \tau_p^{gt} \left[\frac{pE_a \beta \eta}{K_g} \frac{1 - \eta \cos pL_{ps}}{\sin pL_{ps}} \right] \tag{9.3}$$

where

$$p = \sqrt{\frac{2K_g}{E_a r_a \beta}} \quad \beta = \frac{\xi_r - \xi_s}{1 - \eta}$$

and L_{ps} is the distance of the point of transition from the perfectly plastic behavior to the softening behavior.

Residual state (Fig. 9.9d):

$$P_0 = \tau_p^{gt} \left[\frac{pE_a \beta}{K_g} \frac{1 - \eta \cos p\left(L_{ps} - L_{sr}\right)}{\sin p\left(L_{ps} - L_{sr}\right)} + \frac{2\eta}{r_a} L_{sr} \right] \tag{9.4}$$

where L_{sr} is the distance of the point of transition from the softening behavior to the residual plastic behavior.

For estimating the uplift capacity of anchors, one of the formulas presented above can be used depending upon the designer's choice. The design of rockanchors involves not only mechanical considerations but also economic, constructional, and environmental considerations, as well as each country's safety concerns regarding failure. From the mechanical point of view, the design of anchors involves the following:

1) Determination of diameters of bars and boreholes
2) Determination of spacing of anchors
3) Determination of anchorage length

Borehole diameter is governed by the constructional conditions and the existence of a corrosion protection sheath. Diameters of the bars and spacing are governed by the sizes of the bars and the allowable intensity of pull-out load per anchor. From our own experimental program, we found that if the anchor spacing is 30 times the anchor diameter, the interaction between anchors almost disappears. Once the diameters of the borehole and bar and the spacing of anchors are specified, then the problem is to determine the anchorage length for a given load or the pull-out capacity for a given anchorage length.

The long-term behavior of rockanchors under cyclic loading can be evaluated by taking into account the degradation of the shear strength of interfaces as a function of cycle numbers (Aydan et al., 1994). This approach introduces a new concept, based on the principles of mechanics, to the rockanchor design methodology.

The yield strength of interfaces under cyclic loading conditions is taken into account by the procedure described in Subsection 4.6. Figure 9.10 shows the degradation of the shear strength of the grout-tendon interface as a function of cycle number.

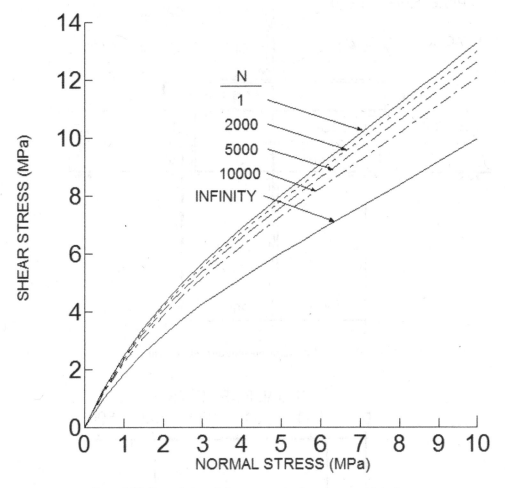

Figure 9.10 Degradation of shear strength of tendon-grout interface.

9.2.1.6 Applications

The presented design method for rockanchor foundations of pylons is applied to *in situ* pull-out tests at two locations, which are typical rock masses found in the Chubu district of Japan. The rock masses were jointed sandstone and granite. Figure 9.11 shows a typical pull-out test setup *in situ*. The axial force and shear stress distributions along a rockanchor installed in granite during the pull-out tests are shown in Figure 9.12.

In situ explorations have shown that the sandstone was classified as weak rock and granite as a medium hard rock. The deformation modulus of the sandstone ranged between 100 MPa and 1000 MPa with an average of 400–500 MPa. The deformation modulus of the granitic site was between 1000 MPa and 6000 MPa with an average of 2000–3000 MPa. The anchorage length of anchors were 30 cm, 100 cm, and 300 cm in the sandstone site and 300 cm in granitic site. 30-cm anchors had a load-bearing capacity of 100–250 MPa, while

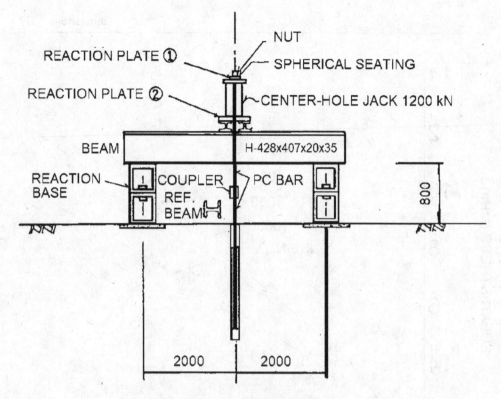

Figure 9.11 A typical pull-out test setup *in situ*.

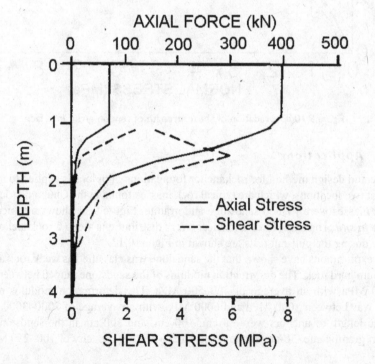

Figure 9.12 Axial force and shear stress distribution along a rockanchor installed in granitic rock mass during an *in situ* pull-out test.

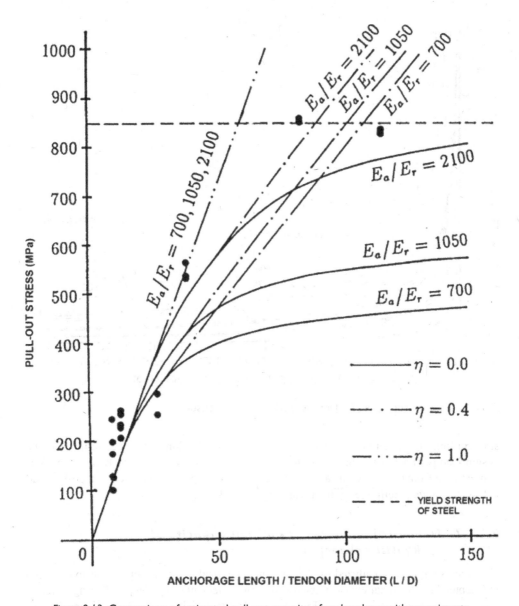

Figure 9.13 Comparison of estimated pull-out capacity of rockanchors with experiments.

100-cm-long anchors failed at a load level between 250–550 MPa. The failure of steel bars caused the failure of the 300-cm-long anchors in both sites. Figure 9.13 shows the load-bearing capacity of rockanchors as a function of the ratio of anchorage length to bar diameter.

Design nomograms were prepared for selected limit states. Figure 9.14 shows an example of the design nomogram, in which the necessary anchorage length of rockanchors were calculated as a function of the elastic modulus of rock masses for various axial force levels, by using the formula for the elastic limit state together with the experimentally obtained shear

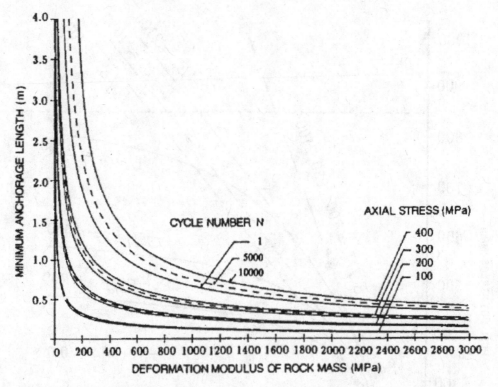

Figure 9.14 Design nomogram for length of rockanchors for elastic limit.

strength formula shown in Figure 4.37. As an actual application of the present method, the design load per anchor has been currently selected as 320 kN and the anchorage length as 3 m. As seen from the figure, the anchors should remain elastic under the current design load for an anchorage length of 3 m in a great range of rock mass conditions.

9.2.1.7   Numerical analyses for pull-out capacity of rockbolts/rockanchors

The first example was analyzed to compare the analytical model for elastic behavior given in Chapter 4 by the linear elastic axisymmetric finite element model by considering different boundary conditions, as shown in Figure 9.15. The material properties and dimensions of bolt and borehole used in the analyses are given in Table 9.1. Figure 9.15 shows distributions of the axial displacement (Fig. 9.15b), the axial stress in the bolt (Fig. 9.15c) and shear stress at the mid-distance of grout annulus (Fig. 9.15d) along the rockbolt calculated by the analytical model (Equation (4.83)), and the finite element analyses. The calculated results are very close to each other, confirming that the analytical model developed with some simplifications of the actual system is quite satisfactory in estimating the stress-strain state in the bolt and its vicinity, as long as the elastic behavior of the system is considered.

The analytical models (Equations (4.90)–(4.100)) developed for the various kinds of elasto-plastic behavior of the bolt-grout interface are also compared with the elasto-plastic

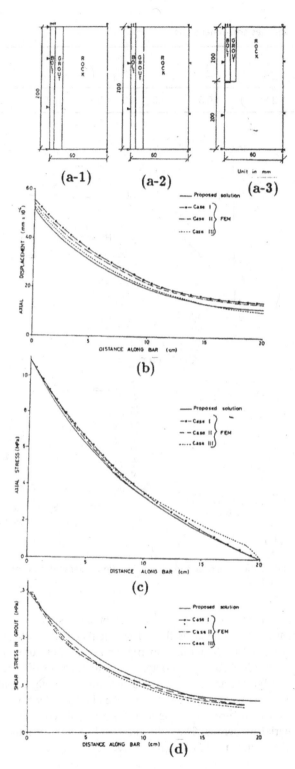

Figure 9.15 Comparison of the analytical model with the finite element method.

Table 9.1 Material properties and dimensions.

E_b GPa	v_b	E_g GPa	v_g	E_r GPa	v_r	r_b mm	r_h mm	L mm
200	0.3	2.8	0.3	3.0	0.3	6.5	25	200

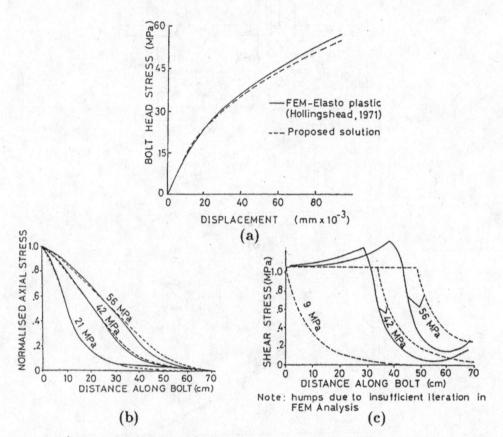

Figure 9.16 Comparison of the analytical model with the finite element method for elastic-perfectly plastic behavior.

axisymmetric finite element models. The considered behaviors are elastic-perfectly plastic and the elastic-softening residual plastic flow. The calculated results for two examples are shown in Figures 9.16 and 9.17. Material properties and dimensions of bolt and borehole are given in Tables 9.2 and 9.3. Figure 9.16 compares the load versus displacement at the bolt head, axial stress distribution, and shear stress distribution in/along the bolt calculated by the axisymmetric finite element analysis by Hollingshead (1971), by considering the elastic-perfectly plastic behavior of grout annulus and those calculated by the analytical model using the same material properties. The solid lines and dotted lines are calculated by the finite element model and analytical model, respectively.

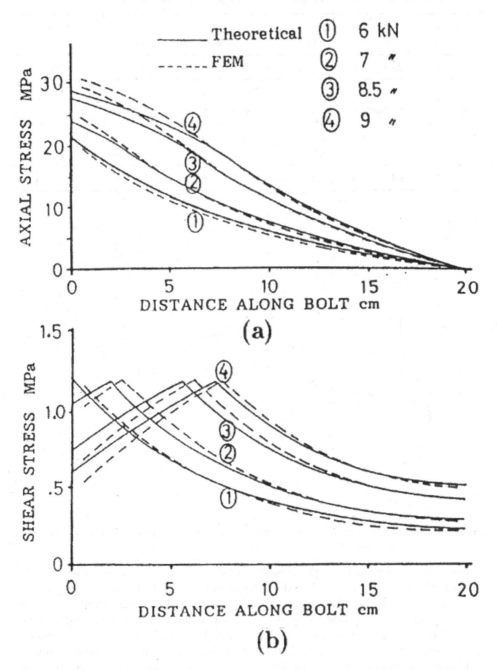

Figure 9.17 Comparison of the analytical model with the finite element method for elastic-softening–residual plastic flow plastic behavior.

Table 9.2 Material properties and dimensions – elastic perfectly plastic behavior.

E_b GPa	v_b	E_g GPa	v_g	E_r GPa	v_r	c_p MPa	η	r_b mm	r_h mm	L mm	γ_r/γ_p
200	0.3	20.4	0.17	35.0	0.2	1.054	1.0	22	28.5	700	–

Table 9.3 Material properties and dimensions – elastic-softening residual plastic flow.

E_b GPa	v_b	E_g GPa	v_g	E_r GPa	v_r	c_p MPa	ϕ (°)	r_b mm	r_h mm	L mm	γ_r/γ_p
210	0.3	3.4	0.3	1.8	0.25	1.2	66	9.5	25	200	2.1

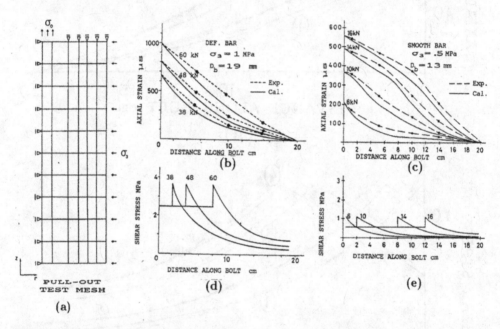

Figure 9.18 Comparison of calculated distributions of stresses in/along the bar with the experimental ones (finite element model).

Figure 9.17 compares the axial stress and shear stress distributions at various levels of load calculated by the analytical model (Equations (4.90)–(4.100)) and the finite element model using the rockbolt element proposed in Chapter 6. As the agreement is very good between results calculated by the finite element model and the analytical model for both cases, shown in Figures 9.16 and 9.17, it can be clearly stated that the analytical model is able to predict the stress-strain state in the bolt and its close vicinity very satisfactorily.

Figure 9.18 compares the distributions of the axial strain and shear stress (solid lines) at various load levels along the bolt calculated by using the proposed finite element model in Chapter 6 with those (dotted lines) measured in pull-out tests of rockbolts of 13 mm in and 19 mm diameter, having a smooth-surface under a confining pressure of 0.5 MPa and a ribbed surface (DEF.BAR) under a confining pressure of 1 MPa, respectively. The predicted axial strain responses are in good agreement with the measured responses, which validate the proposed models.

The finite element analyses to be presented hereafter are the analyses by using the rock-bolt element proposed in Chapter 6.

A series of simulations of pull-out tests was carried out by considering the actual surface configuration of steel bars. Figure 9.19 shows the finite element mesh used in computations,

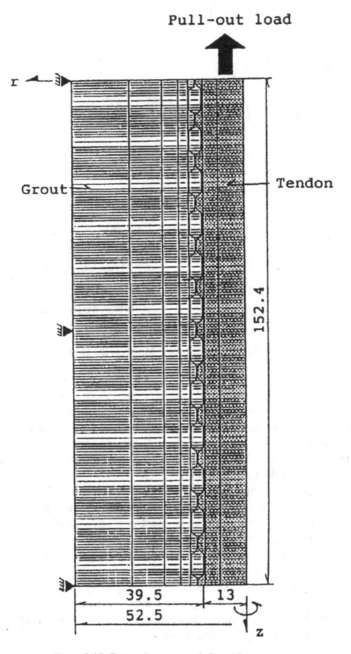

Figure 9.19 Finite element mesh for pull-out tests.

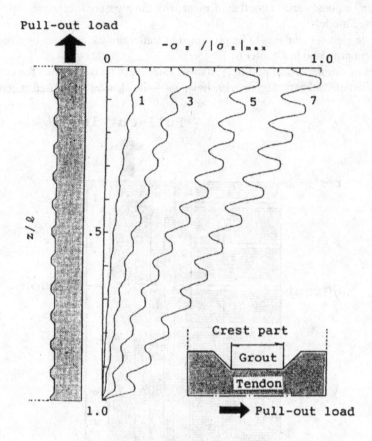

Figure 9.20 Axial stress distribution in the tendon.

and mechanical properties and models of materials used in the analyses are given in Table 9.3. These specific simulations are concerned with the pull-out tests of 150-mm-long rockanchors. Figure 9.20 shows the axial stress distribution in the tendon at various load levels. It is interesting to note that the axial stress distribution is very wavy as a result of the geometry of asperities of the tendon-grout interface.

Figure 9.21 illustrates a typical stress state in the close vicinity of ribs of rockanchors. It is interesting to note that large compressive stresses develop almost perpendicularly to the asperity surface, while the other stress component is always tensile. The tensile stress is maximum just above the top of the rib.

Figure 9.20 shows the distribution of σ_1 and σ_3 along the tendon-grout interface for load levels shown in Figure 9.20. It should be noted that the distributions of the stresses with the consideration of actual asperity configuration are completely different from those with a smooth configuration, which is commonly used in finite element simulations of rockanchors (Hollingshead, 1971; Yap and Rodger, 1984). These analyses demonstrate very clearly the effect of the geometrical dilatancy of interfaces.

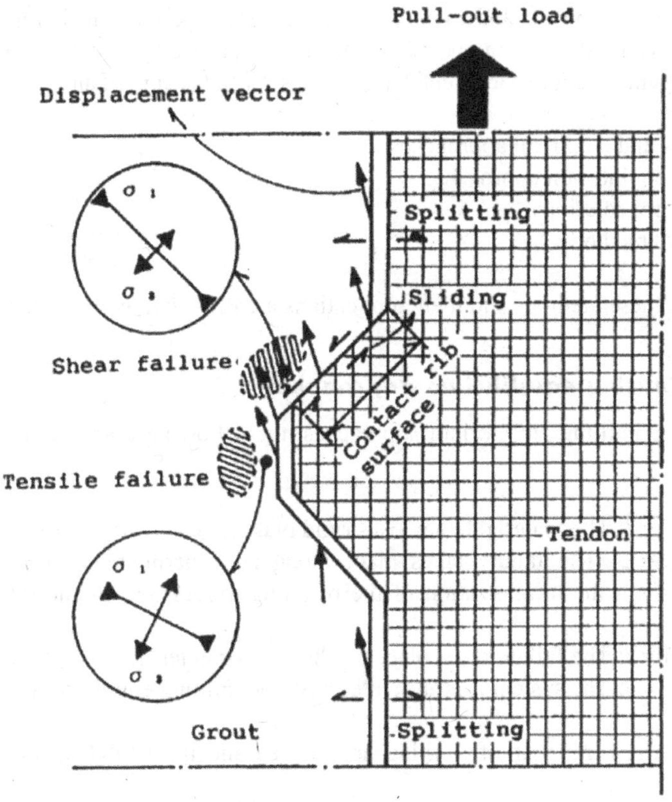

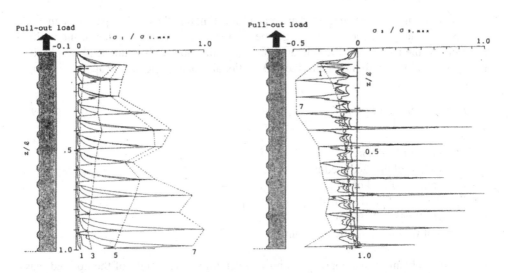

Figure 9.21 Stress state in the close vicinity of an asperity in pull-out out test simulations.

Figure 9 22 Principal stress distributions along the tendon-grout interface.

9.2.2 Design of anchorages

Suspension bridges, towers, and so on generally require large-scale anchorages. The design of anchorages varies depending upon the structure type, as well as practices in each country. There are, in general, four fundamental approaches for the design of anchorages:

- Standards and empirical approaches
- Limiting equilibrium approaches
- Closed-form solutions
- Numerical methods

The details of these methods and their applications are described in this section.

9.2.2.1 Limiting equilibrium approaches

In all available limiting equilibrium approaches, the following assumptions are generally made:

- *Assumption 1*: The material behavior is rigid plastic.
- *Assumption 2*: The failure takes place along an experimentally known surface or obtained from the minimization of the resulting expression for the resistance of the system.
- *Assumption 3*: The failure takes place by shearing or in tension.
- *Assumption 4*: The system is at the state of plastic limiting equilibrium.

In the following, a number of limiting approaches and their fundamental concepts are described.

(1) Dead-weight approach

This method is the earliest suggested approach for estimating the uplift capacity of anchors. The uplift capacity depends upon the geometrical shapes and unit weights of the uplifted medium and the anchor. As the primary element in this approach is the dead weight, it is mainly applicable to only non-cohesive and frictional ground. The uplift capacity is simply given by:

$$F_u = W_a + W_m = \gamma_a V_a + \gamma_m V_m \tag{9.5}$$

where
$W_a =$ weight of anchor
$W_m =$ weight of uplifted medium
$V_a =$ volume of anchor
$V_m =$ volume of uplifted medium
$\gamma_a =$ unit weight of anchor
$\gamma_m =$ unit weight of uplifted medium

The major problem in this approach is how to determine the weight of the uplifted mass. This weight may be determined from empirical relations based on past experiences of uplifting tests. Nevertheless, the estimation of the uplift capacity may be a problem when such empirical relations may not exist for the specific site. A procedure is suggested for determining the weight herein: Let us assume that the anchor has a rectangular base and the uplifted body consists of a center column and side columns, the cross-sections of which have a triangular shape (Fig. 9.23). From the consideration of equilibrium conditions for a side column:

$$\Sigma F_x = N_c - T_\alpha \cos \alpha - N_\alpha \sin \alpha = 0 \tag{9.6a}$$
$$\Sigma F_y = T_c - W_s - T_\alpha \sin \alpha - N_\alpha \cos \alpha = 0 \tag{9.6b}$$

where
W_s = weight of side column
T_α = shear force on plane α
N_α = normal force on plane α
T_c = shear force on plane between center and side columns
N_c = normal force on plane between center and side columns

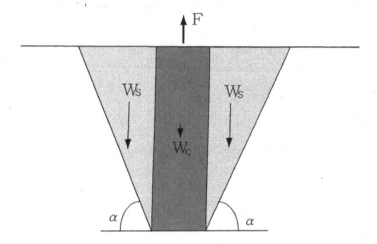

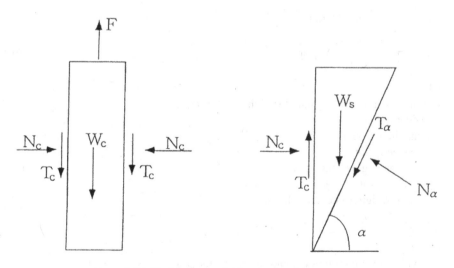

Figure 9.23 Mechanical model for dead-weight approach.

Assuming that the following relationships are held:

$$\frac{T_c}{N_c} = \tan\delta, \quad \frac{T_\alpha}{N_\alpha} = \tan\phi \tag{9.7}$$

where ϕ is friction angle of medium and δ is a shear constant and has a value in between $0°$ and ϕ, yields the following:

$$T_c = W_s \left[1 - \frac{\cos\delta\cos(\alpha+\phi)}{\cos(\alpha+\phi-\delta)} \right] \quad N_\alpha = W_s \frac{\cos\delta\cos\phi}{\cos(\alpha+\phi-\delta)} \tag{9.8}$$

If the dead weight concept is introduced, that is, $T_c = W_s$ on plane α, together with $\cos\delta \neq 0$ and $\cos\phi \neq 0$, one gets the following expression for the value of α for this specific case as:

$$\alpha = 90 - \phi \ T_c = W_s \tag{9.9}$$

The weights of center and side columns for the short and long sides of the rectangle are then given by:

$$W_c = \gamma_{ma} BLH, W_s^L = \gamma_m \frac{LH^2 \tan\phi}{2}, W_s^b = \gamma_m \frac{BH^2 \tan\phi}{2}$$

Then, the total uplift capacity F_u is:

$$F_u = W_c + 2\left(W_s^B + W_s^L\right) \tag{9.10}$$

Let us assume that the length of the short side is related to the length of the long side by:

$$N_s = \frac{B}{L} \tag{9.11}$$

Introducing the concept of the uplift capacity coefficient N_u as:

$$N_u = \frac{F_u}{\gamma_{ma} BLH} \tag{9.12}$$

We have the following relationship:

$$N_u = 1 + \frac{\gamma_m}{\gamma_{ma}} \frac{H}{B} (1 + N_s) \tan\phi \tag{9.13}$$

where
B = minimum width or diameter of the anchor base
H = depth of embedment
L = maximum width of the anchor base
N_s = shape factor
ϕ = friction angle of medium
γ_m = unit weight of medium
γ_{ma} = unit weight of center column
$\gamma_{ma} = \gamma_a n_a + \gamma_m n_m$
$n_a = V_a/V_t$
$n_m = V_m/V_t$
V_a = volumetric fraction of anchor in the center column
V_m = volumetric fraction of uplifted medium in the center column

(2) Ground pressure–shearing resistance–dead weight approach

The approaches, based on the concept of incorporating the resistances resulting from the earth pressure, shearing resistance, and dead weight are proposed in literature. In all proposals, the medium was always assumed to fail along cylindrical, straight, curved truncated conical, or pyramidal surfaces. Although these failure surfaces, in a real sense, are actually observed at the ultimate state, the common assumption is that the shear stress at all points along the surface always obeys the Mohr-Coulomb yield criterion. In other words, the material is assumed to exhibit a perfectly plastic behavior. This assumption may be valid for soils, either frictional or cohesive or both, as they are actually plastic materials from the very beginning. But this assumption cannot be true for rocks, as their behavior is of a brittle nature under the stress state encountered in pull-out conditions. With this in mind, the following proposals may be used for the estimation of the uplift capacity of anchors, particularly installed in soils.

APPROACH 1: FAILURE ALONG A VERTICAL SURFACE

Let us consider a state as depicted in Figure 9.24. The equilibrium equation for an elementary slice is obtained as:

$$\frac{d\sigma_x}{dx} + \frac{2(1+N_s)}{B}\tau = -\gamma_{ma} \tag{9.14}$$

Assume that shear stress τ obeys the Mohr-Coulomb criterion as:

$$\tau = c + \sigma_n \tan \psi \tag{9.15}$$

Normal stress σ_n in this specific case is the horizontal stress σ_h and is related to stress σ_v by the following relationship:

$$\sigma_h = K\sigma_v \tag{9.16}$$

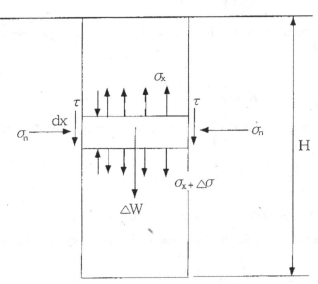

Figure 9.24 Idealized stress state and notations.

Now we will consider two cases, namely, (a) no-arching and (b) with arching.

Case 1: No arching: This is equivalent to assuming that the following relationship holds:

$$\sigma_v = \gamma_m x \tag{9.17}$$

Inserting these relationships into the equilibrium equation and solving the resulting differential equation yields the following:

$$\sigma_x = D - \frac{2(1+N_s)}{B}\left[c_a x + \gamma_m K \tan \psi \frac{x^2}{2} \right] - \gamma_{ma} x \tag{9.18}$$

Integration constant is obtained from the boundary condition:

$$\sigma_x = P_o \; at \; x = 0$$
$$D = P_o$$

Inserting the condition that is $\sigma_x = 0$ at $x = H$ (ground surface) yields the uplift resistance as:

$$P_o = \frac{2(1+N_s)}{B}\left[c_a H + \gamma_m K \tan \psi \frac{H^2}{2} \right] + \gamma_{ma} H \tag{9.19}$$

Case 2: With arching: This is equivalent to assuming that the following relationship holds:

$$\sigma_x = \sigma_v \tag{9.20}$$

The solution of the resulting differential equation is obtained as:

$$\sigma_x = De^{-Cx} - \left[\frac{\gamma_{ma}}{C} + \frac{C_a}{K \tan \phi} \right] \quad C = \frac{2(1+N_s)\tan\phi K}{B} \tag{9.21}$$

Using the same boundary conditions as those given in the previous case, we have the following expression:

$$\sigma_x = P_o e^{-Cx} - \left[\frac{\gamma_{ma}}{C} + \frac{C_a}{K \tan \phi} \right]\left(1 - e^{-Cx} \right) \tag{9.22}$$

Inserting the condition that is $\sigma_x = 0$ at $x = H$ (ground surface) yields the uplift resistance as:

$$P_o = \left[\frac{\gamma_{ma}}{C} + \frac{C_a}{K \tan \phi} \right]\left(e^{C \cdot H} - 1 \right) \tag{9.23}$$

APPROACH 2: FAILURE ALONG AN INCLINED SURFACE

In a similar manner, the equilibrium equation for an elementary slice (Fig. 9.25) is:

$$\frac{d\sigma_x}{dx} + \frac{1}{x}\left[\sigma_x - 2(1+N_s)\left(\tau_\alpha + \sigma_\alpha \frac{1}{\tan\alpha} \right) \right] = -\gamma_{ma} \tag{9.24}$$

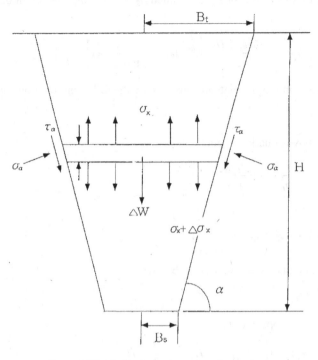

Figure 9.25 Idealized stress state and notations.

Let us assume that the material obeys the Mohr-Coulomb yield criterion at every point on the assumed failure surface as:

$$\tau = c + \sigma_\alpha \tan \phi \tag{9.25}$$

Case 1: No arching: σ_α is related to stress σ_v by the following relationship:

$$\sigma_\alpha = K_\alpha \sigma_v = K_\alpha \gamma_m x \tag{9.26}$$

Solution of the above equation is obtained as:

$$\sigma_x = \frac{D}{x} - 2(1+N_s)\left[c + \frac{\gamma_m x}{2}\left(K_\alpha(\tan\phi - \frac{1}{\tan\alpha}) + 1\right)\right] \tag{9.27}$$

Introducing the boundary condition given below:

$$\sigma_x = P_0 \text{ at } x = H_b$$

yields the integration constant D as:

$$D = \left\{ P_0 + 2(1+N_s)\left[c + \frac{\gamma_m x}{2}\left(K_\alpha(\tan\phi - \frac{1}{\tan\alpha}) + 1\right)\right]\right\} H_b \tag{9.28}$$

The uplift capacity of the anchor P_o is obtained by introducing the condition, that is, $\sigma_x = 0$ at $x = H_t$:

$$P_o = 2(1+N_s)\left[c + \frac{\gamma_m(H_t - H_b)}{2}\left(K_\alpha(\tan\phi - \frac{1}{\tan\alpha}) + 1\right)\right]\left(\frac{H_t}{H_b}\right) \qquad (9.29)$$

Case 2: With arching: σ_α is related to stress σ_v by the following relationship:

$$\sigma_\alpha = K_\alpha \sigma_v = K_\alpha \sigma_x \qquad (9.30)$$

Solution of the above equation is obtained as:

$$\sigma_x = \frac{D}{x^\eta} - c\frac{2(1+N_s)}{\eta} - \frac{\gamma_m x}{\eta + 1} \qquad (9.31)$$

where

$$\eta = 1 + 2(1 + N_s)K_\alpha\left(\tan\phi - \frac{1}{\tan\alpha}\right)$$

Introducing the boundary condition given below:

$$\sigma_x = P_0 \text{ at } x = H_b$$

yields the integration constant D as:

$$D = c\frac{2(1+N_s)}{\eta}\left[\left(\frac{H_t}{H_b}\right)^n - 1\right] + \frac{\gamma_m H_t}{\eta + 1}\left[\left(\frac{H_t}{H_b}\right)^n - \frac{H_b}{H_t}\right] \qquad (9.32)$$

(3) Determination of coefficient K

As noted from the above expression, it is a problem how to determine coefficient K. In literature, there are a number of attempts to determine this coefficient experimentally and theoretically. This coefficient can be determined by the use of Mohr's circle and the transformation law and the sense of movement (Fig. 9.26).

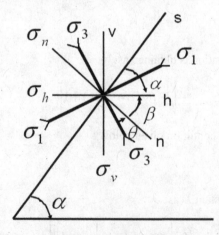

Figure 9.26 Idealized stress state and notations at a material point on the wall.

Let us consider a material point next to the wall which is inclined at an angle $\alpha = 90-\beta$ with respect to the horizontal. The relationship among horizontal, vertical, and shear stress τ_{hv} stress components on the plane can be written as:

$$\begin{Bmatrix} \sigma_h \\ \sigma_v \\ \tau_{hv} \end{Bmatrix} = \begin{bmatrix} \cos^2 \beta & \sin^2 \beta & \sin \beta \cos \beta \\ \sin^2 \beta & \cos^2 \beta & -\sin \beta \cos \beta \\ -\sin \beta \cos \beta & \sin \beta \cos \beta & \cos 2\beta \end{bmatrix} \begin{Bmatrix} \sigma_\alpha \\ \sigma_s \\ \tau_\alpha \end{Bmatrix} \tag{9.33}$$

The relationships among the principal stresses and those on the element next to the wall are:

$$\begin{Bmatrix} \sigma_\alpha \\ \sigma_s \\ \tau_\alpha \end{Bmatrix} = \begin{bmatrix} \cos^2 \theta & \sin^2 \theta & \sin \theta \cos \theta \\ \sin^2 \theta & \cos^2 \theta & -\sin \theta \cos \theta \\ -\sin \theta \cos \theta & \sin \theta \cos \theta & \cos 2\theta \end{bmatrix} \begin{Bmatrix} \sigma_3 \\ \sigma_1 \\ 0 \end{Bmatrix} \tag{9.34}$$

Assuming that the following hold:

$$\sigma_1 = q\sigma_3, \quad \tau_\alpha = \tan \psi \sigma_\alpha, \quad q = \frac{1+\sin \phi}{1-\sin \phi} = \frac{1}{K_a} \tag{9.35}$$

From the above relationships, one can easily write the following relationships:

$$\frac{\sigma_\alpha}{\sigma_1} = \frac{1}{q}(\cos^2 \theta + q \sin^2 \theta); \tag{9.36a}$$

$$\frac{\sigma_s}{\sigma_1} = \frac{1}{q}(\sin^2 \theta + q \cos^2 \theta); \tag{9.36b}$$

$$\frac{\tau_\alpha}{\sigma_1} = \frac{1}{q}(-\sin \theta \cos \theta + q \sin \theta \cos \theta) \tag{9.36c}$$

and

$$\frac{\sigma_h}{\sigma_\alpha} = \cos^2 \beta + \sin^2 \beta \frac{\sigma_s}{\sigma_\alpha} + \sin \beta \cos \beta \frac{\tau_\alpha}{\sigma_\alpha} \tag{9.37a}$$

$$\frac{\sigma_v}{\sigma_\alpha} = \sin^2 \beta + \cos^2 \beta \frac{\sigma_s}{\sigma_\alpha} - \sin \beta \cos \beta \frac{\tau_\alpha}{\sigma_\alpha} \tag{9.37b}$$

$$\frac{\sigma_s}{\sigma_\alpha} = \frac{\sin^2 \theta + q \cos^2 \theta}{\cos^2 \theta + q \sin^2 \theta}, \quad and \quad \frac{\tau_\alpha}{\sigma_\alpha} = \tan \psi \tag{9.37c}$$

Consequently, the relationship between σ_v and σ_α is obtained as:

$$\frac{\sigma_v}{\sigma_\alpha} = \sin^2 \beta + \cos^2 \beta \frac{\sin^2 \theta + q \cos^2 \theta}{\cos^2 \theta + \sin^2 \theta} - \sin \beta \cos \beta \tan \psi \tag{9.38}$$

The above relationship takes the following values for three specific cases:

Active case: $\theta = 45 - \dfrac{\phi}{2}$

$$K = \frac{\sigma_\alpha}{\sigma_v} = \frac{1}{\cos^2\alpha + \sin^2\alpha K_0 - \sin\alpha\cos\alpha\tan\psi}; \quad K_0 = \frac{q^2+1}{2q} \qquad (9.39)$$

Neutral case: $\theta = 45$

$$K = \frac{\sigma_\alpha}{\sigma_v} = \frac{1}{\cos^2\alpha + \sin^2\alpha - \sin\alpha\cos\alpha\tan\psi} \qquad (9.40)$$

Passive case: $\theta = 45 + \dfrac{\phi}{2}$

$$K = \frac{\sigma_\alpha}{\sigma_v} = \frac{1}{\cos^2\alpha + \sin^2\alpha / K_0 - \sin\alpha\cos\alpha\tan\psi} \qquad (9.41)$$

When the walls are vertical, that is, $\alpha = 90$, the above relationships become:

Active case: $\theta = 45 - \dfrac{\phi}{2}$

$$K = \frac{\sigma_\alpha}{\sigma_v} = \frac{1}{K_0} = \frac{2q}{1+q^2} = \frac{1-\sin^2\phi}{1+\sin^2\phi} \qquad (9.42)$$

Neutral case: $\theta = 45$

$$K = \frac{\sigma_\alpha}{\sigma_v} = 1 \qquad (9.43)$$

Passive case: $\theta = 45 + \dfrac{\phi}{2}$

$$K = \frac{\sigma_\alpha}{\sigma_v} = K_0 = \frac{1+q^2}{2q} = \frac{1+\sin^2\phi}{1-\sin^2\phi}; \qquad (9.44)$$

As noted from the above presentation, the selection of the K closely depends upon the assumed state at the wall. The following values may be used for K:

- Backfilling state: The stress state is close to that of the active state. Therefore, the value of K for the active state should be used.
- Rest state: The stress state is close to that of the neutral state as the material consolidates. Therefore, the value of K for the neutral state should be used.
- Pull-out state: The stress state is close to that of the passive state. Therefore, the value of K for the passive state should be used.

(4)   Shear strength only approach

In a cohesive medium, the resistance due to the weight of anchor and anchorage medium is usually of no importance unless the rock is very weak. Foote (1964) suggested that the failure occurs by shearing in a conical shape. Under such circumstances, the normal stresses are usually of a tensile character on the possible failure surface, and the stress distribution will be nonuniform (Fig. 9.27). Let us assume that the shear strength mobilized on the surface at the time of failure is in the following functional form:

$$c = c_0 \left(\frac{x}{l} \right)^n \tag{9.45}$$

Where c_o is the peak cohesion of embedment medium. Considering the equilibrium equations results in:

$$T = (F_u - W)\sin \alpha \qquad N = (F_u - W)\cos \alpha \tag{9.46}$$

Introducing the Mohr-Coulomb criterion in the following form:

$$T = \int_0^s c_0 \left(\frac{x}{l} \right)^n ds - N \tan \phi \tag{9.47}$$

In terms of the embedment depth H, the uplift coefficient N_u for the weightless case is obtained as:

$$N_u = \frac{2}{n+1} \frac{\cos \phi}{\sin \alpha \sin (\alpha + \phi)} \frac{H}{B} (1 + N_s) \tag{9.48}$$

where $N_u = F_u/c_0 BL$ and N_s is shape factor. Minimizing the above expression with respect to α yields the value of α as:

$$\alpha = \frac{\pi}{2} - \frac{\phi}{2} \tag{9.49}$$

(5)   Tensile strength only approach

Contrary to the model proposed by Foote (1964), Evans (1964) suggested that the failure occurs mainly due to tensile breakage. If it is assumed that the tensile failure takes place on a plane perpendicular to tensile force N together with the tensile stress distribution on the failure surface given by:

$$\sigma_n = \sigma_t \left(\frac{x}{l} \right)^n \tag{9.50}$$

then, the statement can be written in the following form:

$$N = \int_0^s \sigma_t \left(\frac{x}{l} \right)^n ds \tag{9.51}$$

In terms of the embedment depth H, the uplift coefficient N_u for the weightless case is obtained as:

$$N_u = \frac{2}{n+1}\frac{1}{\sin\alpha\cos\alpha}\frac{H}{B}(1+N_s) \qquad (9.52)$$

where $N_u = F_u/\sigma_0 BL$ and N_s is shape factor. Minimizing the above expression with respect to α yields the value of α as:

$$\alpha = \frac{\pi}{4}$$

It should be noted that these type of failures are only possible when the pull-out force is applied at the bottom, and there is no possibility of failure along any weakness surface existing within the system.

(6) Closed-Form Solutions (CFM)

There are a few closed-form solutions for estimating the loading capacity of anchorages embedded in rocks. For plate or bulbous anchors, approaches proposed by Vesic (1971) are based upon the expansion of circular or spherical cavities.

The solution proposed by Vesic (1971) involves the parameter of rigidity, which is a function of the elastic modulus of ground and is difficult to know in actual cases. In this subsection, a more practical approach taking into account bulbous anchorages having circular cross-sections to rectangular ones is presented. The following assumptions are made in the derivation of expressions for the estimation of the anchorages.

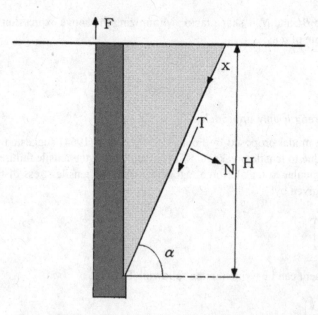

Figure 9.27 Mechanical model for cohesive or tensile failure.

- *Assumption 1*: The problem is axisymmetric and the governing equation has the following form:

$$\frac{d\sigma_r}{dr} + n\frac{\sigma_r - \sigma_\theta}{r} = -\gamma \tag{9.53}$$

  where n is $1 + N_s$.

- *Assumption 2*: Materials obey the following yield criterions in perfectly plastic and residual plastic regions (Fig. 9.28):

  - Cohesive material: Tresca material

    $\sigma_r - \sigma_\theta = 2C^p$ perfectly plastic region
    $\sigma_r - \sigma_\theta = 2C^r$ residual plastic region

  - Frictional material: Coulomb material

    $\sigma_r = q^p\sigma_\theta$ perfectly plastic region
    $\sigma_r = q^r\sigma_\theta$ residual plastic region

where

$$q^p = \frac{1+\sin\phi^p}{1-\sin\phi^p}, \quad q^r = \frac{1+\sin\phi^r}{1-\sin\phi^r}$$

  - Cohesive and frictional material: Mohr-Coulomb material

    $\sigma_r = q^p\sigma_\theta + \sigma_c^p$ perfectly plastic region

    $\sigma_r = q^r\sigma_\theta + \sigma_c^r$ residual plastic region

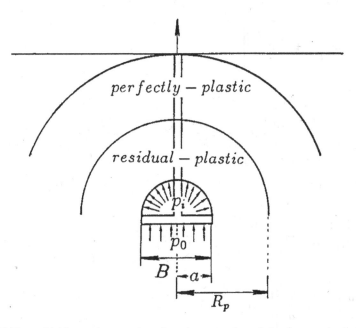

Figure 9.28 The model for anchorages based on the expansion of circular or spherical cavities.

where

$$q^p = \frac{1+\sin\phi^p}{1-\sin\phi^p} \quad q^r = \frac{1+\sin\phi^r}{1-\sin\phi^r}$$

- Tension material

  $\sigma_\theta = -\sigma_t$ at the elastic-plastic boundary

  $\sigma_r > 0 \quad \sigma_\theta = 0$ residual plastic region: no tension region

where σ_t is the tensile strength of the medium.

- *Assumption 3*: Volumetric strain change is zero:

$$\varepsilon_r + n\varepsilon_\theta = 0$$

- *Assumption 4*: The ultimate capacity is assumed to be obtained when the elastic-plastic radius reaches to the ground surface.
- *Assumption 5*: The following relationship holds between the applied plate pressure P_0 and the pressure on the fictitious cavity wall P_i:

$$P_0 = 2\int_0^{\pi/2} P_i \frac{B}{2}\sin\theta d\theta$$

The pull-out capacity coefficient N_u for this kind of material is obtained as

TRESCA MATERIAL:

$$N_u = 4n\left[\ln\left(\frac{H}{R_p}\right) + \eta\left(\ln\frac{R_p}{a}\right)\right] \tag{9.54}$$

where

$$N_u = \frac{P_0 - \gamma_m H}{C^p}; \eta = \frac{C^r}{C^p}; \xi = \frac{\varepsilon_r^r}{\varepsilon_r^p}; \frac{H}{R_p} = \xi^{1/(n+1)}; \frac{R_p}{a} = \frac{D_{H/a}}{\xi^{1/(n+1)}}; D_{H/a} = \frac{H}{a}$$

FRICTIONAL MATERIAL:

$$N_u = 2\left\{\frac{q^p}{(n-1)q^p + q^p} \cdot \left[\left(\frac{H}{R_p}\right)^{n\left(\frac{q^p-1}{q^p}\right)} - \frac{R_p}{H}\right] - \frac{q^r}{(n-1)q^r + q^r}\frac{a}{H}\right.$$

$$\left.\left[\left(\frac{a}{R_p}\right)^{n\left(\frac{q^r-1}{q^r}\right)} - \frac{R_p}{a}\right]\left(\frac{R_p}{a}\right)^{n\left(\frac{q^r-1}{q^r}\right)}\right\} \tag{9.55}$$

where

$$N_u = \frac{P_0}{\gamma_m H}; \ \xi = \frac{\varepsilon_r^r}{\varepsilon_r^p}; \ \frac{H}{R_p} = \xi^{1/(n+1)}; \ \frac{R_p}{a} = \frac{D_{H/a}}{\xi^{1/(n+1)}}; \ D_{H/a} = \frac{H}{a}$$

FRICTIONAL AND COHESIVE MATERIAL:

$$
\begin{aligned}
P_0 = 2\Bigg\{ &\frac{q^p \gamma_m H}{(n-1)q^p + q^p}\left[\left(\frac{H}{R_p}\right)^{n\left(\frac{q^p-1}{q^p}\right)} - \frac{R_p}{H}\right] + \frac{\sigma_c^p}{q^p-1}\left[1-\left(\frac{H}{R_p}\right)^{n\left(\frac{q^p-1}{q^p}\right)}\right] \\
&- \frac{q^r \gamma_m a}{(n-1)q^r + q^r}\left[\left(\frac{a}{R_p}\right)^{n\left(\frac{q^r-1}{q^r}\right)} - \frac{R_p}{a}\right] - \frac{\sigma_c^r}{q^r-1}\left[1-\left(\frac{a}{R_p}\right)^{n\left(\frac{q^p-1}{q^p}\right)}\right]\left(\frac{R_p}{a}\right)^{n\left(\frac{q^r-1}{q^r}\right)}\Bigg\}
\end{aligned}
\tag{9.56}
$$

where

$$N_u = \frac{P_0}{\gamma_m H}; \ \xi = \frac{\varepsilon_r^r}{\varepsilon_r^p}; \ \frac{H}{R_p} = \xi^{1/(n+1)}; \ \frac{R_p}{a} = \frac{D_{H/a}}{\xi^{1/(n+1)}}; \ D_{H/a} = \frac{H}{a}$$

TENSION MATERIAL:

$$N_u = 2\left[\left(\frac{H}{a}\right)^n + \frac{\gamma_m}{\sigma_t}\left(\frac{H}{n+1}\left[\left(\frac{H}{a}\right)^n - \frac{a}{H}\right]\right)\right]
\tag{9.57}$$

where
R_p = radius of plastic zone
a = half width of anchor
n = shape coefficient

9.2.2.2 Examples

The above approaches are applied to a site where three *in situ* pull-out tests on cylindrical steel anchors with a diameter of 800 mm and 6000 mm long and dead weight of 46 kN. The embedment depths were 2130 mm (B-1) and 2490 mm (G-1, G-2) were carried out. Rock mass was andesite with a dominant columnar jointing, and the spacing of discontinuities ranged between 60 mm and 600 mm. The rock mass was classified according to DENKEN, RMQR, RMR, Q-system, and GSI. Table 9.5 gives the values of rock mass classes. The *in situ* p-wave velocity was 3.5 km/s. Table 9.6 gives the mechanical properties of intact rock and estimated rock mass properties.

Anchors were installed using vibro hammers first, and then the surrounding rock mass was grouted. Figure 9.29 shows the displacement and force relations for three pull-out tests. Table 9.7 compares the measured pull-out capacity of anchors with estimated pull-out capacities using the methods explained in the previous subsections. The lower values of cohesion and tensile strength of rock mass were used in computations. The average friction angle of rock mass was assumed to be 39°. While the dead weight-earth pressure (DW-EP) yields the

Table 9.5 The rock classification values of rock mass.

Rock Classification	DENKEN	RQD	RMQR	RMR	GSI	Q-Value
Andesite	CM-CH	50–75	44–56	47–62	42–57	6.67–25.0

Table 9.6 Mechanical properties of intact rock and estimated properties according to different rock mass classifications.

Rock Class	σ_c (MPa)	E (GPa)	c (MPa)	σ_t (MPa)	$\phi(°)$
Intact rock	64	27	8.3	4.3	61
RMQR	7.4–11.6	3.2–4.9	0.96–1.5	0.5–0.78	36–42
RMR	7.0–10.7	8.4–15.0	–	–	–
Q-system	32.3–49.9	16.2–25.1	–	–	–
GSI	2.6–4.4	5.0–9.0	–	–	–
P-wave method	25.9	10.9	–	–	–

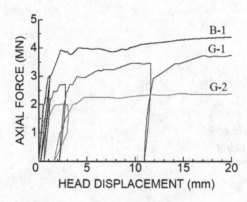

Figure 9.29 Measured force-displacement relation during pull-out tests of anchors installed in andesite.

Table 9.7 Comparison of measured pull-out capacity with estimations from different methods.

Method	Measured		τ_{av} MPa	DW-EP	Tensile	Shear			Aydan(1989)		
	Yield	Ult.			Cone	Cone	CFS		Yield	Ult.	τ_p
B-1	3800	4750	0.887	1978	1307	2510	2097		3260	4530	0.85
G-2	2200	2750	0.439	2311	1558	2992	2292		1945	2819	0.45
G-1	4200	5250	0.839	2311	1558	2992	2292		3640	5328	0.85

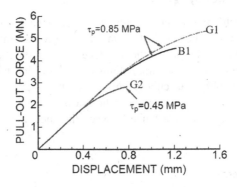

Figure 9.30 Simulation of pull-out tests of anchors installed in andesite.

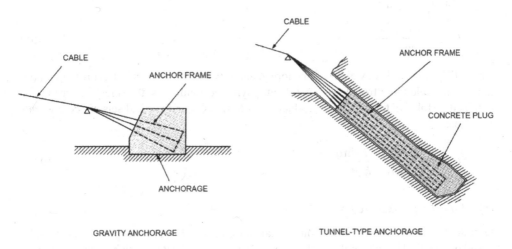

Figure 9.31 Types of anchorages for suspension or cable-stayed bridges.

lowest values, Aydan's method estimate very closely estimates the experimental values for initial yielding and ultimate pull-out resistance. Figure 9.30 shows the computed displacement-force response of the *in situ* pull-out tests using the formulation presented in Chapter 4.

9.2.3   Suspension bridges

Anchorages of suspension bridges may be of two types, namely, gravity anchorage and tunnel-type anchorage as illustrated in Figure 9.31. While gravity anchorage is much more common, tunnel-type anchorages may be preferable in view of environmental, space, and rock mass conditions.

9.2.3.1   Design of anchorages

The design of gravity anchorages is fundamentally similar to the design of gravity dams. The design of the anchorage is checked against sliding with the consideration of the dead weight of anchorage and frictional resistance of the interface between the rock base and anchorage through the introduction of safety factors. Although it is rare, its stability may be checked against rotation, which may be particularly important during earthquakes. The tunnel-type anchorages are sockets into the rock mass, and their design concept is fundamentally based on its dead weight and frictional resistance between the anchorage body and surrounding rock mass.

(a)   Gravity anchorages

The safety factor of anchorages against sliding can be obtained from the force equilibrium shown in Figure 9.32 with the use of frictional resistance, although some cohesive resistance may exist in reality as:

$$SF = \frac{W - T \sin \alpha}{T \cos \alpha} \tag{9.58}$$

where W, T are the dead weight of anchorage and cable force, respectively. α is the inclination of the cable force from horizontal. Under dynamic conditions, the safety factor may be obtained as follows through the introduction of horizontal and vertical seismic coefficients K_h, K_v as:

$$SF_d = \frac{W(1 - K_v) - T \sin \alpha}{T \cos \alpha + K_h W} \tag{9.59}$$

(b)   Tunnel-type anchorages

Despite the use and availability of numerical analyses, the design concept is fundamentally based on the dead weight of the anchorage and frictional resistance with base and sidewalls.

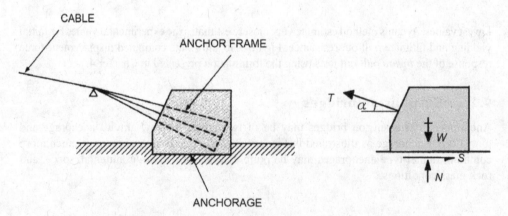

Figure 9.32 Illustration of force system on gravity anchorages.

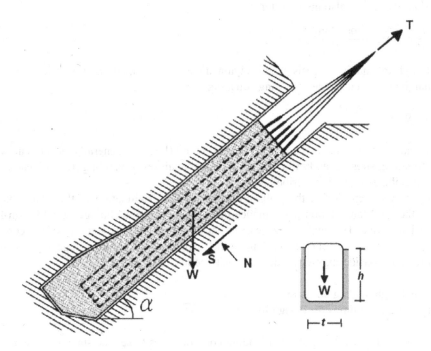

Figure 9.33 An illustration of design concept of tunnel-type anchorages.

Figure 9.33 shows the basic concept of tunnel-type anchorages for typical suspension or cable-stayed bridges.

The force equilibrium of the anchorage shown in Figure 9.33 may be written as:

$$\sum F_s = T - W \sin \alpha = 0 \qquad (9.60a)$$
$$\sum F_n = N - W \cos \alpha = 0 \qquad (9.60b)$$

Where W, N, and T are the dead weight, normal force, and tension acting on the anchorage. The shear resistance of the anchorage is generally assumed to consist of the side resistance and base resistance as given below:

$$S_b = cA_b + N \tan \phi_b; \; S_s = cA_s$$

$$SF = \frac{S_b + 2S_s}{T - W \sin \alpha} \quad \text{or} \quad SF = \frac{c(A_b + 2A_s) + W \cos \alpha \tan \phi}{T - W \sin \alpha} \qquad (9.61)$$

where A_b and A_s are surface areas of the base and sidewalls while c and ϕ are cohesion and frictional resistance of the interfaces between the anchorage and surrounding rock mass. However, a very conservative design concept is generally used, and the resistance is assumed to be frictional as given below:

$$SF = \frac{W \cos \alpha \tan \phi}{T - W \sin \alpha} \quad \text{or} \quad SF = \frac{\cos \alpha \tan \phi}{T / W - \sin \alpha} \qquad (9.62)$$

Equation (9.62) may also be rewritten as:

$$\frac{T}{W} = \sin\alpha + \frac{\cos\alpha\tan\phi}{SF} \tag{9.63}$$

For $SF = 1$, the maximum resistance is obtained when the inclination of the anchorage base is equal to that given by the following equation:

$$\alpha = 90 - \phi \tag{9.64}$$

Numerical methods such as the finite element method (FEM) is generally used to check the local straining, stresses, and safety factors in linear analyses or yielding in nonlinear analyses. A specific example is described herein.

The analyses reported in this subsection are for the examination of the design and to predict the post-construction performance of the tunnel-type anchorage of a 1570-m-long suspension bridge. The analyses are carried out to investigate the stability of the concrete anchorage body and surrounding rock upon the suspension load of the bridge applied on the anchorage by considering two loading intensity conditions using the FEM:

i)   The design loading condition
ii) Loading three times the design loading condition

The purpose of considering the first loading condition was to see the state of stress in concrete and in the surrounding rock and the possibility of any plastification in the anchorage and the rock for given material properties and a yield criterion under the normal loading (i.e. design load) conditions, including earthquake forces.

The second loading condition, on the other hand, was to see the effect of unexpected high loads upon the performance of the anchorage body and the response of the surrounding rock.

For each loading condition, two possible cases are investigated:

i)   Full bonding
ii) No tension slit just under the loading plane

The first case corresponds to an actual situation. The second case, on the other hand, visualizes a tensile crack just under the loading plane in a concrete body. By this, it is intended to see the effect of such a tensile crack on the stability of the concrete anchorage body and surrounding rock.

The rock on the site under consideration is mainly granite. However, there is almost a vertical dioritic volcanic intrusion that seems to have disturbed the surrounding rock, and it is highly weathered. The main portion of the anchorage is situated in granitic rock. The granitic rock is classified as CH class rock in the classification of DENKEN (Ikeda, 1970). The remaining part of the anchorage is located in the dioritic volcanic intrusion (dyke) and is classified as DH class rock in the same classification. The geological cross-section of the rock mass along the anchorage is shown in Figure 9.34. The material properties used in the analyses were determined from *in situ* shear tests and plate-bearing tests in other nearby construction sites with similar geology and are given in Table 9.8. The material properties listed in Table 9.8 are highly conservative and represent the lowest values of the respective tests.

Poisson's ratio for every rock class is assumed to be 0.2.

The elastic modulus, cohesion, and tensile strength is 0.6 times of the undisturbed rock mass.

Table 9.8 The material properties.

Rock Class	Esb MPa	Es MPa	Ed MPa	Cohesion MPa	$\phi(o)$	σ_c MPa	σ_t MPa	γ (kN/m³)
Soil-like	80	75	150	0	35	0	0	19
DH	80–150	180	350	0.1	37	0.4	0.04	20.5
				0.07	32	0.25	0.02	19.0
CL	150–300	330	560	0.5	40	2.14	0.21	22.0
				0.4	40	1.54	0.15	20.0
CM	300–600	600	1200	0.8	40	3.43	0.34	23.5
				0.7	40	3.00	0.30	21.0
CH	600–1200	1280	2400	1.2	45	5.79	0.58	24.5
				1.0	45	4.83	0.48	22.0
CONCRETE		25,000		4.97	45	24	15	24

The anchorage is modeled as an axisymmetric body considering the geological formations shown in Figure 9.35 for FEM model based on the geology shown in Figure 9.34. Besides the modeling of geologic formations mentioned in the previous section, a loosening zone of 2 m wide, which may be caused during the excavation of the anchorage tunnel, was assumed to exist in the finite element model. The material properties of loosening zones are assumed to be 0.6 times those of the respective geological formations. In all analyses reported herein, a finite element program considering an elasto-plastic behavior of rock mass was used. The element type used in the analyses is a four-noded isoparametric element. The dimensions of the analysis domain were taken as two times the total anchorage length vertically and hori-zontally as shown in Figure 9.35.

The gravity in the analyses was taken into account as follows:

$$\gamma^* = \gamma \sin \alpha; \ \sigma_z = \gamma * h; \ \sigma_r = K_o\sigma_z; \ \sigma_\theta = K_o\sigma_z \qquad (9.65)$$

where
γ = unit weight or rock or concrete
h = depth form surface
K_o = lateral initial-stress coefficient
α = inclination of the anchorage axis from horizontal
σ_z = vertical stress
σ_r = radial stress
σ_θ = tangential stress

The value of the lateral stress coefficient K_o, was taken as 1 in view of the ground-stress measurements in the near vicinity of the anchorage site.

The results in terms of the following items:

1) Maximum principal compressive stress in concrete and in rock
2) Maximum principal tensile stress in concrete and in rock
3) Minimum safety factor against shear failure (SFS) for concrete and for rock

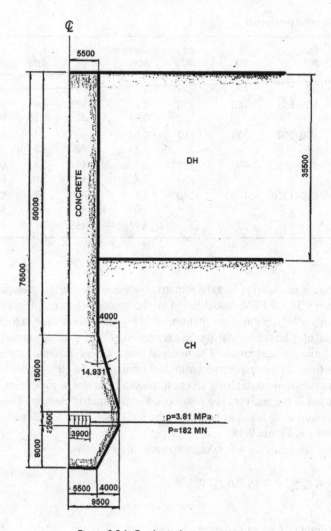

Figure 9.34 Geological cross-section.

Minimum safety factor against shearing (SFS) is defined as:

$$SFS = \frac{c\cos\phi + \dfrac{\sigma_1 + \sigma_3}{2}\sin\phi}{\dfrac{\sigma_1 - \sigma_3}{2}} \qquad (9.66)$$

Note that the sign of the compressive stress is taken as negative (−).

4) Minimum factor of safety against tensile failure (SFT) for concrete and for rock

which is defined as follows:

$$SFT = \frac{\sigma_t}{\sigma_1 \, or \, \sigma_3} \qquad (9.67)$$

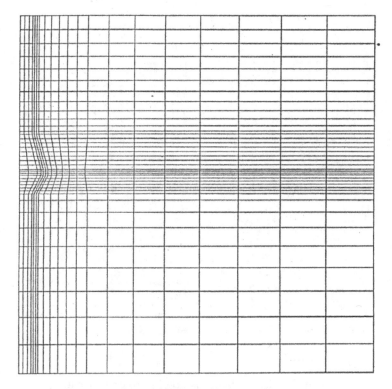

Figure 9.35 Finite element mesh.

First, the calculated results presented for the ordinary design load condition for fully bonded and with no-tension slit cases and are compared and summarized. Figure 9.36 shows principal stress distributions. As noted from stress distributions, the magnitude of compressive principal stresses is higher in the fully bonded case than in the no-tension slit case, as the existence of the tension slit tends to create higher tensile stresses, which result in the reduction in the magnitude of initial *in situ* compressive stresses. As expected, the no-tension slit relieves the concrete block just below the loading plate, which manifests itself as the principal stress directions become more vertical compared with the fully bonded case. The maximum compressive stress and maximum tensile stress are summarized in Table 9.9. When this table is carefully examined, there seems no difference between the two cases, except in the case of concrete. However, as noted from the overall principal stress distribution, there is a remarkable change in the overall distributions.

On the other hand, the comparison of the safety factors may be more relevant, as they will reflect the local changes more clearly. Figures 9.37 and 9.38 show the safety factor distributions against shearing and tensile failures respectively. The minimum safety factors are listed in Table 9.10.

The existence of the no-tension slit has a remarkable effect on the safety factor of concrete against tensile failure, as it results in lower tensile stresses in concrete. In contrast, the safety factors for rock formations tend to decrease in the magnitude as expected. The safety factor in rock formation CH is about 5. The Gauss point at which these values are observed is very near the surface and is next to the anchorage body. The given value for this rock formation corresponds to a very blocky rock mass with a very low value of tensile strength.

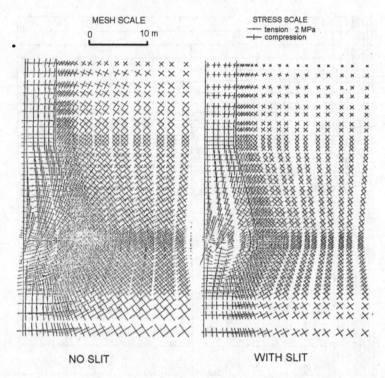

MESH SCALE

0   10 m

STRESS SCALE
--- tension 2 MPa
-+- compression

NO SLIT   WITH SLIT

Figure 9.36 Comparison of principal stress distributions.

Table 9.9 Comparison of maximum compressive and tensile stresses.

Material Type	Fully bonded		No-tension slit	
	Max.Comp.Stress MPa	Max.Ten. Stress MPa	Max. Comp. Stress MPa	Max.Ten. Stress MPa
Concrete	3.59	1.24	6.46	0.35
D_H	0.44	–	0.44	–
C_H	2.34	–	2.34	–
D^*_H	0.44	0.004	0.44	0.004
C^*_H	1.217	–	1.221	–

In conclusion, it may be stated that the stress state at this point is not representative for the overall anchorage body and could have no effect on the stability of the anchorage body.

Next, we summarize and discuss the calculated results for the case of three times the design loading condition. The principal stress distributions and displacement vectors distributions are shown in Figures 9.39 and 9.40 respectively. Table 9.11 and 9.12 compare the maximum compressive and tensile stresses and minimum safety factors in concrete and rock formations.

Although the values of safety factors and maximum compressive and tensile stresses differ from the design load cases, the overall behavior and conclusions are the same. The minimum safety factor distributions in concrete and rock are shown in Figures 9.39 and 9.40.

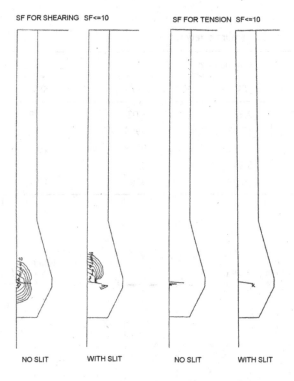

Figure 9.37 Comparison of safety factor distributions (concrete).

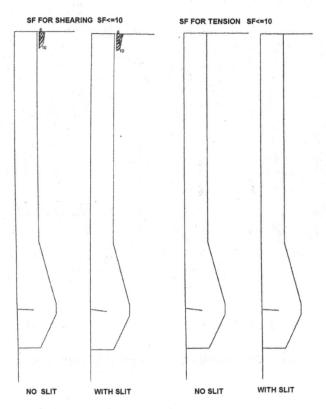

Figure 9.38 Comparison of safety factor distributions (rock).

Table 9.10 Comparison of minimum safety factors.

| Material Type | Fully bonded | | No-tension slit | |
	SFS	SFT	SFS	SFT
Concrete	2.96	1.03	2.69	2.70
D_H	15.43	9999	15.21	9999
C_H	21.12	9999	20.66	9999
D^*_H	5.34	5.51	5.25	5.18
C^*_H	10.29	9999	10.26	9999

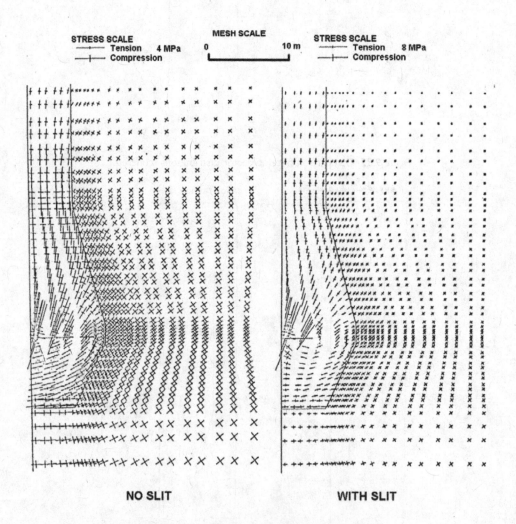

Figure 9.39a Comparison of principal stress distributions.

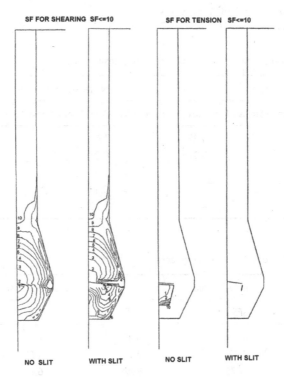

Figure 9.39b Comparison of safety factor distributions (concrete).

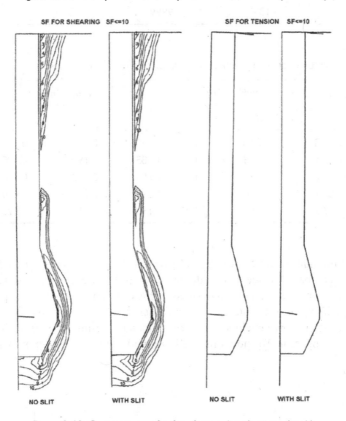

Figure 9.40 Comparison of safety factor distributions (rock).

Table 9.11 Comparison of maximum compressive and tensile stress.

Material Type	Fully bonded		No-tension slit	
	Max. Comp. Stress MPa	Max. Ten. Stress MPa	Max. Comp. Stress MPa	Max. Ten. Stress MPa
Concrete	8.62	5.85	17.24	3.21
D_H	0.44	0.015	0.019	–
C_H	2.34	–	2.34	–
D^*_H	0.44	0.0296	0.44	0.031
C^*_H	1.262	–	1.275	–

Table 9.12 Comparison of minimum safety factors.

Material Type	Fully bonded		No-tension slit	
	SFS	SFT	SFS	SFT
Concrete	0.375	0.23	1.11	0.10
D_H	4.72	2.17	4.65	2.12
C_H	13.76	9999	6.60	9999
D^*_H	1.34	0.54	1.31	0.53
C^*_H	3.13	9999	3.12	9999

Table 9.13 Comparison of stress increment in rock for two loading conditions.

Element No	Aver. Int. Stress	With Int. Stress		$\sigma_1^{3T} / \sigma_1^{1T}$	Without Int. Stress		$\sigma_1^{3T} / \sigma_1^{1T}$
		IT	3T		IT	3T	
324	0.96G	1.16	1.38	1.190	0.194	0.414	2.13
344	0.940	1.13	1.34	1.185	0.190	0.400	2.11
324	0.920	1.16	1.45	1.250	0.240	0.530	2.21
324	0.890	1.10	1.33	1.210	0.210	0.440	2.10
324	0.860	1.07	1.28	1.190	0.210	0.420	2.00

When the minimum compressive stresses given in Tables 9.9 and 9.11 for two loading conditions are compared, it seems that the increase in the intensity of load has almost no effect on the increment of compressive stress in rock. As pointed out in the previous sections, the superimposed initial state of stress causes such an impression. When the superimposed *in situ* stresses are excluded from the calculated results, the effect of such a load increase will be apparent. To show this, we have made such a calculation, which is given in Table 9.13.

The FEM analyses indicate that the anchorage of a suspension bridge based on the pure gravitational and frictional concept is highly safe for the given geometry and material properties, which are determined very conservatively. The anchorage is even safe for loads applied at three times the design load. When the rock has a few discontinuities and has a relatively high tensile strength, the occurrence of tensile cracks in the concrete just under the loading plane does not cause any serious problems. On the other hand, when the rock has a number of discontinuities and low tensile strength, the concrete must be reinforced to behave as a monolithic body to reduce high tensile stresses in rock, which may cause the failure of surrounding rock in the shape of a cone rather than a cylinder.

9.3   FOUNDATIONS UNDER COMPRESSIONS

When the bearing capacity of foundations is likely to be influenced by structural weaknesses, such as bedding planes or major fracture zones, special counter measures are necessary to increase the shear resistance of discontinuities by active as well as passive rockanchors. In addition, steel-concrete piles, which may be viewed as composite cylindrical anchorages, may also be used. This is a common practice during the construction of bridge foundations passing through porous limestone formation in the Ryukyu Archipelago of Japan.

9.3.1   Base foundations

9.3.1.1   Reinforcement of bridge foundations

There may be different modes of foundation failure associated with discontinuous rocks, as illustrated in Figure 9.41. The most likely situations would be planar sliding and flexural toppling. The formulation to consider the effects of rockbolts/rockanchors for the reinforcement of foundations would be similar to those presented in Chapter 8. However, the only difference is the existence of the surcharge load due to bridges. Therefore, no specific formulations would be given herein.

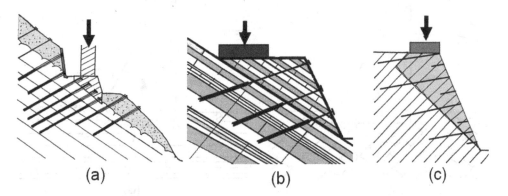

Figure 9.41 Possible failure modes of bridge foundations in layered rock mass: (a) and (b) planar sliding; (c) flexural toppling.

9.3.1.2   Reinforcement of dam foundations

The foundation failure of Malpasset Arch Dam in France in 1959 is one of the most catastrophic events in rock engineering. This event had a profound effect on the design of dams on a rock foundation, with a strong emphasis on the effect of major structural weaknesses existing in rock mass under high water pressures. Figure 9.42 shows some illustrations of possible failure modes of gravity dams. The analyses of the dam foundations can be done through some limiting equilibrium approaches for simple conditions. However, the use of numerical analyses with the consideration of discontinuities in rock mass would be necessary for complex conditions.

(a)   limit equilibrium method for foundation design

A dam subjected to base shearing and reinforced by rockanchors shown in Figure 9.43 is considered. The safety factors of the dam against base shearing and overturning about point O may be easily obtained, respectively, as given below:
Safety factor against base shearing:

$$SF_s = \frac{cL_b + N\tan\phi}{U_s - T_a\cos\alpha} \tag{9.68}$$

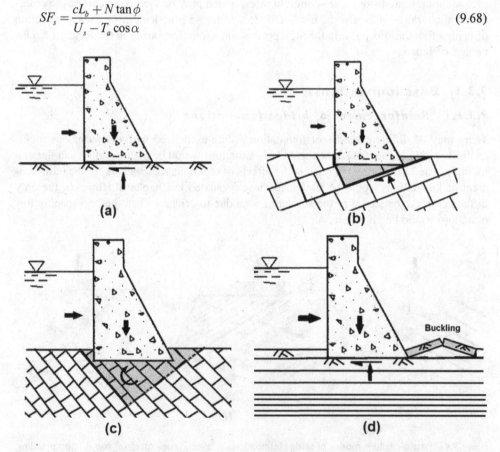

Figure 9.42 Possible failure modes of gravity dams: (a) base shearing, (b) planar sliding along a thor-oughgoing discontinuity plane, (c) flexural or block toppling failure, and (d) buckling failure (modified and re-drawn from US Army Corps of Engineers, 1994).

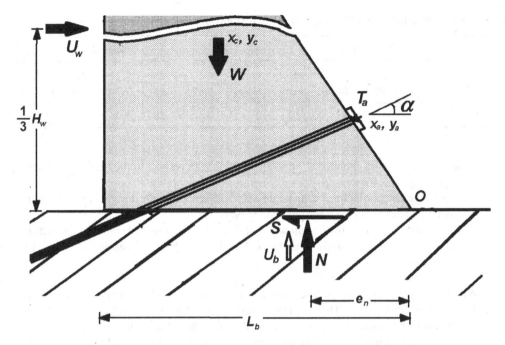

Figure 9.43 Possible force conditions acting on gravity dam foundations.

Safety factor against overturning about point O:

$$SF_m = \frac{W(L_b - x_c) + T_a(y_a \cos\alpha + (L_b - x_a)\sin\alpha)}{\dfrac{H_w}{3}U_s + \dfrac{2}{3}L_bU_b + Ne_n}$$

(9.69)

where

$$N = W + T_a \sin\alpha - U_b$$

(b)   Discrete Finite Element analyses of foundations

A series of analyses using the discrete finite element method (DFEM) on a simple foundation model consisting of lower and upper blocks with a discontinuity plane in between two blocks was carried out by changing the inclination of a rockanchor with respect to the normal of discontinuity plane and its yielding strength. Material properties used in the analyses are given in Table 9.14. The upper block was assumed to be subjected to uniform 10 MN compressive normal load while 7 MN uniformly distributed shear load was applied over the discontinuity plane.

Five different cases were analyzed (Figs. 9.44–9.48):

Case 1: No rockanchor (Fig. 9.44)
Case 2: Rockanchor inclined at an angle of −45° with respect to the normal of discontinuity (Fig. 9.45)

Table 9.14 Properties of blocks. discontinuity, and rockanchor used in analyses.

Blocks		Discontinuity						Rockanchor		
λ	μ	E_n	G_s	h	σ_t	c	ϕ	σ_t	A_b	G_g
(MPa)	(MPa)	(MPa)	(MPa)	mm	(MPa)	(MPa)	(°)	(MPa)	(m²)	(GPa)
100	100	5	1	10	0	0	25	200	0.1	3

Case 3: Rockanchor inclined at an angle of +45° with respect to the normal of discontinuity (Fig. 9.46)

Case 4: Rockanchor inclined at an angle of 0° with respect to the normal of discontinuity and rockanchor behaves elastically (Fig. 9.47)

Case 5: Rockanchor inclined at an angle of 0° with respect to the normal of discontinuity and rockanchor yields (Fig. 9.48). Tensile strength of rockanchor is reduced to 100 MPa.

When the friction angle of the discontinuity plane was set to 40°, the behavior was elastic and no relative slip occurred. The limiting friction angle for elastic behavior is 35°, and if the friction angle of the discontinuity plane is less than 35°, the relative slip is likely.

The friction angle of the discontinuity plane was reduced from 40° to 25°. The computation results for this case (Case 1) are shown in Figure 9.44. The nonlinear analysis was based on the secant method. As noted from Figure 9.44, the relative slip of the upper block accelerates after each computation step. The installation of a rockanchor at different inclinations with respect to the normal of the discontinuity plane restricts the relative slip of the upper block. Particularly, the effect of the rockanchor is largest among all cases when its inclination is +45°. This is in accordance with theoretical considerations presented in Chapter 4. When the relative slip of the discontinuity plane is considered, the largest relative slip occurs

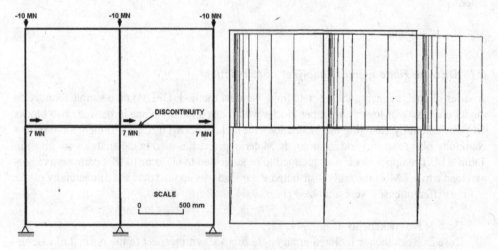

Figure 9.44 Illustration of the DFEM model and deformed configurations (Case 1).

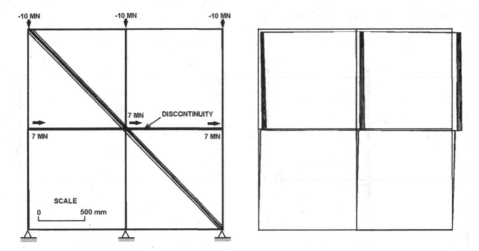

Figure 9.45 Illustration of the DFEM model and deformed configurations (Case 2).

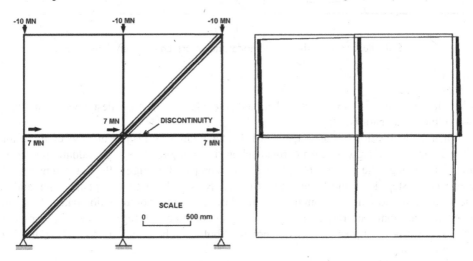

Figure 9.46 Illustration of the DFEM model and deformed configurations (Case 3).

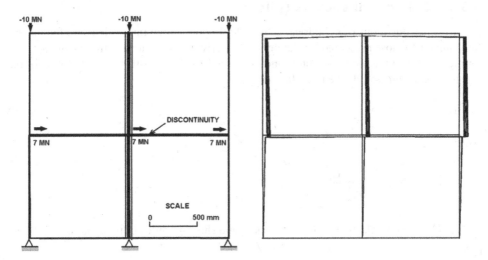

Figure 9.47 Illustration of the DFEM model and deformed configurations (Case 4).

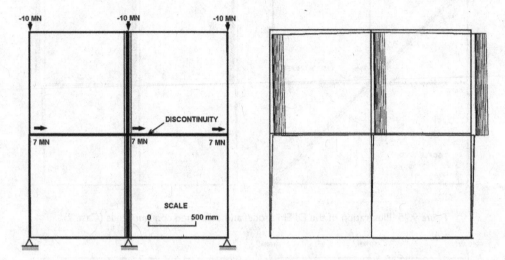

Figure 9.48 Illustration of the DFEM model and deformed configurations (Case 5).

when the inclination is −45°. On the other hand, the relative slip is smallest when the inclination of the rockanchor is +45°.

When the inclination of the rockanchor is 0°, the relative slip is just in between those for Case 2 and Case 3. However, if the rockanchor yields, the relative slip gradually increases after its yielding. The computational results are compared in Figure 9.49 as a function of computation step. It should be noted that the analysis is the pseudo-dynamic type, and if the behavior of the analyzed domain remains elastic, no further deformation takes place and it remains constant with respect to the increase of computation step number. This series of analyses clearly illustrates that rockanchors/rockbolts can be quite effective in reinforcing the foundations.

9.3.2 Cylindrical sockets (piles)

As governing equations of cylindrical sockets to support bridge foundations are fundamentally similar to those presented in Chapter 4, the derivations of the governing equations are omitted herein. If the socket and surrounding ground behaves elastically, the solution of the resulting equation would take the following form:

$$w_b = A_1 e^{\alpha z} + A_2 e^{-\alpha z}. \tag{9.70}$$

where

$$\alpha = \sqrt{\frac{2K_r}{E_b r_b}}, \; K_r = \frac{G_r}{r_b \ln(r_0 / r_b)}$$

r_b, E_b, G_r are radius, elastic modulus of socket, and shear modulus of rock mass.

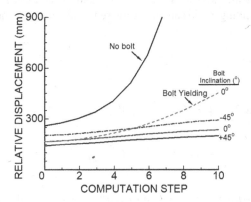

Figure 9.49 Relative slip response of the upper block for different inclinations of rockanchor.

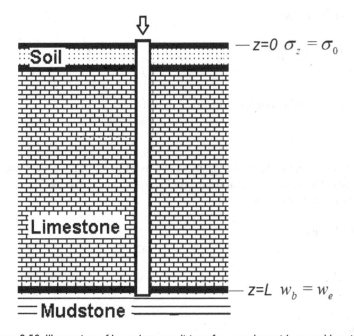

Figure 9.50 Illustration of boundary conditions for a socket with an end bearing.

Let us introduce the following boundary conditions for a socket with an end-bearing added (see Figure 9.50):

$$\sigma_z = \sigma_0 \text{ at } z = 0 \tag{9.71a}$$
$$w_b = w_e \text{ at } z = L \tag{9.71b}$$

Thus, integration constants A_1, A_2 of Equation (9.70) can be obtained as given below:

$$A_1 = \frac{1}{e^{\alpha L} + e^{-\alpha L}}\left(w_e - \frac{\sigma_0}{E_b \alpha}e^{-\alpha L}\right), \quad A_2 = \frac{1}{e^{\alpha L} + e^{-\alpha L}}\left(w_e + \frac{\sigma_0}{E_b \alpha}e^{\alpha L}\right) \tag{9.72}$$

The axial displacement and axial stress of the socket and shear stress at the interface between the socket and surrounding rock can be expressed using the integration constants given above as follows:

Axial displacement:

$$w_b = \frac{1}{e^{\alpha L} + e^{-\alpha L}}\left[w_e\left(e^{\alpha z} + e^{-\alpha z}\right) + \frac{\sigma_0}{E_b \alpha}\left(e^{\alpha(L-z)} - e^{-\alpha(L-z)}\right)\right] \tag{9.73a}$$

Axial stress:

$$\sigma_b = \frac{1}{e^{\alpha L} + e^{-\alpha L}}\left(-E_b \alpha w_e\left(e^{\alpha z} - e^{-\alpha z}\right) + \sigma_0\left(e^{\alpha(L-z)} + e^{-\alpha(L-z)}\right)\right) \tag{9.73b}$$

Interface shear stress:

$$\tau_b = \frac{K_r}{e^{\alpha L} + e^{-\alpha L}}\left[w_e\left(e^{\alpha z} + e^{-\alpha z}\right) + \frac{\sigma_0}{E_b \alpha}\left(e^{\alpha(L-z)} - e^{-\alpha(L-z)}\right)\right] \tag{9.73c}$$

If the end of the socket is rigidly supported so that:

$$w_e = 0 \tag{9.74}$$

For this particular case, the axial displacement and axial stress of the socket and shear stress at the interface between the socket and surrounding rock can be expressed using the integration constants given above as follows:

Axial displacement:

$$w_e = \frac{\sigma_0}{E_b \alpha}\frac{e^{\alpha(L-Z)} - e^{-\alpha(L-Z)}}{e^{\alpha L} + e^{-\alpha L}} \tag{9.75a}$$

Axial stress:

$$\sigma_b = \sigma_0\frac{e^{\alpha(L-Z)} + e^{-\alpha(L-Z)}}{e^{\alpha L} + e^{-\alpha L}} \tag{9.75b}$$

Interface shear stress:

$$\tau_b = \frac{\sigma_0 K_r}{E_b \alpha}\frac{e^{\alpha(L-Z)} - e^{-\alpha(L-Z)}}{e^{\alpha L} + e^{-\alpha L}} \tag{9.75c}$$

However, it is very unlikely that the ends of the sockets would be rigidly supported. Therefore, the displacement of the surrounding ground beneath the socket tip may be approximated

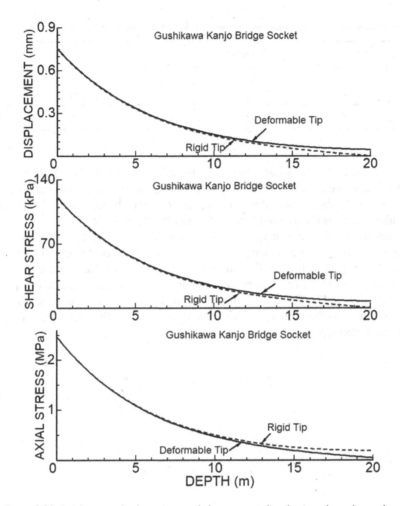

Figure 9.51 Axial stress, displacement, and shear stress distribution along the socket.

using the following equations proposed by Timoshenko and Goodier (1951, 1970), Aydan et al. (2008), and Aydan (2016):

Timoshenko and Goodier (1951, 1970):

$$w_e = \frac{\pi(1+v_r^2)}{2E_r} r_b P_e \qquad (9.76)$$

Aydan et al. (2008) and Aydan (2016):

$$w_e = \frac{(1+v_r)}{E_r} r_b P_e \qquad (9.77)$$

Table 9.15 Properties of blocks. discontinuity and rockanchor used in analyses.

Rock mass		Socket			
E_r (MPa)	v_r	E_s (GPa)	r_s (m)	L m	σ_o (MPa)
720	0.25	20	0.6	20	2.476

where E_r, v_r are elastic modulus and Poisson's ratio of rock mass and p_e is the pressure acting at the tip of the socket. p_e is fundamentally is unknown. One needs to obtain it, by requiring the continuity of displacement of the socket tip and rock mass beneath at $z = L$ using an iterative technique, such as the Runga-Kutta or Newton-Raphson method.

The method described above was applied to a reinforced concrete socket for a bridge foundation. The socket was 20 m long and had a diameter of 1.2 m. Table 9.15 gives material properties. Figure 9.51 shows the distributions of axial stress, displacement, and shear stress for two cases: Case 1: the end was rigidly supported and Case 2: the socket displacement obeys to Equation (9.77). The dotted lines correspond to Case 1 while continuous lines correspond to Case 2. As the socket is relatively long, the stress at the socket tip is only 7.53% of that at the socket top.

Chapter 10

Dynamics of rock reinforcement and rock support

10.1 INTRODUCTION

Dynamic problems, such as rockburst, earthquakes, and blasting, cause dynamic loads on rock support and rock reinforcement. As mentioned in previous chapters, rockbolts and rockanchors are commonly used as principal support members in underground and surface excavations. They are generally made of steel bar or cables, which are resistant against corrosion. These support members may be subjected to earthquake loading, vibrations induced by turbines, vehicle traffic, and long-term corrosion. Figure 10.1 shows the cable rockanchors used at a rock slope ruptured during the 2008 Iwata-Miyagi earthquake. Figure 10.2 shows

Figure 10.1 Rockanchors ruptured by the 2008 Iwate-Miyagi earthquake.

Figure 10.2 The state of rockanchors and rockbolts at underground excavations that experienced rockburst and spalling problems.

the state of rockanchors and rockbolts at underground excavations that experienced rock-burst and spalling problems. This chapter presents theoretical, numerical, and experimental studies on rockbolts and rockanchors under shaking and impulsive loading, in particular.

10.2   DYNAMIC RESPONSE OF POINT-ANCHORED ROCKBOLT MODEL UNDER IMPULSIVE LOAD

It is very rare to see any discussions or experimental results on the load-displacement-time or stress-strain-time responses. The author devised an experimental setup, which consists of an acrylic bar attached to a strain gauge and fixed to a support at the top. The diameter and length of the acrylic bar were 8 mm and 200 mm, respectively. An object with a given weight was instan-taneously applied to the lower end of the bar. The strain response was monitored using WE7000 dynamic data acquisition system with a sampling interval of 10 ms. Figure 10.3 illustrates the experimental setup. In the first stage, 500 gf was applied and then the load was increased by 1543 gf. Figure 10.4 shows the strain variation with time. As noted from the figure, the strain

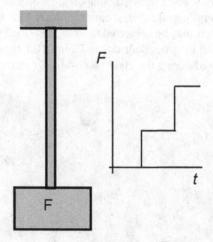

Figure 10.3 Schematic illustration of the experimental setup.

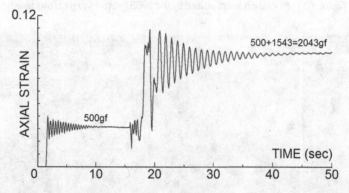

Figure 10.4 Strain response of an acrylic bar subjected to the weight of an object.

fluctuates and becomes asymptotic to the static strain level for the applied stress level. Although the experiment is quite simple, it clearly shows that the loading of samples and structures, as well as the excavation of rock engineering structures, should be treated as a dynamic phenomenon.

A numerical experiment on the response of a 2-m-long point anchored steel rockbolt with a diameter of 25 mm was analyzed using a dynamic finite element method developed by the author (Aydan, 1994). The model was subjected to an impulsive rock load of 25 kN, and the numerical model is fundamentally similar to the experimental setup shown in Figure 10.3 except for the size of rockbolt and magnitude of the load. The rockbolt was assumed to be visco-elastic of the Voigt-Kelvin type with a viscosity coefficient of 10 MPa s. Figure 10.5 shows the displacement responses of the steel rockbolt at selected distances from the fixed end. After about 10 ms, the displacement responses of the steel rockbolt converge to those obtained from the static solution. If the rockbolt is purely elastic, the head of the rockbolt response would show a cyclic displacement behavior, as shown in Figure 10.6. As noted in the experiment, every material in nature has some viscous resistance, so the displacement of rockbolts eventually converges to those measured under static condition.

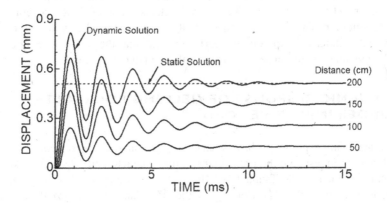

Figure 10.5 Displacement responses of the rockbolt at selected distances from the fixed end.

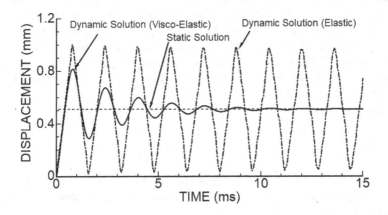

Figure 10.6 Comparison of displacement responses of the rockbolt head for elastic and visco-elastic condition.

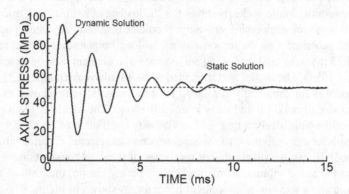

Figure 10.7 Comparison of axial stress of the rockbolt head for dynamic and static conditions.

Figure 10.7 shows the response of the axial stress in the rockbolt for the displacement response shown in Figures 10.5 and 10.7. The axial stress in the rockbolt is several times that under static condition. This result is fundamentally similar to that shown in Figure 10.4. The experiment and numerical experiment clearly illustrate that the rockbolts may experience loads under dynamic condition several times of those under static condition.

10.3   DYNAMIC RESPONSE OF YIELDING ROCKBOLTS UNDER IMPULSIVE LOAD

In recent years, many experiments have been carried out on yielding rockbolts under laboratory conditions to investigate their shock absorption characteristics against rockburst (e.g. Roberts and Brummer, 1988; Tannant and Buss, 1994; Ortlepp, 1992; Stacey and Ortlepp, 1994; Gaudreau *et al.*, 2004). The yielding rockbolts fundamentally utilize the displacement hardening frictional resistance while the tendon itself behaves elastically or elasto-strain hardening plastically under shock loads. The macroscopic behavior of yielding rockbolt can be modeled through an elastic-displacement hardening plastic model. The displacement capacity of yielding rockbolts of is at least 20 times that of typical grouted rebar. The pull-out response of a yielding rockbolt can be modeled as shown in Figure 10.8.

When a yielding rockbolt is subjected to a shock load of dropping mass (m), it can be modeled as shown in Figure 10.9. Before yielding, the overall response would be visco-elastic and elasto-visco-plastic after yielding. The governing equation of motion illustrated in Figure 10.9 may be given as:

Visco-elastic response:

$$m\frac{d^2u}{dt^2}+c_e\frac{du}{dt}+k_eu=mg \tag{10.1}$$

Where m, c_e, k_e, g, and u are mass of dropping mass, damping coefficient and stiffness under visco-elastic regime, and displacement of yielding rockbolt. The equation above is subjected to the following initial condition given by:

$$u=u_o \quad \text{and} \quad v=v_o \quad \text{at } t=0.$$

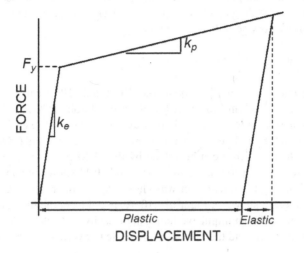

Figure 10.8 Static pull-out response of a yielding rockbolt.

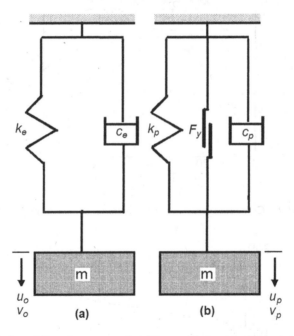

Figure 10.9 Mechanical modeling of response of yielding rockbolt subjected to a shock load due to a dropping mass: (a) visco-elastic regime and (b) elasto-visco-plastic regime.

As for elasto-visco-plastic regime, one can write the following equation of motion:

$$m\frac{d^2u}{dt^2} + c_p\frac{du}{dt} + k_p u + F_y = mg \tag{10.2}$$

Where c_p, k_p, and F_y are damping coefficient and stiffness under elasto-visco-plastic regime and yield resistance of yielding rockbolt. Equation (10.2) is subjected to the following condition at the onset of elasto-visco-plastic behavior:

$$u = u_p \quad \text{and} \quad v = v_p \quad \text{at } t = t_y$$

The non-homogenous differential Equations (10.1) and (10.2) can be easily solved to obtain the responses of yielding rockbolt for given conditions in visco-elastic and elasto-visco-plastic regimes (see a textbook such as Kreyszig, 1983). These equations can also be solved through numerical methods, such as the finite difference method (FDM). As an example, Equations (10.1) and (10.2) can be solved by the FDM by considering an experiment reported by Gaudreau et al. (2004). Figures 10.10 and 10.11 show the computed responses. The time interval during the integration was chosen as 0.2 microseconds, and parameters used in computations are shown in Figure 10.10. As noted from the figures, the response of the yielding rockbolt under an impulsive load can be easily evaluated through the mechanical model described above, and the responses are close to overdamped condition.

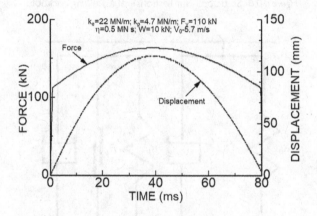

Figure 10.10 Computed force and displacement response of yielding rockbolt.

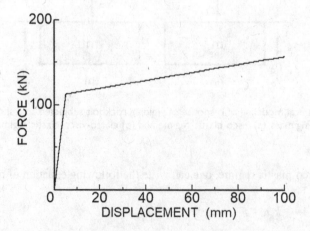

Figure 10.11 Computed force and displacement response of yielding rockbolt.

10.4   TURBINE INDUCED VIBRATIONS IN AN UNDERGROUND POWERHOUSE

Aydan *et al.* (2008, 2012) performed some investigations and measurements at a pumped-storage scheme consisting of two reservoirs and an underground powerhouse constructed about 40 years ago. The underground powerhouse is 55 m long, 22 m wide, and 39 m high and has two turbines. The maximum water level variation may reach 45 m at full capacity. Vibrations induced by turbines in an underground powerhouse were measured at several locations in relation to the assessment of the behavior of rockbolts and rockanchors during operation. The measurements were carried out at penstock, draft shaft, turbine top, a rockbolt above the penstock (Figs. 10.12 and 10.13). Figures 10.14–10.17 show the records of measured vibrations at the draft shaft, turbine top,

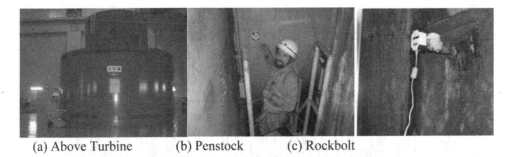

(a) Above Turbine   (b) Penstock   (c) Rockbolt

Figure 10.12 Views of locations of vibration measurements.

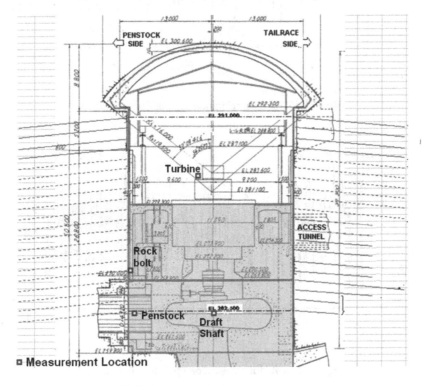

Figure 10.13 Locations of vibration measurements.

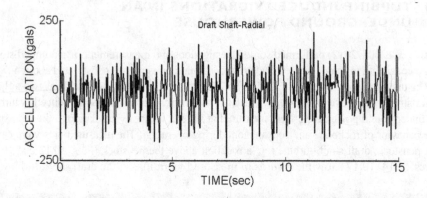

Figure 10.14 Radial acceleration records at the draft shaft.

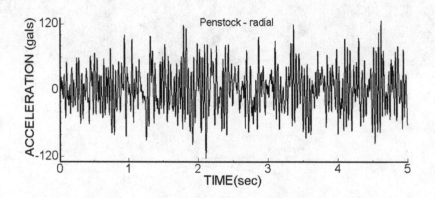

Figure 10.15 Radial acceleration records at the penstock.

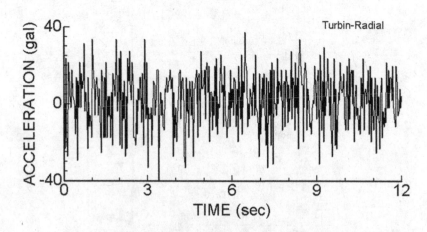

Figure 10.16 Radial acceleration records at the turbine.

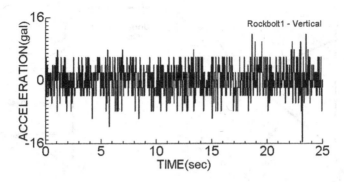

Figure 10.17 Radial acceleration records at the rockbolt above the penstock.

and penstock during operations. The maximum accelerations range between 120 gals and 250 gals. It should be noted that these vibrations are transmitted to the surrounding rock mass. Furthermore, the vibrations are higher when the water flow is stopped or started in relation to the operation of the powerhouse turbines of this pumped-storage scheme.

10.5   DYNAMIC BEHAVIOR OF ROCKBOLTS AND ROCKANCHORS SUBJECTED TO SHAKING

Owada and Aydan (2005) and Owada *et al.* (2004) carried out some model experiments on the development of axial forces in rockanchors and grouted rockbolts, stabilizing the potentially unstable blocks in sidewalls of the underground openings using shaking tables by considering the situation illustrated in Figure 10.18. The model rockbolts tested were rockanchors and fully grouted rockbolts. The experimental results are described in the following two subsections.

10.5.1   Model tests on rockanchors restraining potentially unstable rock block at sidewall of underground openings

The physical model and instrumentation used in experiments are illustrated in Figure 10.19. Two laser displacement sensors produced by KEYENCE (LB-01) together with a data acquisition and logger system were used to measure the displacement of the model and also the motion of the shaking table. The displacement of the shaking table was derivedto obtain the base acceleration. If the sampling interval is 10 ms, the acceleration obtained by derivation of the displacement response is almost the same as the acceleration response measured directly using the accelerometers (AR-10TF) produced by TOKYO-SOKKI. In addition to displacement response measurements, acoustic emissions were also measured by AE-Tester (Human Data Co. & NF Sensor) near the potential sliding plane. The shaking table, which is capable of imposing acceleration up to 1200 gals, is used. The frequency of imposed acceleration waves can be adjusted as desired.

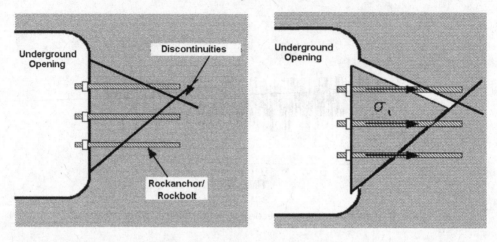

Figure 10.18 Possible situation of a potentially unstable rock blocks at sidewalls of underground openings.

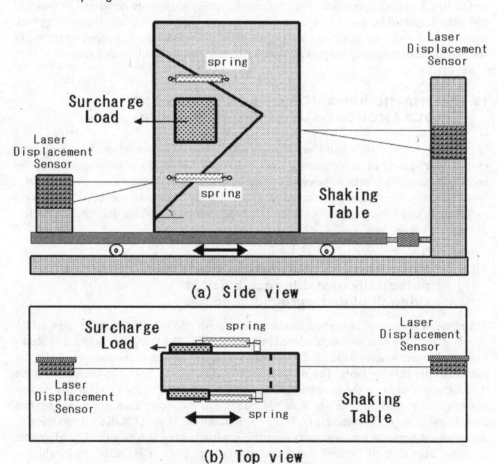

(a) Side view

(b) Top view

Figure 10.19 The experimental setup for model tests.

The block was 274 mm high, 137 mm wide, and 37 mm thick. A plane of discontinuity with an inclination of 58° was introduced into the block. The friction angle of the saw-cut surface was measured by tilting test, and its value ranged between 24° and 26°. Two springs with a length of 30 mm as the model rockanchors or rockanchors were fixed at the both sides of the block so that the springs cross the potential sliding surface as shown in Figure 10.19.

The displacement-load relation of springs was obtained by an additional loading test. The following relationship between displacement (mm) and load (gf) of the spring was established:

$$T = 57\delta^{0.5} \tag{10.3}$$

The total weight of the potentially unstable block was 242.4 gf and the block was unstable without springs due to the high inclination angle of the sliding plane. From the limit equilibrium analyses, the horizontal pull-out force (T) at the time of sliding under the dead weight and surcharge load can be obtained as:

$$\frac{T}{W_t} = \tan(\alpha \pm \phi) \tag{10.4}$$

Where α and ϕ are inclination and friction angle of discontinuity plane, respectively. Plus sign is for upward motion and minus sign is for downward motion. Equation (10.4) implies that to move the block upward, the force will be almost 10 times the load for downward motion. In other words, the base acceleration, which may cause the downward sliding of the block, may be incapable of causing the upward sliding and the block will remain stationary under such loading situations. If the deformation of the sliding block itself is negligible, the axial force on the spring may be estimated from the rigid motion of the block. The extension of the spring as a function of the block motion can be shown to be:

$$\delta_s = \ell - \ell_o = \sqrt{\left(\ell_o + \delta_h\right)^2 + \delta_v^2} - \ell_o \tag{10.5}$$

Where l_o, δ_h, and δ_v are the initial length of springs and horizontal and vertical movement of the sliding block. The force in the spring can then be obtained by inserting the spring extension value from Equation (10.5) into Equation (10.3).

Four shaking experiments were carried out on the models described above. The responses measured in an experiment numbered 4 are shown in Figure 10.20. The maximum base acceleration was 590 gals.

As noted from the figure, the anchor force increased after each slip event in a step-like fashion and becomes asymptotic to a certain value thereafter. Furthermore, the block does not return to its original position. This implies that the axial force becomes higher than the applied pre-stress level following each slip event. It was interesting to note that the axial force on the rockanchor at the end of shaking was more than twice that of the initial stage. Finally, the bolt force becomes sufficient to stop the sliding of the block, if it does not fail. The experiments clearly demonstrate that the bolts and anchors used as a part of the support system may experience greater load levels than that at the time of their installation after each passage of dynamic loads. This outcome has important implications, in that support systems may fail during their service lives not simply due to their deterioration but also to dynamic loads resulting from different causes, as mentioned in the introduction.

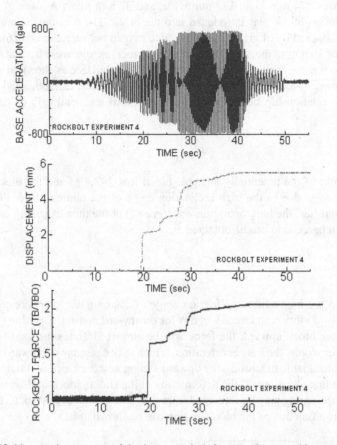

Figure 10.20 Measured responses of displacement, bolt force, and imposed base acceleration.

10.5.2   Model tests on rockanchors restraining potentially unstable rock block in roof of underground openings

The physical model and instrumentation used in experiments are rearranged to investigate the response of potentially unstable rock blocks in the roof, as illustrated in Figure 10.21. Two laser displacement sensors produced by KEYENCE (LB-01) together with a data acquisition and logger system as illustrated in Figure 10.21 were used to measure the displacement of the model and also the motion of the shaking table. The base acceleration of the shaking table was obtained from the derivation of time-displacement response. In addition to displacement response measurements, acoustic emissions were also measured by AE-Tester (Human Data Co. & NF Sensor) near the potential failure planes. The shaking table was the same one as used in side-wall experiments.

Figure 10.22 shows the dynamic response of the block in the roof during the experiments. The block is not symmetric and is slightly eccentric. As noted from the figure, the load variations in rockbolts are not symmetric. Furthermore, some residual displacement of the block occurs. This definitely induces some residual loads following shaking. In other words, if the

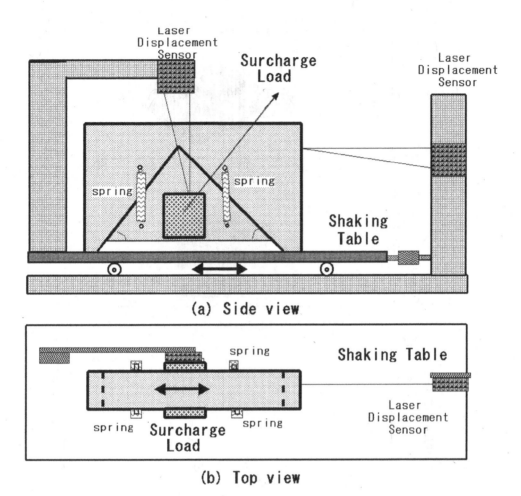

Figure 10.21 Experimental setup for dynamic response of potentially unstable block in the roof.

block does not return to its original position after shaking terminated, there would be some residual forces in the rockbolts, which may accumulate in the rockbolts so that their eventual rupturing may be induced.

10.6   PLANAR SLIDING OF ROCK SLOPE MODELS

The same shaking table was used by changing the configuration of the model to study the planar sliding of rock slopes. Figure 10.23 shows a typical experimental setup. The block was 137 mm high, 137 mm wide and 37 mm thick. A plane of discontinuity with an inclination of 58° was introduced into the block. The friction angle of the saw-cut surface was measured by a tilting test and its value ranged between 24° and 26°. Two springs with a length of 30 mm as the model rockbolts and rockanchors were fixed at both sides of the block, so that

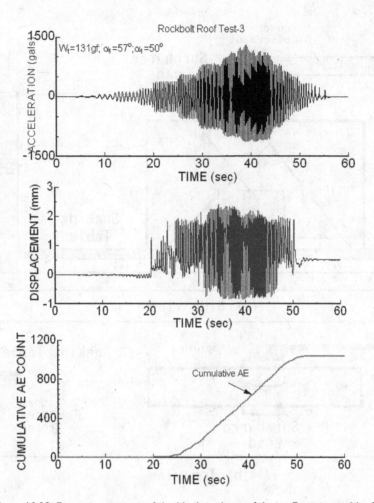

Figure 10.22 Dynamic response of the block in the roof during Experiment No 3.

the springs cross the potential sliding surface as shown in Figure 10.23. The characteristics of rockbolt models are the same as those described in previous subsections.

The dead weight of the potentially unstable block was 69.7 gf and a surcharge load of 172.7 gf was imposed on the potentially unstable block. The block was unstable without springs due to the high inclination angle of the sliding plane. From the limit equilibrium analyses, the horizontal pull-out forces at the time of sliding under the dead weight and surcharge load can be obtained from Equation 10.4.

More than four experiments were performed. As the results are quite similar to each other, one example is described herein. The model conditions are fundamentally the same. The only difference is the level and pattern of the acceleration form imposed on the models. Figure 10.24 shows the acceleration responses. The maximum base acceleration was 590 gal in Rockbolt Experiment 4. As a result, the ultimate displacement, ultimate bolt force and

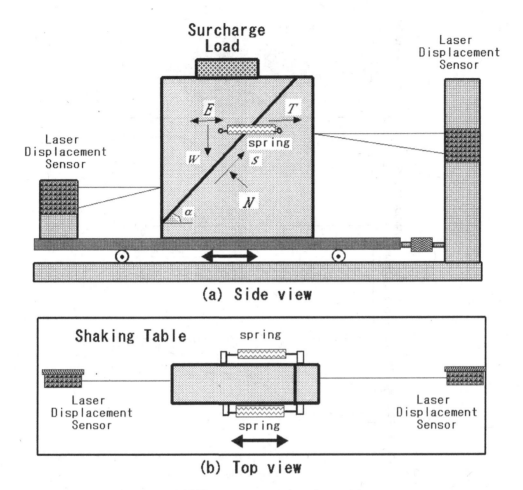

Figure 10.23 The experimental setup for model tests.

cumulative AE become higher for the model subjected to higher acceleration levels. During each increase of acceleration level, the displacement and bolt force and cumulative AE increase simultaneously. As the acceleration level becomes stationary, the responses of all parameters tend to become stationary. This is a quite natural result, as the bolt force on the sliding plane becomes larger after each cycle which increases the resistance of the plane against sliding. Finally, the bolt force becomes sufficient to terminate the sliding of the block, if it does not fail. This simple yet very meaningful experiment clearly demonstrates that the bolts and anchors used as a part of the support system may experience load levels greater than that at the time of their installation after each passage of dynamic loads. This outcome has important implications, in that support systems may fail during their service lives not simply due to their deterioration but also to cyclic loads resulting from different causes, as mentioned in the introduction.

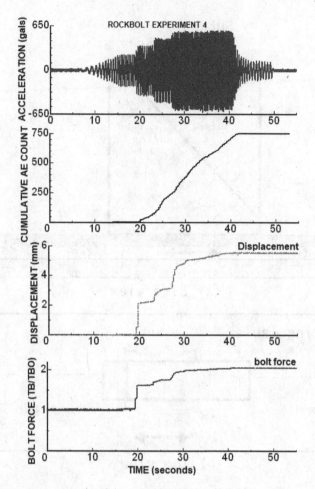

Figure 10.24 Measured responses of displacement, bolt force, cumulative AE, and imposed base acceleration.

Next, a series of layered rock slope models utilizing Ryukyu limestone was carried out to investigate the effect of fully grouted rockbolt models. Figure 10.25 shows an overall view of the rockslope model on a 1D horizontal shaking table, while Figure 10.26 shows the instrumentation of a bolted rock slope model. A series of experiments were also carried out to see the effect of bolting on vibration characteristics of the bolted and unbolted rock slope models. In this section, the reinforcement effect of the rockbolts on the failure of rockslopes with and without rockbolts are described herein.

Figure 10.27 shows views of model rock slopes with/without rockbolts. In these particular experiments, the length of rockbolts were investigated. As noted from Figures 10.27b and 10.27c, the unstable body of the rock slope model behaves as if a monolithic body if the rockbolts are not anchored in the stable part. On the other hand, the movement of the potentially unstable body is restrained despite some relative small displacements, as the rockbolts are not pre-stressed (passive type). Figures 10.28–10.30

show the applied base acceleration, responses of acceleration, and displacement of the potentially unstable body. As pointed out, when the rockbolts are short, there is no fundamental difference in terms of acceleration value at the failure. However, the movement of the potentially unstable body is greatly restrained despite some small relative slips among layers.

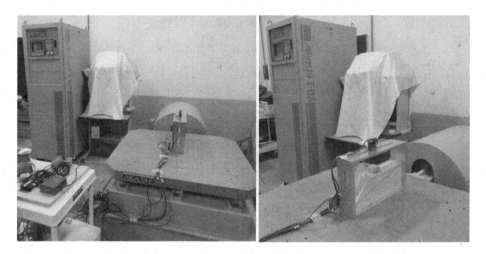

Figure 10.25 Overall views of the rock slope model on a shaking table.

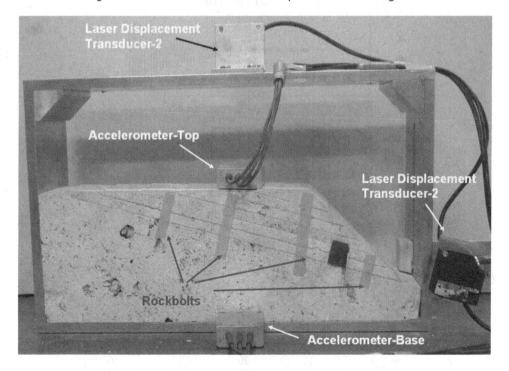

Figure 10.26 Instrumentation of a bolted rock slope model.

BEFORE SHAKING   AFTER SHAKING
(a) UNBOLTED

(b) SHORT BOLTS

(c) LONG BOLTS

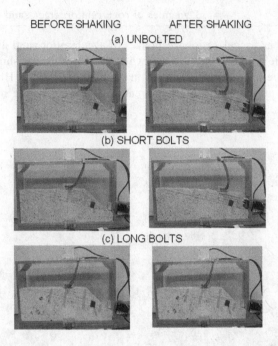

Figure 10.27 Views of layered rock slope model with fully grouted rockbolt models.

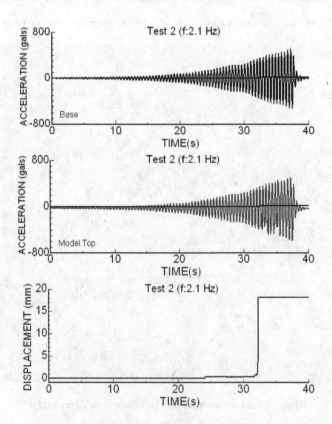

Figure 10.28 Measured acceleration and displacement response (unbolted).

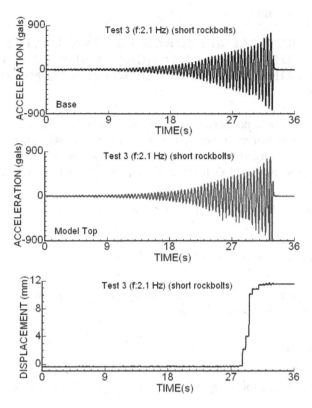

Figure 10.29 Measured acceleration and displacement response (short rockbolts).

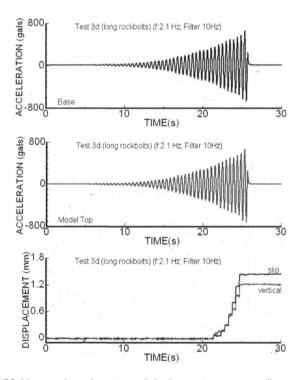

Figure 10.30 Measured acceleration and displacement responses (long rockbolts).

10.7   A THEORETICAL APPROACH FOR EVALUATING AXIAL FORCES IN ROCKANCHORS SUBJECTED TO SHAKING AND ITS APPLICATIONS TO MODEL TESTS

A theoretical approach was developed by Owada and Aydan (2005) for evaluating the axial forces in rockanchors in their experiments presented in the previous subsections. This approach is described in this subsection and applied to experiments. The following limiting equilibrium equations of the potentially unstable block for s and n directions can be written as follows (Fig. 10.31):

$$\sum F_s = W_t \sin\alpha + E\cos\alpha - T\cos(\alpha - \beta) - S = m\frac{d^2s}{dt^2} \tag{10.6a}$$

$$\sum F_n = W_t \cos\alpha + E\sin\alpha - T\sin(\alpha - \beta) - N = m\frac{d^2n}{dt^2} \tag{10.6b}$$

Let us assume that the inertia force for n-direction during sliding is negligible and the resistance of the failure plane is purely frictional as given below:

$$\left|\frac{S}{N}\right| = \tan(\phi) \tag{10.7}$$

One can easily obtain the following equation for the rigid body motion of the sliding rock mass body:

$$m\frac{d^2s}{dt^2} = A \tag{10.8}$$

Where
$$A = W_t(\sin\alpha - \cos\alpha \tan\phi) + E(\cos\alpha + \sin\alpha \tan\phi) - T(\cos(\alpha - \beta) + \sin(\alpha - \beta)\tan\phi)$$

Since the shaking force E will be proportional to the mass of the sliding body, it can be related to ground shaking in the following form:

$$E = \frac{a_g(t)}{g}W_t \tag{10.9}$$

Where a_g and g are ground acceleration and gravitational acceleration, respectively.

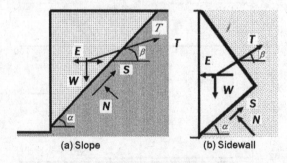

(a) Slope   (b) Sidewall

Figure 10.31 Mechanical model.

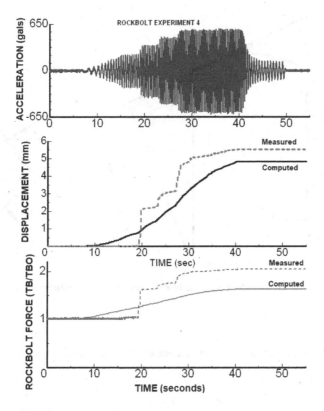

Figure 10.32 Comparison of computed responses with measured responses.

The theoretical approach is applied to model tests shown in Figure 10.24 by selecting a friction angle of 26° on the basis of friction tests using a tilting machine. The computed results are shown in Figure 10.32. As noted from the figure, the computed results are quite similar to experimental results both quantitatively and qualitatively. However, the computations indicate that the yielding should start earlier than the measured results show. The discrepancy may result from the complexity of actual frictional behavior of a sliding surface. Nevertheless, the theoretical model is capable of modeling the dynamic response of the support system.

10.8 APPLICATION OF THE THEORETICAL APPROACH TO ROCKANCHORS OF AN UNDERGROUND POWERHOUSE SUBJECTED TO TURBINE-INDUCED SHAKING

Rock mass generally contains geological discontinuities, and these discontinuities play a major role in the local instabilities of underground openings (i.e. Aydan, 1989; Kawamoto *et al.*, 1991). The potentially unstable blocks were identified around an underground cavern shown in Figure 10.33 based on *in situ* investigations, as well as on geological investigations during the construction phase (Aydan, 2016; Aydan *et al.*, 2012).

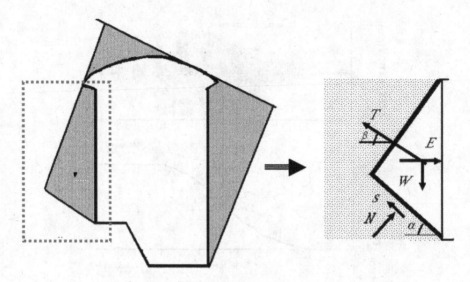

Figure 10.33 Identified unstable blocks and force equilibrium conditions.

The axial force T of rockanchors is also computed from a relationship similar to Equation 10.3, given below:

$$T = C\delta^b \tag{10.10}$$

The displacement of rockanchors induced by the relative displacement during each relative motion due to sliding was computed from geometrical considerations. It should be also noted that there is no sliding if the rockanchor force attains a certain level. Furthermore, the oscillations of anchor force due to visco-elastic response during each increment of axial force were neglected in computations.

Figure 10.34 shows the computational results for the block shown in Figure 10.33 for two different situations using the induced radial vibration record at the penstock. In the first case, no rockanchors were considered, while rockanchors were assumed to be installed in the second case. The inclination of discontinuity was set to 35° and its friction coefficient was determined from tilting tests as 35.5°. When no rockanchors are installed, the block tends to slide downward when vibrations induced are sufficient to cause sliding. However, as the relative sliding movements of the block are restricted, the amount of sliding becomes less and the anchor force tends to become asymptotic to a certain level for the given amplitude of vibrations. This computational result also implied that the axial forces may increase during their service life from forces resulting from vibrations near a cavern and/or due to occasional earthquakes. These increases may also lead to the rupture of rockanchors in the long term, as well as the reduction of cross-sections of rockanchors due to corrosion.

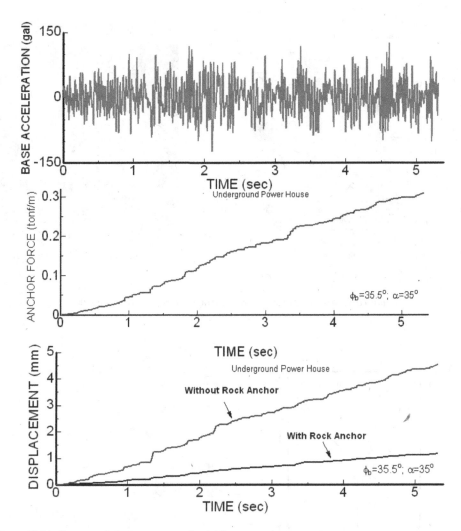

Figure 10.34 Computed displacement and axial force responses.

10.9   MODEL TESTS ON FULLY GROUTED ROCKBOLTS RESTRAINING A POTENTIALLY UNSTABLE ROCK BLOCK AGAINST SLIDING

Owada and Aydan (2005) and Owada *et al.* (2004) also carried out some model experiments on the development of axial forces on grouted rockbolts stabilizing the potentially unstable blocks in sidewalls of underground openings or rock slopes subjected to planar sliding, as shown in Figure 10.35 using shaking tables. An acrylic bolt equipped with three strain gauges shown in Figure 10.36 was used, and the inclination of the bolt was varied to 45°,

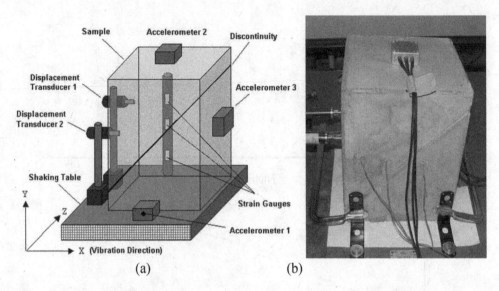

Figure 10.35 An illustration of experimental setup (a) and its view (b).

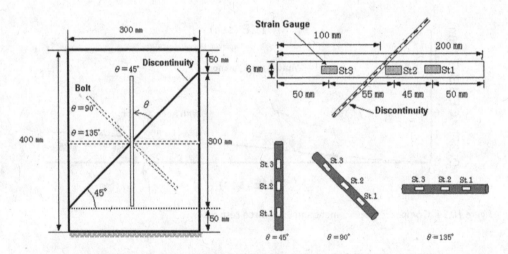

Figure 10.36 A drawing of an instrumented sample and position of strain gauges.

90°, and 135° with respect to the sliding plane, which is inclined at an angle of 45° to the horizontal. Strain gauge number 2 (St2) is set next to the discontinuity plane. The dynamic response of the acrylic bolt under impulsive gravity loading has been already shown in Figure 10.4. The samples with different orientations of model rockbolts were also tested under static loading condition, and the experimental results are shown in Figure 10.36 and Figure 10.38.

Figures 10.39 to 10.42 show the applied base acceleration and strains in the bolt and displacement of the potentially unstable block. As noted from the figures, some residual strains

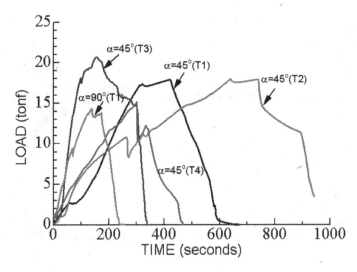

Figure 10.37 The load response of samples with a bolt with different orientations.

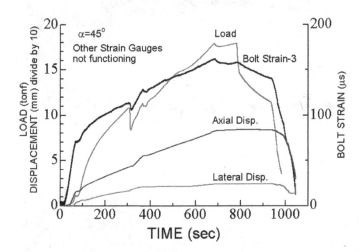

Figure 10.38 Load, strain, and displacement responses of a sample with a bolt installed at an angle of 45°.

occur in the bolts following the termination of shaking. Similarly, the block is displaced permanently. Strain at gauge 2 is always largest, as expected. Although the measured strain levels are small compared with their yield strain level, the permanent straining results from the permanent displacement of the potentially unstable block.

As shown in Figure 10.42, the amplitude of acceleration is increased stepwise up to 450 gals. When the acceleration was less than 200 gals, the straining of the bolt and displacement of the potentially unstable block was small. However, strains and permanent displacement become larger after each acceleration level increment.

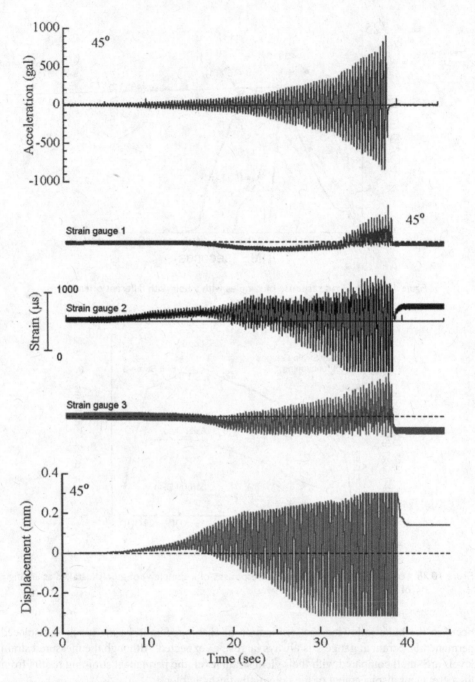

Figure 10.39 The responses of bolt strains and displacement of unstable block in relation to applied base acceleration (bolt angle 45°).

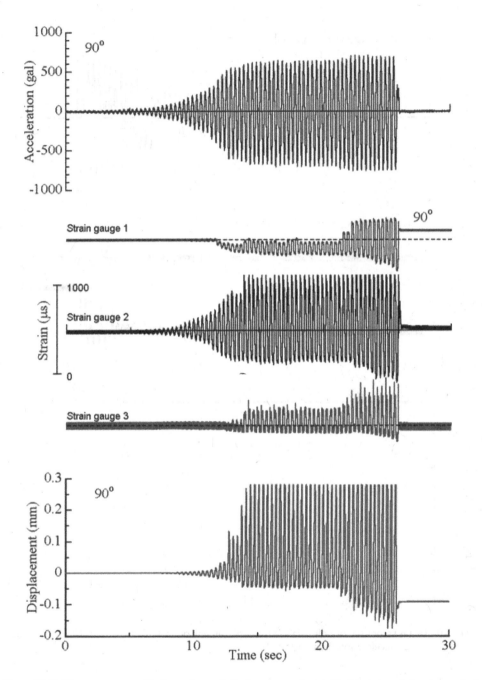

Figure 10.40 The responses of bolt strains and displacement of unstable block in relation to applied base acceleration (bolt angle 90°).

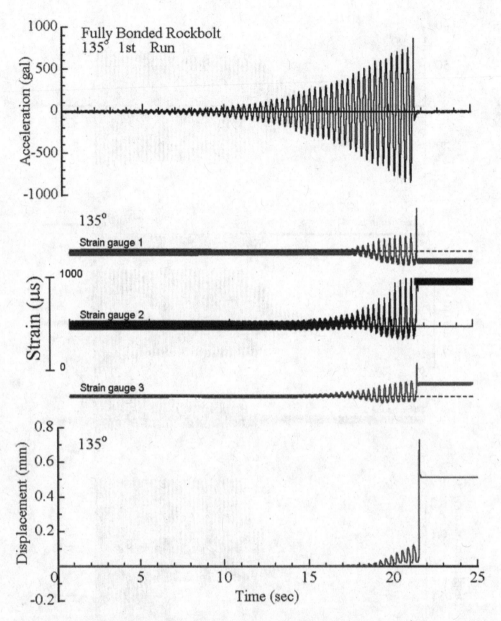

Figure 10.41 The responses of bolt strains and displacement of unstable block in relation to applied base acceleration (bolt angle 45°).

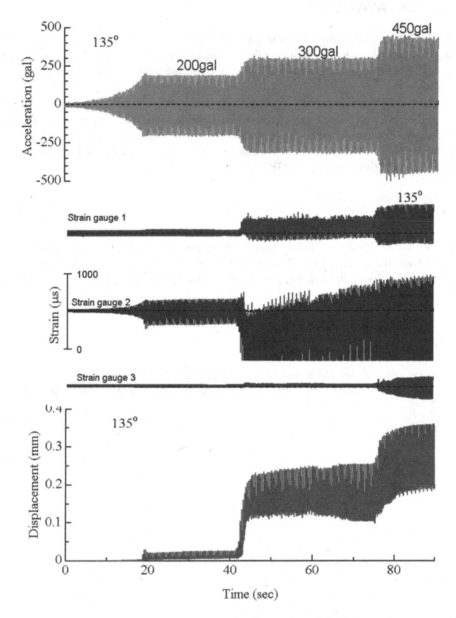

Figure 10.42 The responses of bolt strains and displacement of unstable block in relation to applied base acceleration (bolt angle 135°).

10.10   EXCAVATIONS

In this section, several examples for the dynamic response of surrounding rock and rockbolts during excavations are given. Although the assumed excavation geometries may be simple, the examples should be sufficient to illustrate the fundamental aspects of rock excavation dynamics.

Tunnel excavation is generally done through drilling-blasting or mechanically such as TBM and/or excavators. The most critical situation on stress state is due to the drilling-blasting type excavation, since the excavation force is applied almost impulsively. The dynamic response of circular tunnels during excavations under hydrostatic *in situ* stress condition can be given as (Fig. 10.43):

$$\frac{\partial \sigma_r}{\partial r} + \frac{\sigma_\theta - \sigma_r}{r} = \rho \frac{\partial^2 u}{\partial t^2} \tag{10.11}$$

Where σ_r, σ_θ, u, ρ, and r are radial, tangential stresses, radial displacement, density, and distance from the center of the circular cavity, respectively. Let us assume that the surrounding rock behaves in a visco-elastic manner of the Kelvin-Voigt type, which is specifically given as:

$$\begin{Bmatrix} \sigma_r \\ \sigma_\theta \end{Bmatrix} = \begin{bmatrix} D_1 & D_2 \\ D_2 & D_1 \end{bmatrix} \begin{Bmatrix} \varepsilon_r \\ \varepsilon_\theta \end{Bmatrix} + \begin{bmatrix} C_1 & C_2 \\ C_2 & C_1 \end{bmatrix} \begin{Bmatrix} \dot{\varepsilon}_r \\ \dot{\varepsilon}_\theta \end{Bmatrix} \tag{10.12}$$

Where ε_r, ε_θ and $\dot{\varepsilon}_r$, $\dot{\varepsilon}_\theta$ are radial and tangential strain and strain rates, respectively.

The strain and strain rates are related to the radial displacement in the following form:

$$\varepsilon_r = \frac{\partial u}{\partial r}; \varepsilon_\theta = \frac{u}{r} \quad \text{and} \quad \dot{\varepsilon}_r = \frac{\partial \varepsilon_r}{\partial t}; \dot{\varepsilon}_\theta \frac{\partial \varepsilon_\theta}{\partial t} \tag{10.13}$$

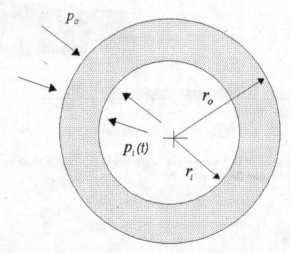

Figure 10.43 Illustration of dynamic excavation of a circular opening under hydrostatic *in situ* stress condition.

Eringen (1961) developed a closed-form solution for Equation (10.11) under blasting loads. In order to deal with more complex boundary and initial conditions and material behavior, Equation (10.11) is preferred to be solved using dynamic finite element code. The discretized finite element form of Equation (10.11) together with the constitutive law given by Equation (10.12) takes the following form:

$$\mathbf{M\ddot{U}} + \mathbf{C\dot{U}} + \mathbf{KU} = \mathbf{F} \tag{10.14}$$

Equation (10.14) has to be discretized in time domain and the resulting equation would take the following form:

$$\mathbf{K^*U}_{n+1} = \mathbf{F^*}_{n+1} \tag{10.15}$$

The specific form of matrices in Equation (10.15) may change depending upon the method adopted in the discretization procedure in time domain. For example, if the central difference technique is employed, the final forms would be the same those given in Subsection 6.6.2. A finite element code has been developed by the author and used in the examples presented in this subsection.

10.10.1   Unbolted circular openings

In this subsection, the dynamic response of a circular tunnel under the impulsive application of excavation force is presented. The results were initially reported in the publication by Aydan (2011) and mentioned in Chapter 3.

Figure 10.44 shows the responses of displacement, velocity, and acceleration of the tunnel surface with a radius of 5 m. As noted from the figure, the sudden application of the excavation force, in other words, the sudden release of ground pressure, results in 1.6 times the static ground displacement at the tunnel perimeter, and shaking disappears almost at 2 seconds. As time progresses, it becomes asymptotic to the static value and velocity and acceleration disappear.

The resulting tangential and radial stress components near the tunnel perimeter (25 cm from the opening surface) are plotted in Figure 10.45 as a function of time. It is of great interest that the tangential stress is greater than that under static condition. Furthermore, very

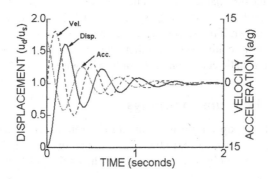

Figure 10.44 Responses of displacement, velocity, and acceleration of the tunnel surface.

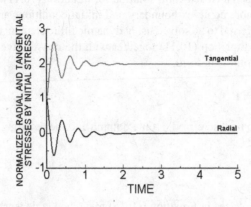

Figure 10.45 Responses of radial and tangential stress components near the tunnel surface (25 cm away from the perimeter of the opening).

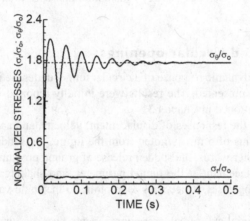

Figure 10.46 Variation of radial and tangential stresses in rock mass at a distance of 25 cm from the opening perimeter for bolted case.

high radial stress of tensile character occurs near the tunnel perimeter. This implies that the tunnel may be subjected to a transient stress state, which is quite different than that under static conditions. However, if the surrounding rock behaves elastically, stresses will become asymptotic to their static equivalents. In other words, the surrounding rock may become plastic even though the static condition may imply otherwise.

10.10.2 Bolted circular openings

The circular underground opening was assumed to be reinforced by rockbolts of 4 m long with a 25-mm diameter installed in a hole with a radius of 36 mm. The spacing of rock-bolts was assumed to be 1 m by 1 m. Figure 10.46 shows the radial and tangential stresses

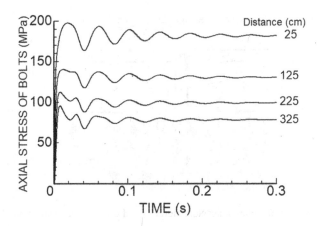

Figure 10.47 Variation of axial stress in rockbolts at a distance of 25 cm from the opening perimeter for bolted case.

normalized by the *in situ* stress at a distance of 25 cm from the opening surface. It is interesting to note that the radial and tangential stresses are less than those of the unsupported case, which indicates the reinforcement effect of rockbolts through the internal pressure as well as shear interaction between rockbolts and surrounding ground.

Figure 10.47 shows the variation of axial stress in the rockbolts at selected distances from the head of the rockbolts. The overall response of axial stress in rockbolts is quite similar to the changes of the stresses in rock mass. One important observation is that the axial stress in rockbolts would fluctuate and the rockbolts would experience larger axial stresses than those under static condition.

10.11   DYNAMIC RESPONSE OF ROCKBOLTS AND STEEL RIBS DURING BLASTING

Blasting induces vibrations on the support and reinforcement system. Examples of vibrations induced in rockbolts and steel ribs were recently measured by Aydan *et al.* (2016) using portable accelerometers QV3-OAM-SYC. The accelerometers can be synchronized and set to the triggering mode and are capable of recording pre-trigger waves for a period of 1.2 s. The trigger threshold and the period of each record can be set to any level and chosen time as desired. Vibrations in a rockbolt and a steel rib of the main tunnel at the Tarutoge site were induced by a blasting operation at the evacuation tunnel at 21:28 on February 28, 2014. The amount of explosive was 71.2 kgf with 10 rounds with a delay of 0.4–0.5 s. The threshold value for triggering was set to 10 gals. Accelerometer S17 was fixed on the steel-rib while accelerometer S18 was fixed onto a rockbolt plate and they were about 27 m away from the blasting location (Fig. 10.48). Figure 10.49 shows the acceleration records measured by accelerometers S17 and S18. When acceleration responses are compared, the vibration

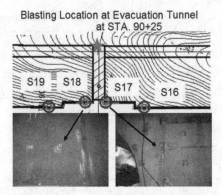

Figure 10.48 Locations of accelerometers fixed on steel ribs and rockbolts.

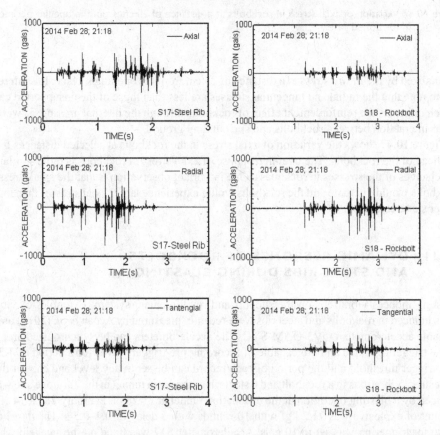

Figure 10.49 Acceleration responses of axial, radial, and tangential components measured in a steel rib and a rockbolt at Tarutoge Tunnel.

levels induced by blasting in steel ribs are much higher than those in rockbolts. While the radial components are close to each other, distinguishable differences are observed in axial and tangential components. The results imply that blasting operations may cause more damage on steel ribs than on rockbolts.

Chapter 11

Corrosion, degradation, and nondestructive testing

11.1 INTRODUCTION

The mechanism, site examples, and techniques for evaluating corrosion in steel and iron materials in relation to the long-term performance and degradation of reinforcement and support systems are presented in the first part of this chapter. Some theoretical models are developed for the axial and traverse dynamic nondestructive tests, and numerical simulations are carried out based on these fundamental equations by considering possible conditions *in situ*. Specifically, the bonding quality, the existence of corroded parts or couplers, and pre-stress are simulated through numerical experiments. The results of the numerical simulations can be used to interpret the dynamic responses to be measured in nondestructive dynamic tests *in situ*. Furthermore, a theoretical method is described for simulating lift-off experiments, which are classified as destructive testing. Several applications of this procedure to practice are given together with measurements obtained from some nondestructive tests under laboratory and *in situ* conditions.

11.2 CORROSION AND ITS ASSESSMENT

11.2.1 The principle of iron corrosion

Iron corrosion generally occurs due to chemical or electrochemical action of water in the presence of oxygen and hydrogen (Fig. 11.1). Two chemical reactions play important roles in this process. The following reaction makes iron dissolve:

$$2H^+ + 2e \rightarrow H_2 \tag{11.1}$$

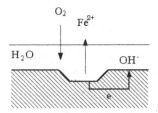

Figure 11.1 Illustration of electrochemical cell action.

Then, it results in an oxidation half reaction given by:

$$Fe \rightarrow Fe^{2+} + 2e \tag{11.2}$$

These two reactions result in the following equation:

$$Fe + 2H^+ \rightarrow Fe^{2+} + H_2 \tag{11.3}$$

In the first stage, iron is oxidized to iron ions. When the pH is greater than 4–5 and less than 9–10, oxygen from the air is reduced to hydroxide ions, OH^-, through the following action:

$$\frac{1}{2}O_2 + H_2O + 2e \rightarrow OH^- \tag{11.4}$$

The above reactions finally result in the formation of Fe^{2+} and $2OH^-$ given as:

$$Fe + \frac{1}{2}O_2 + H_2O \rightarrow Fe^{2+} + 2OH^- \tag{11.5}$$

$Fe(OH)_2$ is formed and the corrosion process is completed as given below:

$$Fe(OH)_2 + \frac{1}{4}O_2 + \frac{1}{2}H_2O \rightarrow Fe(OH)_3 \tag{11.6}$$

This process results in rust, which is a flaky, reddish-brown solid.

11.2.2 Factors controlling corrosion rate

Corrosion rate is said to be dependent on several factors, such as environment, pH concentration of dissolved oxygen, and temperature. Their effects are as follows:

i)   Effect of pH

In an acidic environment, the effect of pH is largest and the corrosion rate is about 0.05 mm/year when pH ranges between 4 and 10 at a temperature of 22°C. If pH becomes greater than 10, corrosion rate decreases.

ii) Effect of dissolved oxygen

Dissolved oxygen accelerates corrosion as cathode element or decelerating action through the formation of an oxidized protective layer. If the concentration of dissolved oxygen is about 10–15 ppm, the corrosion rate is largest under room temperature. However, if the temperature becomes higher and concentration increases, the rate of corrosion decreases.

iii) Effect of temperature

If temperature increases, corrosion rate increases. Nevertheless, the concentration of dissolved oxygen increases by the increase of temperature, and this results in the reduction of its cathode action.

The effects of various factors are summarized in Tables 11.1 to 11.4. In underground opening environments, the action of groundwater on rockbolts and rockanchors is about 0.03 mm/ year. However, pH may increase this rate to 0.05 mm/year. The temperature of underground

Table 11.1 Corrosion rate of iron in relation to environment (μm/year).

Condition	Air			Water			Soil		
	Ind.	Urban	Non-urban	River	Sea	Drinking	High	Med.	Low
Unprotected	1000	500	500	500	1000	100	1000	300	50
Protected	100	50	300	300	200	150	150	100	30

Table 11.2 Effect of pH on corrosion rate (mm/year).

pH	2	3	4	5	6	7	8	9	10	11	12	13	14
Rate	0.35	0.11	0.05	0.05	0.05	0.05	0.05	0.05	0.05	0.03	0.02	0.01	0.03

Table 11.3 Effect of dissolved oxygen.

Dissolved oxygen (ml/l)	5	10	15	20	25
Rate (mmd)	45	65	55	15	10

Table 11.4 Effect of temperature on corrosion rate.

Temperature		30	40	50	60	70	80	90	100	110	120
Rate	sealed	0.25	0.29	0.33	0.37	0.41	0.45	0.49	0.53	0.57	0.61
	unsealed	0.24	0.27	0.3	0.33	0.36	0.39	0.36	0.00	–	–

openings is generally less than 30°C, and the general rate of corrosion is expected to be about 0.024 mm/year.

11.2.3   Experiments on corrosion rate of rockbolts

A series of experiments are carried out on steel bars of smooth rockbolts. Figure 11.2 shows a typical rockbolt sample. The surface of smooth bars is covered with a protective layer. A 10-mm-wide strip of this protective layer was removed to see the effects of the protection layer.

Water sampled from five different locations was used. Tap water, seawater in Orido Bay, Koseto, hot spring water, underground waters from the Mitake abandoned mine and a research gallery in the Tono underground research laboratory at about 120 m below the ground. Samples of rockbolts were immersed in water as illustrated in Figure 11.3. Three sets of five samples were fully immersed, while three sets of five samples were immersed up to half of the height.

Figure 11.4 shows the variation of weight of samples for a period of 18 months. The weight of samples corresponds to the weight after cleaning. Views of samples before the tests, half-immersed samples, and fully immersed samples are shown in Figure 11.4. The corrosion was high when samples were half-immersed in water.

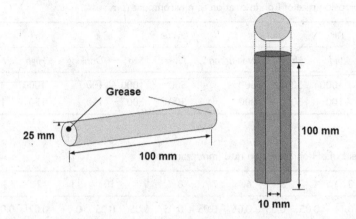

Figure 11.2 Illustration of rockbolt sample for corrosion rate experiments.

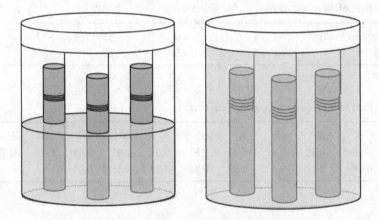

Figure 11.3 Illustration of immersion condition of samples.

Figure 11.4 Views of half-immersed samples before and after immersion and cleaning.

To evaluate the corrosion state of samples at given durations, the electrical resistivity, elastic-wave velocity, weight, and dimensions of each sample were measured. The values averaged for measurements on three samples for each water group.

Figure 11.5 shows an experimental setup to measure the electrical resistivity of samples. Measurement results for non-corroded and corroded samples are shown in Figures 11.6 and 11.7.

Although there is a distinct difference between the values of electrical resistivity of non-corroded and corroded samples, the values are greatly scattered due to contact conditions after each measurement. Although the electrical resistivity has a high potential for sensitive measurements, the values are highly scattered by measurements at certain time intervals, such as days and months. Figures 11.8 and 11.9 show the variation of weight and elastic wave velocity of samples. The largest variations occurred on samples immersed in hot spring water sampled from Koseto in Shizuoka City.

Figure 11.5 Experimental setup to measure electrical resistivity of samples.

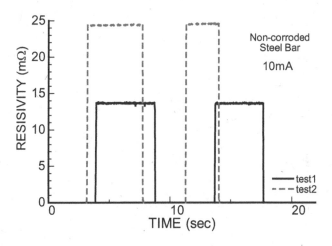

Figure 11.6 Electrical resistivity of non-corroded bar.

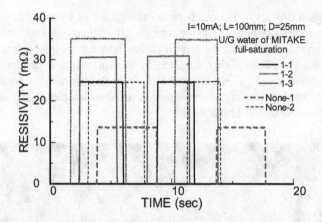

Figure 11.7 Comparison of electrical resistivity of corroded bars with those of non-corroded bar.

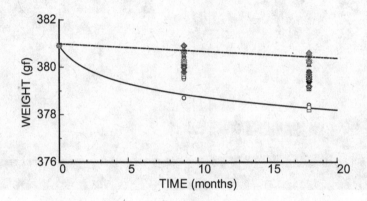

Figure 11.8 Variation of weight of sample with elapsed time.

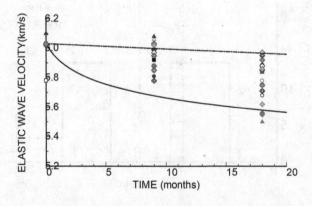

Figure 11.9 Variation of elastic wave velocity of samples with elapsed time.

11.2.4　Observations of iron bolts at Koseto hot spring discharge site

As mentioned above, some observations and measurements were carried out at Koseto hot spring water discharge site, since the largest variations of measured quantities occurred in samples immersed in water from this site. Figure 11.10 shows the layout of the site and locations of measured bolts. Measurements were carried out on diameter and length of bolts on the top, front, and back sides of the platform. Figures 11.11–11.13 show the locations of the bolts and their number. Tables 11.5–11.8 give the diameter and length of the bolts at each respective location. Figures 11.14–11.17 show the diameter

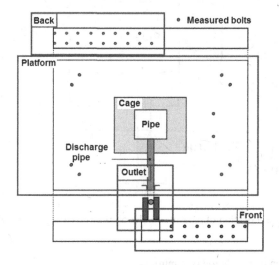

Figure 11.10　Layout of the Koseto hot spring sites and locations of bolts.

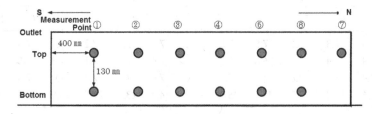

Figure 11.11　Location of measured bolts on the front side of the platform.

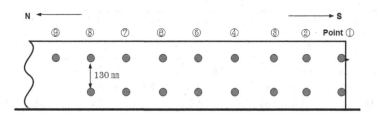

Figure 11.12　Location of measured bolts on the back side of the platform.

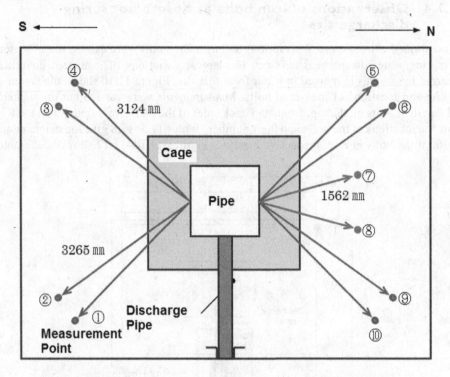

Figure 11.13 Location of measured bolts on the top surface of the platform.

Table 11.5 Corrosion on front side (units in mm).

	P-1		P-2		P-3		P-4		P-5		P-6		P-7
	T	B	T	B	T	B					T	B	T
D (Hor)	7.3	2.8	8.0	7.8	7.8	8.0	7.2	7.4	7.2	7.4	7.4	7.3	7.2
D (Ver)	7.4	3.4	7.5	6.8	7.9	8.0	7.6	7.6	7.1	7.3	7.5	7.8	7.3
Average	7.4	3.1	7.8	7.3	7.9	8.0	7.4	7.5	7.2	7.4	7.5	7.6	7.3
Length	23.6	15.3	22.5	23.1	25.0	23.3	24.4	23.3	22.5	24.8	22.2	20.8	25.1

Table 11.6 Corrosion on back side (units in mm).

	P-1		P-2		P-3		P-4		P-5		P-6		P-7		P-8		P-9	
	T	B	T	B	T	B					T	B	T	T	T	B	T	B
D-H	6.9	7.6	7.9	7.7	6.5	6.9	7.7	7.8	7	6.6	7.4	7	7.3	6.4	7.2	7.6	7.9	7.9
D-V	7.8	8	7.5	7.5	6.8	6.8	7.7	7	6.6	7.4	7.7	7.4	6.9	7.3	7.7	7.5	8	7.4
Av.	7.35	7.8	7.7	7.6	6.65	6.85	7.7	7.4	6.8	7	7.55	7.2	7.1	6.85	7.45	7.55	7.95	7.65
L	25.1	22.9	22	22.3	23.4	22.3	21.8	22.3	24.6	22.1	21.4	22.6	24.4	22.4	23.6	22.5	23.9	21.8

Table 11.7 Corrosion on the top surface of the platform (units in mm).

	P-1	P-2	P-3	P-4	P-5	P-6	P-7	P-8	P-9	P-10
D-H	18.9	18.5	18.2	18.8	18.5	18.2	17	16.9	19	18.7
D-V	18.5	18.5	18.3	19	18.8	18.8	18.1	16	18.6	18.8
Av.	18.7	18.5	18.25	18.9	18.65	18.5	17.55	16.45	18.8	18.75
L	55.2	55.1	50.2	47.5	54.6	46.6	42.7	32	46.6	43.5

Table 11.8 Corrosion of the bolt at outlet (units in mm).

Location	P-1	P-2
Diameter (EW)	10.80	15.20
Diameter (V)	15.30	16.80
Diameter (Average)	13.05	16.00

Table 11.9 Corrosion of the plate at outlet (units in mm).

Location	P-3	P-4
Thickness	8.60	5.50

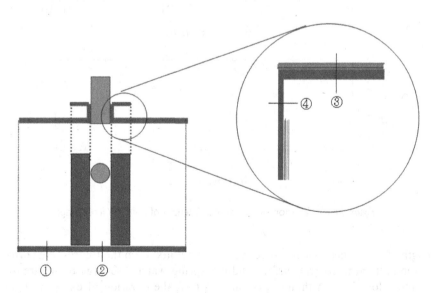

Figure 11.14 Measurement locations at the outlet.

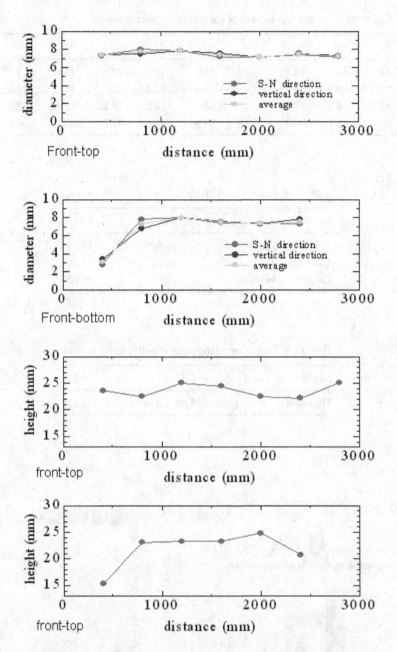

Figure 11.15 Variation of diameter and length of bolts on front side.

and length of the bolts on each respective side. As noted from the figures, the corrosion rate of the bolts near ground surface and hot spring water discharge outlet are higher than at other locations. Although pH value is 6–7, the duration of exposure is about 10 years.

Figure 11.16 Views of the site and the sample.

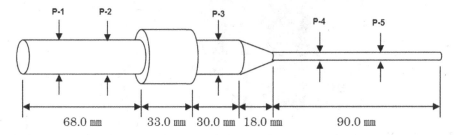

Figure 11.17 Position of measurement points.

Table 11.10 Measurements.

	Diameter (mm)		
	P-1	P-2	P-3
Before removal	16.74	16.92	15.93

Table 11.11 Measurements.

	Diameter (mm)	
	P-4	P-5
After removal (30 years)	9.85	10.93

11.2.5   Corrosion of iron at Ikejima Seashore

Figure 11.16 shows views of the site and the state of the sample. Thirty years have elapsed since the construction. Points 1, 2, and 3 shown in Figure 11.17 correspond to original views before the removing the corroded parts. Points 4 and 5 are in the state that corroded parts have fallen. Results are measurements given in Tables 11.10 and 11.11. The corrosion rate is extremely high at this location.

11.2.6   Corrosion of deformed bar at Tekkehamam hot spring site

The corrosion state of a deformed bar at the Tekkehamam hot spring site is measured (Fig. 11.18). The total duration was 4 years, and the temperature of the monitoring hut where the bar was located ranges between 80°C and 90°C near the hot spring discharge, while the temperature ranges between 34°C and 44°C at the upper part of the monitoring hut. The diameter of the bar was measured at several locations, as illustrated in Figure 11.19.

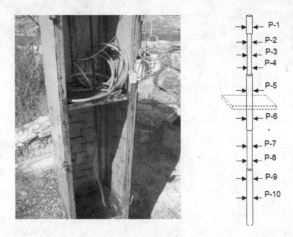

Figure 11.18 Views of the deformed bar in the monitoring hut and locations of measurements.

Figure 11.19 Several views of the mine.

The diameter of points numbered 1, 4, 6, 9, and 10 correspond to the diameter before removing the corroded part, and 2, 3, 4, 7, and 8 correspond to the diameter after the removal of corroded parts. The diameter of the bars measured at various locations are given in Tables 11.12 and 11.13. As noted from the tables, the corrosion rate is extremely high.

11.2.7   Corrosion of an iron bar at Moyeuvre abandoned iron mine and its investigation by X-ray CT scanning technique

Moyeuvre, which provided iron for the famous Eiffel tower, began being mined in 1881 and was finally closed in 1930 (Fig. 11.19). The iron ore consists of three layers, and its overall thickness is about 15 m. The overburden ranges between 5 m and 120 m. The iron ore is sandwiched between marl layers. The mining method was room and pillar. A bolt used for fixing the rails was sampled from the mine (Fig. 11.20).

Table 11.12 Diameter of the bar at various locations.

	Diameter (mm)				
	P-1	P-5	P-6	P-9	P-10
Before removal	16.74	16.92	15.93	14.69	15.23

Table 11.13 Diameter of the bar at various locations.

	Diameter (mm)				
	P-2	P-3	P-4	P-7	P-8
After removal (4 years elapsed)	9.85	10.93	10.11	12.51	12.82

Figure 11.20 Sample iron bolt.

The corrosion state of the bolt could be investigated using X-ray CT scanning technique. Despite the unsuitability of steel for X-ray CT scanning, the images clearly displayed corrosion of the bolt. Figure 11.21 shows several X-ray CT scan images of the bolt. As noted from the figure, the corrosion depth was generally less than 1 mm and some of corroded parts were missing due to handling and transportation. The surface of the bolt was severely undulating. Figure 11.22 compares the actual image with the scanned image. The rusty surface of the bolt could not be clearly shown in the X-ray CT scanning images.

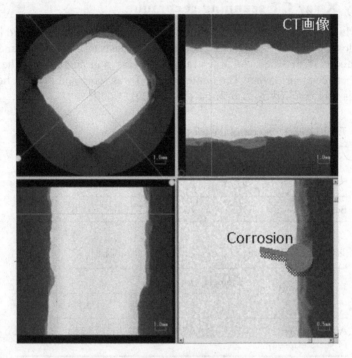

Figure 11.21 X-ray CT scanning images of corroded bolt.

Figure 11.22 Comparison of the actual and X-ray CT scanning images of the corroded bolt.

11.2.8   Simulation of corrosion

The governing equation of corrosion can be associated with mass conservation law if ϕ represents the corrosion state as a scalar function as given below:

$$\frac{d\phi}{dt} = -\nabla \cdot \mathbf{q} + Q \tag{11.7}$$

where ϕ may be related to a product of corrosion. The corrosion rate may also be given in analogy to Fick's law as:

$$\mathbf{q} = -\lambda \nabla \phi \tag{11.8}$$

where λ is corrosion flux. On the other hand, the momentum conservation law with the negligence of inertia component can be given in incremental form as follows:

$$\nabla \cdot \dot{\sigma} = 0 \tag{11.9}$$

It should be also noted that deformability tensor, density, and corrosion rate may be assumed to be dependent upon the corrosion potential:

$$\mathbf{D} = \mathbf{D}(\phi), \; \rho = \rho(\phi), \; \lambda = \lambda(\phi) \tag{11.10}$$

However, it should be noted that if such a concept is introduced, the necessary experiments should be carried out. Equations (11.7) and (11.9) can be solved using numerical methods such as the finite element method. If the finite element method is used, the governing equation can be represented as given below:

$$[M]\{\dot{\theta}\} + [H]\{\theta\} = \{Q\} \tag{11.11}$$

where

$$[M] = \int [N]^T [N] dV; \; [H] = \lambda \int [B]^T [B] dV; \; \{Q\} = \int [\bar{N}]^T \{q_n\} d\Gamma$$
$$[K]\{\dot{U}\} = \{\dot{F}\} \tag{11.12}$$

where

$$[K] = \int_V [B]^T [D][B] dV; \{\dot{F}\} = \int_V [B]^T [D]\{\dot{e}_s\} dV + \int_S [\bar{N}]^T \{i\} dS$$

Figures 11.23 and 11.24 show an example of solutions by assuming that $\lambda = 0.001$ mm²/year. The diameter of the rockbolts is 25 mm.

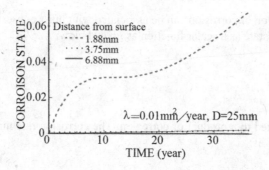

Figure 11.23 Simulation of corrosion for a rockbolt with a diameter of 25 mm.

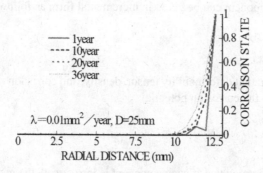

Figure 11.24 Simulation of corrosion for a rockbolt with a diameter of 25 mm.

11.2.9 Effect of corrosion on the physico-mechanical properties of tendon

Steel is commonly used as tendon material in rockbolts and rockanchors. Some samples were also gathered from existing powerhouses, a bridge in Shimizu built about 70 years ago, and those presented in the previous subsection. Table 11.14 gives some physico-mechanical properties of tendon materials collected from different sites. As noted from the table, physico-mechanical properties of tendons are strongly influenced by the corrosion process.

In particular, the wave-velocity may be used for evaluating the state of corrosion of materials as explained in previous sections (Fig. 11.25). The corrosion state may be related to the wave velocity of tendon materials before and after exposure through the following equation:

$$\alpha = 1 - \left(\frac{v_p^*}{v_p^o} \right)^2 \qquad (11.13)$$

where α is interpreted as corrosion ratio.

Table 11.14 Physical and mechanical parameters of various steel bars and cables.

Material	Unit Weight (kN/m³)	P-wave velocity (km/s)	S-wave velocity (km/s)	Elastic Modulus (GPa)	Tensile Strength (MPa)
PC steel bar	75.6	6.05	3.23	200–210	
PC wire cable	69.6	4.80	–	165–170	
Deformed bar	70–74	5.6–5.9	3.0–3.2	200	
Steel	77–80	5.95	3.23	200	
Iron	76–78	5.91	3.21	200	
Ikejima-corroded smooth bar		4.83			
Shimizu Bridge, smooth bar (70 years)	75.4–78.5			196–213	425–458
Iron bolt (Moyeuvre Mine)	–	5.26	–	–	–
Theoretical estimations	75.6	5.97	3.19	200	

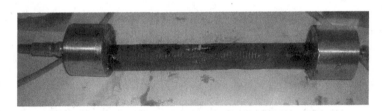

Figure 11.25 Elastic wave velocity measurement of steel rockanchor from Mazegawa underground power station.

Table 11.15 Corrosion of iron depending upon environment.

	Koseto	Ikejima	Tekkehamam	Moyeuvre
Corrosion (%)	7–13	41	20–35	
Corrosion rate (mm/year)	0.15–0.50	0.22–0.33	0.32–0.61	0.1

Table 11.16 Corrosion rate of Swellex rockbolts (from Atlas Copco).

	Mine A Canada	Mine B Canada	Tunnel Japan	Mine C Australia	Mine D Australia
pH	7.8	3.4	6.7	7.9	3.0
Corrosion rate (mm/year)	0.1–0.5	0.1–0.5	0.02–0.1	0.05–0.2	0.1–0.5

In many countries, the corrosion rate of steel/iron is always a major point of interest. Atlas Copco investigated the corrosion rate of Swellex rockbolts under different environments as given in Table 11.15. From this table, it is noted that the corrosion rate is about 0.1–0.2 mm/year.

11.2.10   Estimation of failure time of tendons

In this section, the failure time of pre-stressed tendons subjected to corrosion and creep is assessed. The long-term strength of materials can be expressed using the following equation (i.e. Aydan and Nawrocki, 1998):

$$\frac{\sigma_{il}}{\sigma_{ts}} = 1 - b \ln\left(\frac{t}{t_s}\right) \tag{11.14}$$

Where t_s is short-term testing period (5–20 minutes). Coefficient b is an empirical constant and its value for steel ranges between 0.02 and 0.04.

As many tendons have circular cross-sections, the corrosion starts from the outer surface and progresses inward. As a result, the initial cross-section area (A_o) of the tendon would decrease as time progresses. Thus, one can write the following relationship for the cross-section area for any time normalized by the initial cross-section area:

$$\frac{A}{A_o} = \left(\frac{r}{r_0}\right)^2 \tag{11.15}$$

Let us assume that the decrease of the radius of the bar due to corrosion may be estimated from the following equation:

$$r = r_o - \beta t \tag{11.16}$$

If an initial pre-stress acts on the tendon, the current stress on the tendon may be related to the pre-stress in the following form provided that the force remains the same:

$$\frac{\sigma}{\sigma_o} = \left(\frac{r_o}{r}\right)^2 \tag{11.17}$$

The tendon will rupture when the stress given by Equation (11.17) becomes equal to the strength given by Equation (11.14). If we denote the failure time by t_f, it would be specifically obtained as follows:

$$t_f = t_s \exp\left(\frac{1}{b}\left[1 - \frac{\sigma}{\sigma_{ts}}\left(\frac{r_o}{r_o - \beta t}\right)^2\right]\right) \tag{11.18}$$

As the surface area of cable-type anchors is larger than the bar-type rockanchor, Equation (11.18) implies that the failure time of cable anchors would be shorter than that of the bar-type anchors. Table 11.17 is a summary of corrosion rates of steel used in underground excavations in various countries. Figures 11.26 and 11.27 show the estimation of failure time of bar-type rockanchors of 27 mm and cable-type anchors with 6-mm steel wires for an initial pre-stress value of 1/3 of their ultimate load-bearing capacity.

Table 11.17 Corrosion rates of steel in underground excavations.

	Mine-A Canada	Mine-B Canada	Tunnel Japan	M-UGPH Japan	Mine-E France	Mine-D Australia	Mine-E Australia
pH	7.8	3.4	6.7	6.7	–	7.9	3.0
Rate (mm/year)	0.1–0.5	0.1–0.5	0.02–0.1	0.025	0.1	0.05–0.2	0.1–0.5

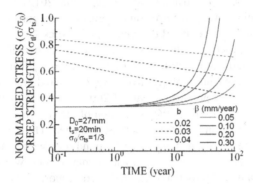

Figure 11.26 Estimation of failure time of bar-type rockanchor.

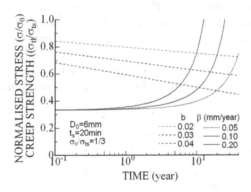

Figure 11.27 Estimation of failure time of cable-type rockanchor.

When the anchor ruptures, the anchor head would be thrown and its ejection velocity could be estimated from a similar procedure described for the ejection velocity of rock fragments during rockburst. The ejection velocity (unit is m/s) can be given as

$$v = 0.0255\sigma_{ct} \qquad (11.19)$$

The failure stress (σ_{ct}) would be a fraction of its original tensile stresses. It is very likely to be in the range of 0.6–0.8 times the short-term tensile strength. As the tensile strength of high-grade steel bars is about 1 GPa, the ejection velocity of anchor head could likely be in the range of 15–20 m/s.

11.3   EFFECT OF DEGRADATION OF SUPPORT SYSTEM

The support system of the caverns consists of rockbolts, rockanchors, and arch concrete. The rockanchors installed on sidewalls are expected to perform their functions during the service life of the cavern. The rockanchors may consist of either high-strength steel bars and/or high-strength steel wires. The underground environment is always humid and groundwater may have corrosive dissolved chemical substances. Steel is known as a corrosion-resistant material. Site investigations by the authors as well as by other researchers indicate that rockbolts and rockanchors may undergo corrosion (Aydan *et al.*, 2008, 2012). Cable-type rockbolts or rockanchors are much more prone to corrosion in view of *in situ* pull-out experiments. Theoretical estimations for the service life of different types of rockbolts and rockanchors under pre-stress based on the experimentally measured corrosion rate confirm this experimental finding. The last computational example is concerned with the effect of the degradation of the support system. The excavation of the cavern was simulated first, and rockanchors with a 5-m-anchorage length and a pre-stress value of 200–300 kN are assumed to be installed. Since blasting technique was used for the excavation of the cavern, it is likely that the rock mass adjacent to the excavation boundary might be damaged. The computation was carried out for the following three cases, namely, Case 1: sound rock with supports; Case 2: sound rock with degraded supports; and Case 3: rock with damaged zone, whose deformation modulus is 1/10 of the undamaged rock mass and degraded support. Figure 11.28 illustrates the pattern of rockanchors, the extent of the blasting-induced damage zone, and selected points for comparisons of displacement. Table 11.18 gives the computed displacements at selected points. As expected, the degradation of the support system increases the displacement field around the cavern. Nevertheless, the amplitude of the increase is not very large. When there is a blasting-induced damage zone around the cavern, the amplitude of the displacement field becomes much larger. In any case, no plastic zone developed around the cavern. However, it should be noted that these type of continuum analyses are valid for global stability of the caverns. Rock mass always has discontinuities in different forms. These pre-existing discontinuities may constitute unstable rock blocks around the cavern. The size of these blocks may be quite large when the cavern height and width are more than 40 m and 25 m, respectively. In such cases, the stabilization of such blocks by a support system may become the prime concern (Aydan, 1989; Kawamoto *et al.*, 1991).

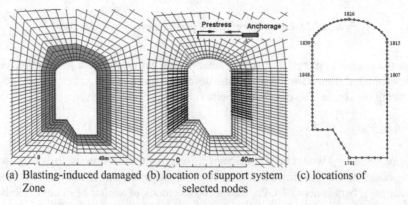

(a) Blasting-induced damaged   (b) location of support system   (c) locations of
    Zone   selected nodes

Figure 11.28 Illustration of blasting-induced damage zone, locations of support system, and selected nodes.

Table 11.18 Horizontal and vertical displacement of selected points.

Node No.	CASE 1		CASE 2		CASE 3	
	U (mm)	V (mm)	U (mm)	V (mm)	U (mm)	V (mm)
1781	4.966	−0.343	4.946	−0330	12.32	−13.33
1807	−13.89	−12.86	−14.03	−12.89	−39.19	−18.40
1815	−8.947	−15.63	−9.049	−15.65	−27.51	−24.69
1826	0.006	−22.15	0.008	−22.17	1.597	−49.14
1839	7.587	−16.12	8.647	−16.08	27.54	−25.03
1848	13.03	11.25	13.24	−11.29	38.84	−15.65

11.4   NONDESTRUCTIVE TESTING FOR SOUNDNESS EVALUATION

The geoenvironmental conditions imposed on rockbolts and rockanchors may be adverse and may cause corrosion over a short period of time. While rockbolts are generally made of steel bars, rockanchors may be either high-strength steel bar or cables consisting of several wires. As the surface area of cable rockanchors are much larger, they are much more prone to corrosion compared with steel bars.

The maintenance of existing rock engineering structures, such as underground power-houses, dams, and slopes, requires the reassessment of support members. Although *in situ* pull-out tests and lift-off tests would be the best options for assessing the soundness and acting stress of the rockbolts and rockanchors, performing such tests is costly and labor inten-sive. Furthermore, such tests may require the reinstallation of rockbolts and rockanchors if they are pulled out or broken during the tests. Therefore, the use of nondestructive tests to assess their soundness and/or acting stresses is desirable with a main concept as illustrated in Figure 11.29. Despite several attempts to assess the state of support members, it is difficult to say that such techniques are satisfactory for practical purposes. Furthermore, the structural part of rockbolts and rockanchors used during the nondestructive tests is quite limited com-pared with the overall size of the support system.

The soundness investigation involves the state of corrosion of steel bars, cracks in tendon and grouting material, debonding of interfaces, and acting stresses. There have been sev-eral attempts to develop such systems for nondestructive evaluations of the soundness of rockanchors and rockbolts (i.e. Tannant *et al.*, 1995; Aydan *et al.*, 2005, 2008; Ivanović and Neilson, 2008; Beard and Lowe, 2003). Due to very limited exposure of rockbolts and rock-anchors embedded in rock masses, it is necessary to utilize numerical techniques to capture the fundamental features of the expected responses, which may be used to interpret measured responses from *in situ* nondestructive tests. With this in mind, some theoretical models have been developed for the axial and traverse dynamic tests, and then numerical simulations are carried out based on these fundamental equations by considering possible conditions *in situ*. Specifically, the bonding quality, the existence of corroded parts or couplers, and pre-stress are simulated through numerical experiments. The results of the numerical simulations can be used to interpret the dynamic responses to be measured in nondestructive dynamic tests *in*

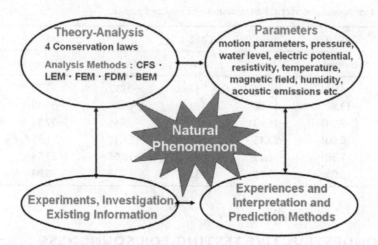

Figure 11.29 The main concept for developing nondestructive assessment of soundness of rockbolts and rockanchors.

situ. Several applications of this procedure to practice are given together with measurements obtained from some nondestructive tests under laboratory and *in situ* conditions.

11.4.1   Impact waves for nondestructive testing of rockbolts and rockanchors

Aydan (Aydan, 1989; Aydan *et al.*, 1985, 1986, 1987, 1988; Kawamoto *et al.*, 1994) studied the static response of rockbolts and rockanchors and developed some theoretical and numerical models. However, dynamic responses for use in nondestructive tests require dynamic equilibrium equations for axial and traverse responses and their numerical representations and solutions. These dynamic equilibrium equations and their numerical representations are given in the next subsections.

11.4.1.1   Mechanical models

By modification of the static equilibrium equations developed for rockbolts and rockanchors by Aydan (1989), the equation of motion for the axial responses of rockbolts and rockanchors, together with the consideration of inertia component including mass proportional damping, can be written in the following form (Fig. 11.30):

$$\rho \frac{\partial^2 u_b}{\partial t^2} + h_a^* \frac{\partial u_b}{\partial t} = \frac{\partial \sigma}{\partial x} + \frac{2}{r_b}\tau_b \tag{11.20}$$

Where ρ, u_b, σ, r_b, h_a^*, and τ_b are density, axial displacement, axial stress and radius of tendon, axial damping ratio, and shear stress along the tendon-grout interface, respectively.

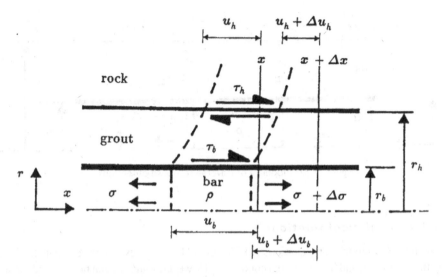

Figure 11.30 Modeling of dynamic axial response of tendons.

Constitutive relationships for axial stress and shear stress may be given in the following forms, respectively:

$$\sigma = E_b \varepsilon + \eta \dot{\varepsilon}; \varepsilon = \frac{\partial u_b}{\partial x}; \dot{\varepsilon} = \frac{\partial \varepsilon}{\partial t} = \frac{\partial}{\partial t}\left(\frac{\partial u_b}{\partial x}\right)$$ (11.21a)

$$\tau_b = K_g u_b$$ (11.21b)

The effect of surrounding grouting annulus and rock mass is taken into account trough shear stiffness K_g, which was originally derived by Aydan et al. (1985, 1986, 1987, 1988; see also Chapter 4) and Aydan (1989), and it is specifically given as:

$$K_g = \frac{G_g}{r_b \ln(r_h/r_b)} \cdot \left(\frac{X-1}{X}\right) \text{ and } X = \frac{G_r}{G_G} \frac{\ln(r_h/r_b)}{\ln(r_0/r_h)} + 1$$ (11.22)

Where G_g, G_r, r_h, and r_o are shear modulus of grouting material and surrounding rock, radius of hole, and radius of influence.

As for traverse response of the free part of the tendon assuming the existence of the axial tensile force, the following equation may be written (Fig. 11.31):

$$\rho \frac{\partial^2 w_b}{\partial t^2} + h_t^* \frac{\partial w_b}{\partial t} = \sigma_p \frac{\partial^2 w_b}{\partial x^2}$$ (11.23)

Where w_b, σ_p, and h_t^* are traverse displacement, applied pre-stress, and traverse damping ratio of tendon.

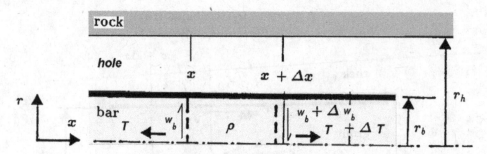

Figure 11.31 Modeling of traverse response of tendons.

11.4.1.2 Analytical solutions

The analytical solutions for Equations (11.20) and (11.23) are extremely difficult for given constitutive laws, boundary, and initial conditions. However, solutions can be obtained for simple cases, which may be useful for interpreting results of site investigations. Equation (11.20) may be reduced to the following form by omitting the effect of damping and interaction with surrounding rock as:

$$\frac{\partial^2 u_b}{\partial t^2} = V_p^2 \frac{\partial^2 u_b}{\partial x^2} \tag{11.24}$$

where

$$V_p = \sqrt{\frac{E_b}{\rho}}$$

The general solution of partial differential Equation (11.24) may be given as:

$$u_b = h(x - V_p t) + H(x + V_p t) \tag{11.25}$$

For a very simple situation, the solution may be given as follows:

$$u_b = A \sin \frac{2\pi}{L} \left(x \pm V_p t \right) \tag{11.26}$$

where L is tendon length. Thus, the Eigen values of the tendon may be obtained as follows:

$$f_P = n \frac{1}{2L} V_p, n = 1,2,3, \tag{11.27}$$

Similarly, the Eigen values of traverse vibration of the tendon under a given pre-stress may be obtained as follows:

$$f_T = n\frac{1}{2L}V_T, n = 1,2,3,$$

(11.28)

where

$$V_T = \sqrt{\frac{\sigma_o}{\rho}}$$

11.4.1.3   Finite element formulation

(a) Weak form formulation

The integral form of Equation (11.20) may be written as follows:

$$\int \delta u_b \left(\rho \frac{\partial^2 u_b}{\partial t^2} + h_a^* \frac{\partial u_b}{\partial t} \right) dx = \int \delta u_b \frac{\partial \sigma}{\partial x} dx + \int \delta u_b \frac{2}{r_b} \tau_b dx$$

(11.29)

Introducing the following identity into Equation (11.29):

$$\delta u_b \frac{\partial \sigma}{\partial x} = \frac{\partial}{\partial x}(\delta u_b \sigma) - \frac{\partial \delta u_b}{\partial x}\sigma$$

(11.30)

one can easily obtain the weak form of Equation (11.20) as follows:

$$\int \delta u_b \left(\rho \frac{\partial^2 u_b}{\partial t^2} + h_a^* \frac{\partial u_b}{\partial t} \right) dx + \int \frac{\partial \delta u_b}{\partial x}\sigma dx - \int \delta u_b \frac{2}{r_b}\tau_b dx = \int \delta u_b \sigma_n \big|_{x=a}^{x=b}$$

(11.31)

(b) Finite element formulation

Displacement field of rockbolts and rockanchors is discretized in space in a classical finite element form as:

$$u_b = [N]\{U_b\}$$

(11.32)

With the use of Equation (13.32), one can write the following:

$$a_b = \frac{\partial^2 u_b}{\partial t^2} = [N](\{\ddot{u}_b\}); \; v_b = \frac{\partial u_b}{\partial t} = [N]\{\dot{u}_b\}$$

(11.33)

$$\varepsilon_b = \frac{\partial u_b}{\partial x} = \frac{\partial}{\partial x}([N(x)]\{u_b\}) = [B]\{u_b\}; \; \varepsilon_b = \frac{\partial \dot{u}_b}{\partial x} = \frac{\partial}{\partial x}([N(x)]\{\dot{u}_b\}) = [B]\{\dot{u}_b\}$$

(11.34)

If Equations (11.32) to (11.34) are inserted into Equation (11.31), one can easily obtain the following expression:

$$[M_e]\{\ddot{u}_b\}+[C_e]\{\dot{u}_b\}+[K_e]\{u_b\}=\{f_e\}$$ (11.35)

where

$$[M_e]=\int \rho[N]^T[N]dx;\ [C_e]=\int h_a^*[B]^T[B]dx+\int \eta[B]^T[B]dxh;$$
$$[K_e]=\int E[B]^T[B]dx-\int \frac{2Kg}{r_b}[N]^T[N]dx$$

Similarly, the finite element form of Equation (11.23) may be written as follows:

$$[M_e^t]\{\ddot{w}_b\}+[C_e^t]\{\dot{w}_b\}+[T_e^t]\{w_b\}=\{t_e^t\}$$ (11.36)

where

$$[M_e^t]=\int \rho[N]^T[N]dx;\ [C_e^t]=\int h_t^*[N]^T[N]dx;\ [T_e]=\int \sigma_p[B]^T[B]dx$$

11.4.1.4 Properties of rockbolts/rockanchors, grouting material, and interfaces

(a) Properties of rockbolts/rockanchors

Aydan and his group have been performing destructive and nondestructive tests on non-corroded and corroded iron and steel bars (Aydan, 1989; Aydan *et al.*, 2005, 2012, 2016 and some unpublished reports). Various relevant parameters are given in Table 11.14. This table also includes the parameters of steel bars sampled from 70-year-old reinforced concrete structures. Figure 11.32 shows the strain-stress responses of widely used steel bars and cable anchors under tension. In this figure, steel bar (WC) corresponds to a steel bar with a section of areal reduction to represent the effect of corrosion.

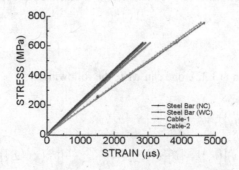

Figure 11.32 Strain-stress response of steel bars and cable anchors.

(b) Properties of grouting material and interfaces

The overall behavior of rockbolt/rockanchors is also influenced by the characteristics of grouting material. Depending upon the composition, material properties of grout may differ. The properties of some grouting material tested by the author are given in Table 11.19.

The load-bearing capacity of rockbolts and rockanchors are strongly influenced by the shear strength characteristics of interfaces. These interfaces are tendon-grout (T-G), grout-protection sheath (G-S) and grout-rock (G-R). The short-term, creep, and cyclic characteristics of interfaces have been studied by Aydan (1989) and Aydan *et al.* (1985; 2016), and some of the experimental results are shown in Chapter 4. Figure 11.33 compares the shear strength of various interfaces with that of grouting material. As noted from the figure, the surface morphology of interfaces greatly influences their shear strength, which is much less than that of the grouting material.

Table 11.19 Physico-mechanical properties of grouting material (28 days).

Grout Type	Unit Weight (kN/m3)	Elastic Modulus (GPa)	Poisson's ratio	UCS (MPa)	Tensile Strength (MPa)
M-T	20.0–20.4	14.7–18.4	0.17–0.25	19–21.9	1.5–1.8
M-M	19.1	11.6	–	53.1	5.54
M-R	20.2–20.6	13.7–17.3	0.17–0.25	19–22	2.3–2.7

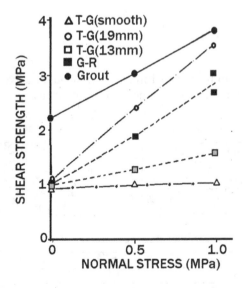

Figure 11.33 Comparison of shear strength of interfaces with that of grouting material (from Aydan, 1989).

11.4.1.5   Evaluation of corrosion of rockbolts and rockanchors

The evaluation of corrosion from the measured responses is one of the most important items in the interpretation of nondestructive test investigations. This definitely requires some mechanical models for interpretation. The mechanical properties of the corroded part of the steel are much less than those of steel. Furthermore, the corrosion may be limited to a certain zone where the anti-corrosive protection may be damaged due to either relative motions at rock discontinuities or chemical attacks of corrosive elements in the groundwater. If the corrosion is assumed to be taking place uniformly around the steel bar for a certain length, as illustrated in Figure 11.34, the equivalent elastic modulus (E_b^*), shear modulus (G_b^*), and density (ρ_b^*) of the tendon may be obtained using the micro-structure theory (Aydan *et al.*, 1996, 2005, 2012) as follows:

$$\frac{E_b^*}{E_b} = \frac{(1-\alpha)+\alpha E_c/E_b}{\lambda+(1-\lambda)\big((1-\alpha)+\alpha E_c/E_b\big)}, \quad \frac{G_b^*}{G_b} = \frac{(1-\alpha)+\alpha G_c/G_b}{\lambda+(1-\lambda)\big((1-\alpha)+\alpha G_c/G_b\big)}, \quad (11.37a)$$

$$\frac{\rho_b^*}{\rho_b} = (1-\lambda)+\lambda\bigg[(1-\alpha)+\alpha\frac{\rho_c}{\rho_b}\bigg], \quad \lambda=\frac{l_c}{L}, \quad \alpha=\frac{A_c}{A_b} \qquad (11.37b)$$

Where E_c, G_c, and ρ_c are properties of the corroded part. A_c is corrosion area. If the properties of the corroded part are negligible, then the above equations take the following form:

$$\frac{E_b^*}{E_b} = \frac{(1-\alpha)}{\lambda+(1-\lambda)(1-\alpha)}, \quad \frac{G_b^*}{G_b} = \frac{(1-\alpha)}{\lambda+(1-\lambda)(1-\alpha)} \qquad (11.38)$$

If corrosion is uniformly distributed over the total length of the tendon, then one may write the following equation:

$$\alpha = 1-\left(\frac{v_p^*}{v_p^o}\right)^2 \qquad (11.39)$$

The dynamic parameters to be obtained from *in situ* nondestructive tests may be used to obtain the dimensions of corrosion using the models presented above.

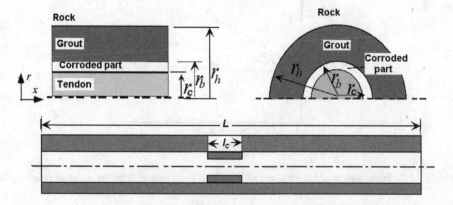

Figure 11.34 Geometrical illustration for the evaluation of the effect of the corroded part.

11.4.1.6. Numerical analyses

Some numerical analyses were already reported by Aydan *et al.* (2008) for 10-m-long rock-anchors. Numerical experiments presented in this section are performed to clarify the dynamic responses of rockanchors and rockbolts under impact waves for various conditions encountered in actual situations and in relation to some experiments carried out in the laboratory. Specifically, the following conditions are considered together with that of damping effects.

Case 1: Unbonded and non-corroded bar
Case 2: Bonded and non-corroded bar
Case 3: Unbonded bar with corrosion
Case 4: Bonded bar with corrosion
Case 5: Unbonded and non-corroded bar under pre-stress
Case 6: Unbonded and non-corroded bar under variable pre-stress
Case 7: Unbonded bar with corrosion under pre-stress

Rockanchors were assumed to be 1200-m-long with a 26-mm diameter and elastic. When rockanchors are bonded, they are bonded along their entire length. The shear modulus of grouting material and rock is assumed to be 2 GPa and 0.5 GPa, respectively. The effect of the bonding is taken into account according to Aydan's model for rockbolts. Figures 11.35 and 11.36 shows the axial dynamic response of rockanchors for Case 1 and Case 2 with and

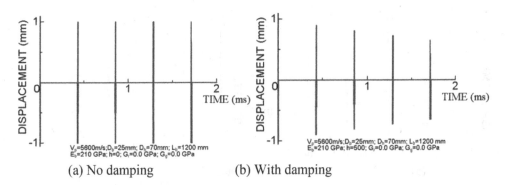

(a) No damping   (b) With damping

Figure 11.35 Axial dynamic displacement responses of unbonded rockanchors with and without damping (Case 1).

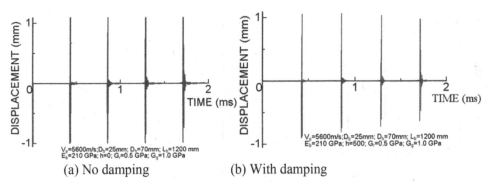

(a) No damping   (b) With damping

Figure 11.36 Axial dynamic displacement responses of bonded rockanchors with and without damping (Case 2).

without damping. As noted from the figures, the wave travels according to p-wave velocity and the bonding has little influence on the arrival time response of the reflected wave. Nevertheless, some noise-like responses are noted following the fundamental wave, and the amplitude of the reflected waves differs. The amplitude of waves decreases with time if damping is introduced, which was also discussed in Chapter 10.

Next, two numerical experiments were concerned with the effect of corrosion on the dynamic response of unbonded and bonded rockanchors (Cases 3 and 4). Figure 11.37 shows the computed dynamic displacement responses for corroded tendon at the middle with a cross-sectional reduction of 20%. Compared with the responses in previous cases, there are some reflected waves before the arrival of the reflected main shock. Furthermore, the amplitude of the reflected main shocks is no longer the same, and the noise-like waves are noted following the main shock. The amplitude and duration of these noise-like responses become larger as time passes. However, the results from these numerical analyses indicate that it is possible to locate the corrosion location. The amplitude of the reflection from the corroded part depends upon the geometrical and mechanical characteristics of the corroded part.

The effect of the pre-stress in rockanchors on the traverse response of rockanchors is investigated (Case 5). In the numerical tests, the pre-stress values are varied. Figure 11.38 shows the dynamic displacement responses for a pre-stress of 500 MPa under undamped and damped conditions. It should be noted that the travel time of the waves entirely depends

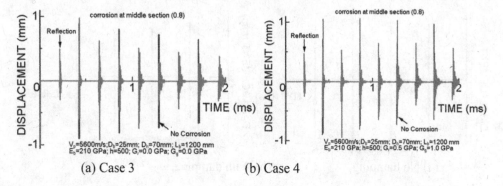

(a) Case 3   (b) Case 4

Figure 11.37 Axial dynamic displacement responses of unbonded and bonded rockanchors with corrosion.

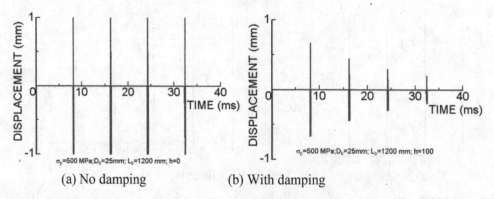

(a) No damping   (b) With damping

Figure 11.38 The effect of pre-stress on dynamic traverse responses of pre-stressed tendons (Case 5).

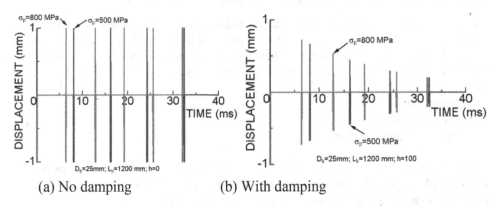

(a) No damping   (b) With damping

Figure 11.39 The effect of pre-stress variation on dynamic traverse responses of pre-stressed tendons (Case 6).

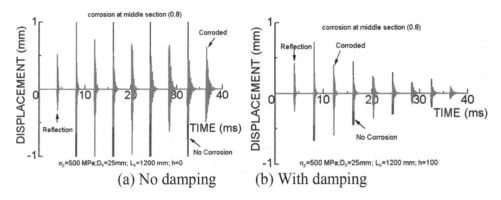

(a) No damping   (b) With damping

Figure 11.40 The effect of corrosion on axial wave responses of pre-stressed tendons (Case 7).

upon the pre-stress value. Figure 11.39 shows the dynamic traverse response of the bar under different values of pre-stress. As also reported by Aydan et al. (2008), the travel time of the reflected traverse wave becomes shorter as the value of pre-stress increases. These results indicate that the traverse wave responses should provide valuable information on the pre-stress state of actual rockanchors.

The final example is concerned with the effect of the pre-stressed tendon with a corroded part in the middle section (Case 7). Figure 11.40 shows the dynamic displacement responses for non-corroded and corroded tendons subjected to the same pre-stress. The corroded part is assumed to have 80% of the original cross-section. When there is no corrosion, the travel time of the reflected displacement wave is 8 ms. Since the behavior of the tendon is assumed to be elastic, the reflected wave arrives at each 8-ms interval. However, when it is corroded at the middle section, it is noted that a wave is reflected before the arrival of the main wave. Furthermore, the amplitude of the reflected main shocks are no longer the same and the noise-like waves are noted following the main shock. The amplitude and duration of these noise-like responses become larger as time passes. However, the results from these numerical analyses indicate that it is possible to locate the corrosion location. The amplitude of the reflection from the corroded part is associated with the geometrical and mechanical characteristics of the corroded part.

11.4.1.7   Identification of reflected waves from records

As the application and observation of rockbolts and rockanchors for nondestructive testing is very small, the utilization of reflected waves is the essential item in the evaluation of soundness of the rockbolts and rockanchors. There are different techniques to identify the arrival time of reflected waves for the first fundamental mode. These may be categorized as

i)   Manual pick-up
ii) Structural function
iii) Auto-correlation function

To illustrate these approaches, nondestructive tests are carried out on steel bars having different lengths (2580, 3530, 6040, and 9000 mm). The actual records for steel bars for four different lengths under shock waves are shown in Figure 11.41.

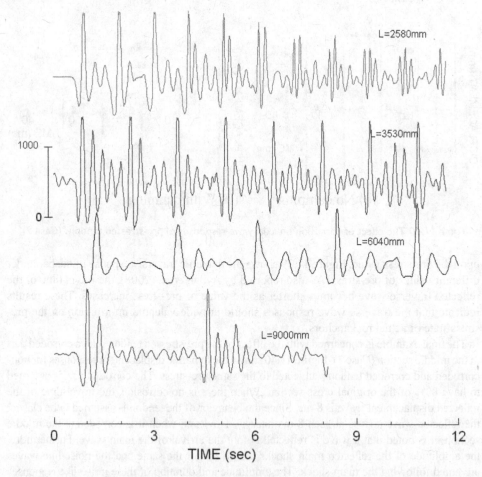

Figure 11.41 Acceleration waves for steel bars with different length under shock wave (unit is gals).

The arrival time of reflected waves can be manually picked up from the records if they distinctly appear in the records. For this purpose, a "peak-to-peak" approach should be appropriate. The structural function (SF) used in wave processing is given in the following form:

$$SF = \frac{1}{T} \int_{t=0}^{t=T} (\phi(t) - \phi(t+\tau))^2 \, dt$$

$$(11.40)$$

Similarly, an auto-correlation function (ACF) is given by:

$$ACF = \frac{1}{T} \int_{t=0}^{t=T} \phi(t)\phi(t+\tau) \, dt$$

$$(11.41)$$

Where ϕ, t, and τ are observation function, time, and time lag, respectively. For a periodic function with given constant amplitude, when the time lag coincides with the periodicity of the wave, the values of SF and ACF would be minimum and maximum. The applications of Equations (11.40) and (11.41) to the records shown in Figure 11.41 are shown in Figure 11.42. It is interesting to note that both SF and ACF approaches yield exactly the same results.

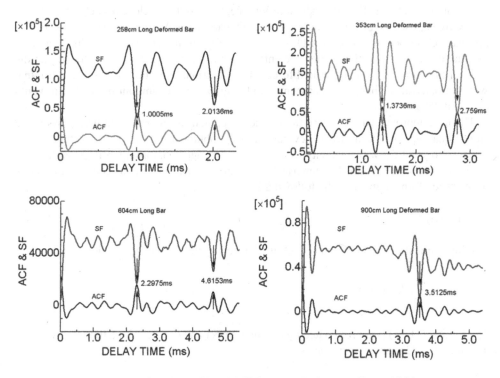

Figure 11.42 Computed SF and ACF for records shown in Figure 11.36.

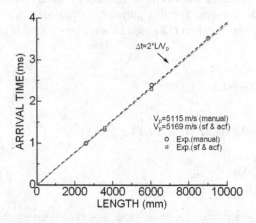

Figure 11.43 The relation between bar length and arrival time of reflected waves determined from three different approaches.

The results of arrival time of reflected waves obtained from three different approaches are plotted in Figure 11.43. The manually picked-up arrival times yielded that the velocity of the bar is about 5115 m/s while the SF and ACF approaches yielded that the velocity of bars is about 5169 m/s. Despite a slight difference, the results are close to each other. As the bars were exposed to the atmosphere for several years, some rusting of bars was noted. The approach described in Subsection 11.3.1.5 was applied to the computed results by assuming that corrosion took place along the entire length of the bars, and it was found that the ratio of the corrosion was estimated to be 14–25% depending upon the assumed original wave velocity of the bars. If it is assumed to be 5900 m/s, the amount of corrosion would be about 25%.

The next approach would to use the frequency content of the recorded waves. As noted from Equation (11.26), the frequency content to be obtained from Fourier spectra analysis would also yield the wave velocity of the tendon. Fourier transformation is generally used to simulate real wave forms by numerical approximate functions. Let us consider the actual acceleration form is given by the following function:

$$a = a(t) \tag{11.42}$$

Fourier transform of Function (11.42) is replaced by the following function:

$$a(t) = \sum_{k=0}^{\infty} \left[A_k \cos(kt) + B_k \sin(kt) \right] \tag{11.43}$$

If Equation (11.43) is approximated by a finite number (N) of data with a time interval (Δt), Equation (11.43) can be rewritten as:

$$\tilde{a}(t) = \frac{A_o}{2} + \sum_{k=1}^{N/2-1} \left[A_k \cos\left(2\pi f_k t\right) + B_k \sin\left(2\pi f_k t\right) \right] + \frac{A_{N/2}}{2} \cos 2\pi f_{N/2} t \tag{11.44}$$

where

$$f_k = \frac{k}{N\Delta t} \tag{11.45}$$

If time (t) is represented by $m\Delta t$, coefficients A_k and B_k in Equation (11.43) would be expressed as:

$$A_k = \frac{2}{N}\sum_{m=0}^{N-1} a_m \cos\frac{2\pi km}{N} \quad k = 0,1,2,\ldots,\frac{N}{2}-1,\frac{N}{2} \tag{11.46a}$$

$$B_k = \frac{2}{N}\sum_{m=0}^{N-1} a_m \sin\frac{2\pi km}{N} \quad k = 1,2,\ldots,\frac{N}{2}-1 \tag{11.46b}$$

For kth frequency, the maximum amplitude, phase angle, and power would be obtained as:

$$C_k = \sqrt{A_k^2 + B_k^2}; \quad \phi_k = \tan^{-1}\left(-\frac{B_k}{A_k}\right); \quad P_k = \frac{C_k^2}{2} \tag{11.47}$$

It should be noted that the amplitude of Fourier coefficient C_k is multiplied by half of the period of the record (T/2) in Fast Fourier Transformation (FFT). Therefore, Fourier spectra explained above differ from the FFT spectra in the value of amplitudes and its unit.

11.4.1.8 Applications to actual measurements under laboratory conditions

Several bar-type tendons (1200 mm in length, 36 mm in diameter) and cable-type tendons (1300 mm in length and six wires with a diameter of 6 mm) with/without artificial corrosion under bonded and unbonded conditions have been prepared (Figs. 11.44 and 11.45), and the responses of the tendons under single impact waves induced by an impact hammer or special

Figure 11.44 Samples with bar-type and cable-type tendons embedded in rock and a typical experimental setup.

Schmidt-hammer-like device (ponchi) were measured. Three different sensors were used, and two of them had a center hole for inducing impact waves on tendons (Fig. 11.46). The waves can be recorded as displacement, velocity, or acceleration, and the device for monitoring and recording consists of an amplifier and a small hand-held type computer (Fig. 11.47). A single person can perform the measurements, and the Fourier spectra of recorded data can be stored in the small hand-held computer; they can also be visualized in the measurement spot. In the following, results obtained would only be given without reference to the sensor or hammer unless it is mentioned. Most of the results and details of experiments described in this section can be found in an invited lecture by Aydan (2012) presented at the EITAC Annual Convention and Aydan (2017).

Figure 11.48 shows the wave responses of a 1200-mm steel bar induced by an impact hammer and its numerical simulation using the numerical method described previously. The wave velocity inferred from acceleration records directly was about 5520 m/s. The Fourier spectra of wave records induced by the impact hammer and a special Schmidt-hammer-like device (ponchi) are shown in Figure 11.49. The frequency content interval is about 2300 and the inferred velocity of the steel bar was 5520 m/s. These results are consistent with each other.

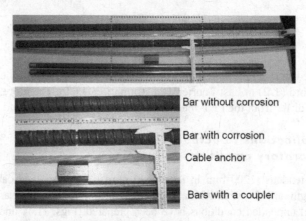

Bar without corrosion

Bar with corrosion

Cable anchor

Bars with a coupler

Figure 11.45 Bar-type and cable-type tendons with/without artificial corrosion used in experiments.

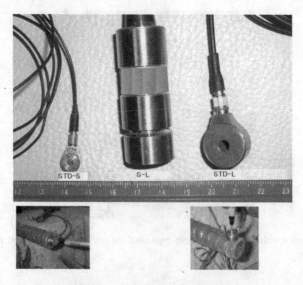

Figure 11.46 Views of sensors.

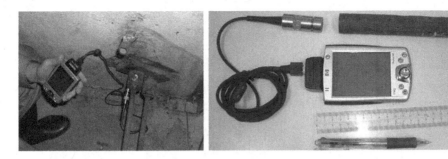

Figure 11.47 Views of PC-pocket-type sampling and recording device.

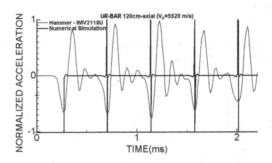

Figure 11.48 Impact hammer-induced wave response and its numerical simulation for a 1200-m-long steel bar.

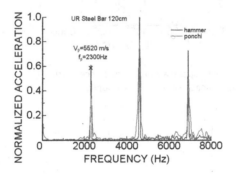

Figure 11.49 Normalized Fourier spectra of recorded acceleration records induced by impact hammer and a special Schmidt-hammer-like device.

Figures 11.50 and 11.51 show the effect of a coupler on the steel bar with a length of 1000 mm and the acceleration response of single 1000-mm bar also shown in the figures. As noted from the figure, two reflections occur in the 2-m-long coupled bar. The main reflection is from the other end of the coupled bar and the secondary reflection is due to the coupler.

Figure 11.52 shows the Fourier spectra of a 1200-mm bar with and without artificially induced area reduction to simulate corrosion. As expected from the numerical analysis, high-frequency content would be generated by the partially reflected waves from the artificial corrosion zone. This feature is clearly observed in the computed Fourier spectra. Figure 11.53

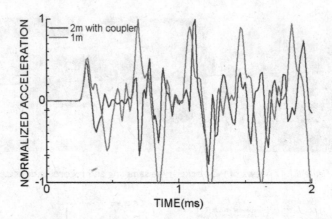

Figure 11.50 Impact hammer-induced wave responses of one and two 1-m-long two bars connected to each other with a coupler.

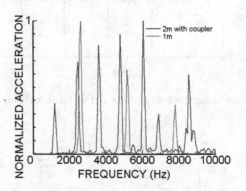

Figure 11.51 Normalized Fourier spectra of recorded acceleration records of one and two 1-m-long bars connected to each other with a coupler.

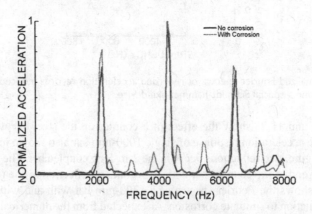

Figure 11.52 Normalized Fourier spectra of recorded acceleration records for bars with or without corrosion.

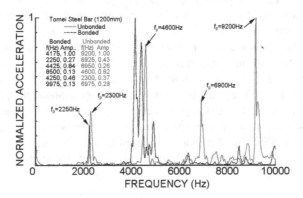

Figure 11.53 Normalized Fourier spectra of recorded acceleration records of bonded and unbonded bars.

Figure 11.54 Views of some *in situ* nondestructive tests examples.

shows the effect of bonding of the bar to the surrounding rock. Although the frequency of the bonded bar is slightly smaller than that of the unbonded rockanchor, the Fourier spectra for the first mode are quite close to each other. Nevertheless, the frequency content starts to change after the second or higher modes.

11.4.1.9 Some applications to rockbolts and rockanchors in in situ

The monitoring and recording system described in the previous subsection is utilized for nondestructive testing of rockbolts, bar-type rockanchors, and cable-type rockanchors installed in various rock masses (Fig. 11.54). Some of these measurements are shown in Figures 11.55 to 11.58 together with numerical simulations for the evaluation of *in situ* measurements. As noted from the figures, the existence of grouting causes the leakage of waves into the surrounding medium so that the amplitude of reflected waves is drastically reduced.

Figure 11.59 shows the traverse acceleration responses of a cable-type rockanchor with a 30 m length during a lift-off test. The numbers in the figure corresponds to axial forces (60, 164, 268, 372, 476, and 580 kN) applied to the rockanchor. The arrival times during the increase of axial loads are expected to be shorter. This feature is well observed when the axial load was over 268 kN. There are different causes for explaining some of discrepancies between measured and anticipated responses. The most difficult aspect associated with cable-type anchors is that the induction of impact wave and measurement location differs due to the inherent structure of the cable anchors.

One important aspect for maintaining rock engineering structures is to measure the initial parameters of the structure in question at the time of construction. Figure 11.60 shows an impact experiment on a 4-m-long rockbolt at a tunnel excavated in Shimajiri mudstone on Okinawa Island (Japan) before and after installation. The measurement is done on the rock-bolt installed about one week after the construction. As noted from the figure, the amplitude of the acceleration wave of grouted rockbolts drastically decreases compared with that of the rockbolt before installation. In other words, the amplitude of acceleration of the non-grouted rockbolt gradually decreases while the impact energy is distributed into the surrounding

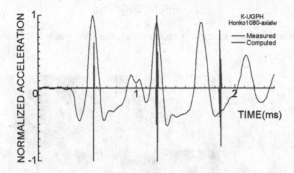

Figure 11.55 Measured and computed acceleration responses of a rockbolt at K-UGPH.

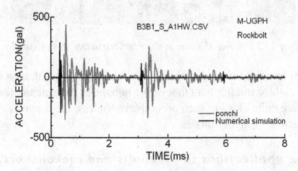

Figure 11.56 Measured and computed acceleration responses of a rockbolt at M-UGPH.

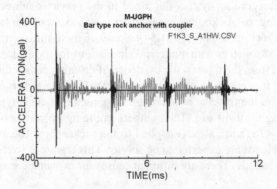

Figure 11.57 Measured and computed acceleration responses of a bar-type rockanchor with a coupler at M-UGPH.

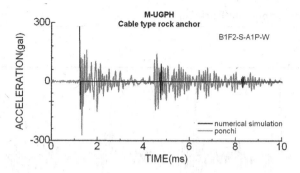

Figure 11.58 Measured and computed acceleration responses of a cable-type rockanchor at M-UGPH.

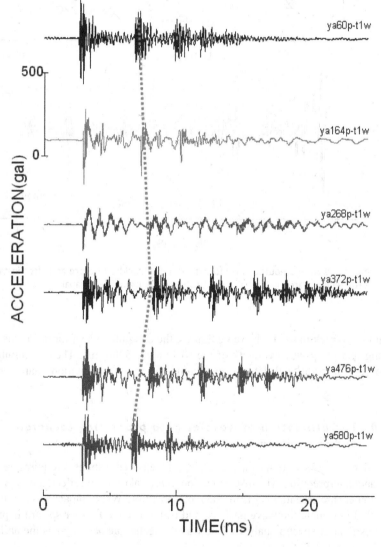

Figure 11.59 Measured traverse acceleration responses of a cable-type rockanchor during a lift-off test at N-HPP.

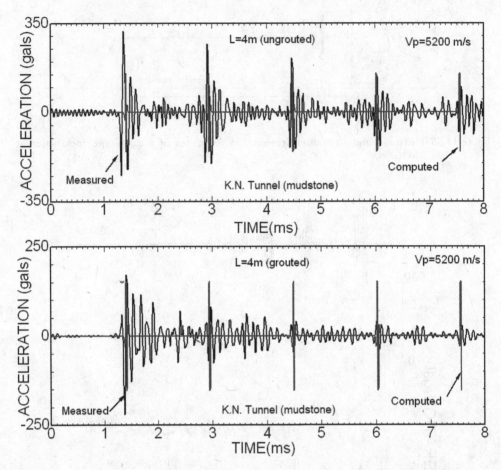

Figure 11.60 Measured and computed wave forms for a 4-m rockbolt before and after installation at a tunnel excavated in Shimajiri mudstone on Okinawa Island, Japan.

grout annulus and rock mass. In the same figures, the computed wave forms are also plotted by assuming that the p-wave velocity of the rockbolt is 5200 m/s. These computed wave forms also provide additional data for evaluating the wave propagation characteristics of rockbolts.

11.4.1.10 The utilization of wavelet data processing technique and some issues

As described in previous sections, the processing of measured responses requires tremendous efforts for data interpretation. The discrimination of arrival times of reflected waves and their frequency content are essential items for data interpretation. When impact waves are utilized, it is desirable to evaluate the wave packets, and using discrete Fourier spectra is preferable to continuous Fourier spectra analyses. The wavelet technique actually has the ability to do

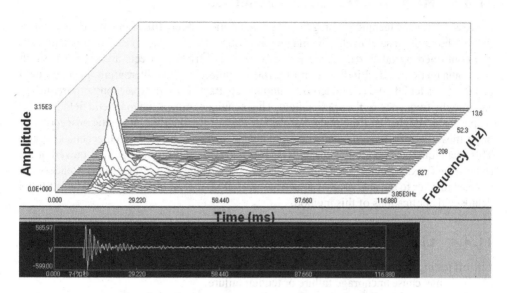

Figure 11.61 A wavelet processing of acceleration response.

such analyses, which can evaluate the data for each arrival time of wave packets reflected from the end as well as some structural weaknesses in the rockanchor systems. Figure 11.61 shows an example of using the wavelet technique for one of experiments reported in Subsection 11.3.1.6. However, it should be also noted that the wavelet data processing technique also results in a vast amount of processed data, which may sometimes be confusing during data interpretation.

As pointed out previously, the application of impact waves on rockbolts and rockanchors and the monitoring responses of reflected waves are carried out at the rockbolt/rockanchor head, which has a very limited physical space. Therefore, it is desirable to utilize a sensor with a central hole to induce impact waves and easy to mount to the head of rockbolt/rockanchors in real applications.

11.4.2   Guided ultrasonic wave method

The guided ultrasonic wave method is an NDT method using the vibration characteristics or propagation velocity characteristics of the ultrasonic wave varying with the stress of the pre-stressing tendon (i.e. Beard and Lowe, 2003; Chaki and Bourse, 2009; Buys *et al.*, 2009). A short-duration Gaussian windowed sine burst is used to excite a guided wave in the bolt from the free end. The wave is then reflected from the other end and from any major defects. From the reflection arrival time and knowledge of the wave velocity dispersion curves, the positions of the defects or the bolt length can be calculated. Although this method is widely adopted for the inspection of cracks or corrosion in pipelines, it presents limitations when applied on bonded pre-stressing tendons. The guided ultrasonic wave experiences loss when penetrating surrounding medium. Readers are advised to refer to the articles on the details and applications of this method mentioned above.

11.4.3   Magneto-elastic sensor method

A stress-measuring technique using the magnetostriction effect utilizes magnetic field distortion by the strain of steel itself. The magnetostriction effect was discovered by Joule in 1842 and confirmed by others (i.e. Villari, 1865). Magnetostriction effects are such that when a ferromagnetic material in the form of a rod is subjected to an alternating magnetic field parallel to its length, the rod undergoes alternate contractions and expansions at a frequency equal to the frequency of the applied field. The technique utilizes the magnetic field induced by an electromagnet and the magnetostriction effect, in which the magnetic susceptibility or permeability of a ferromagnetic material changes when subjected to a mechanical stress. This technique was applied to the nuts of rockanchors used in underground powerhouses to infer their axial forces (i.e. Laguerre *et al.*, 2002; Akutagawa *et al.*, 2008). Readers are advised to refer to the articles by Laguerre *et al.* (2002) and Akutagawa *et al.* (2008) for the details and applications of this method.

11.4.4   Lift-off testing technique

Lift-off testing is also used to check the state of rockanchors and the level of the axial forces. This test may cause anchorage failure or tendon failure.

11.5   CONCLUSIONS

The assessment of support systems such as rockbolts and rockanchors are important for maintaining existing rock engineering structures, particularly for dynamic loads that involve blasting, vehicle traffic, and vibrations induced by power generation. As time goes by, support systems may deteriorate as a result of corrosion, micro cracking, and motions along rock discontinuities in rock masses.

In the first part of this chapter, several experimental setups are used to clarify the response of rockbolts and rockanchors through model tests subjected to dynamic loads through shaking tables. The experiments described in this chapter clearly demonstrated that the dynamic loads may cause additional stretching on support systems utilizing rockbolts and rockanchors, due to the permanent deformation along rock discontinuities as a result of active sliding and active toppling type of motions. A theoretical model for the model tests was presented and the computed results are compared with the measured responses. The computed results are quite similar to the measured results both qualitatively and quantitatively, despite some discrepancies. However, these discrepancies could be resulting from the complexity of actual frictional behavior of discontinuities compared with the assumed frictional behavior in computations. In addition, the fully grouted rockbolts were tested in the model tests under dynamic loads. The experiments clearly showed that the strain of rockbolts crossing discontinuities could be very high and that the dynamic loads may cause additional stretching of rockbolts and rockanchors due to the permanent deformation along rock discontinuities, as a result of active sliding and active toppling type of motions. These results are in accordance with those from the shaking table tests of mechanically anchored rockbolts and rockanchors.

Nondestructive tests utilizing dynamic responses of support members are preferable to destructive tests. These tests should generally provide information on the constructional conditions, corrosion, and axial force conditions. In the first part of this study, some theoretical and finite element models for the axial and traverse dynamic tests of rockbolt and rock-anchors are presented. Then, results of numerical tests are explained by considering possible conditions *in situ*, such as the bonding quality, the existence of corroded parts, and pre-stress. The numerical experiments indicate that it is possible to evaluate the length of the rockbolts and rockanchors, the value of the applied pre-stress, and the location of corrosion. A nondestructive system was developed using impacts waves, and various typical situations were experimentally investigated. The nondestructive testing technique based on impact waves is promising. Nevertheless, there is still some room for further advancement by upgrading sensors and data-acquisition systems.

In addition to nondestructive testing, lift-off tests provide direct information on the soundness of rockanchors and their axial load. Nevertheless, this test is expensive and labor-intensive and replacement of rockanchors is necessary if they rupture. Lift-off tests on rockanchors at an underground power station constructed about 40 years ago were carried out to check the soundness of rockanchors and the value of pre-stress applied at the time of construction. The responses of the bar-type anchors are quite close to their original state and their deterioration is almost none, despite 40 years elapsing since their installation. On the other hand, the responses of cable-type rockanchors are remarkably different from their original state and permanent displacements occurred after each cycle of loading. These results also imply that the cable-type rockanchors consisting of wires are quite vulnerable to deterioration with time and they may not be qualified as permanent supporting systems.

Chapter 12

Conclusions

This book provided a fundamental, complete treatise on rock reinforcement and rock support under static, dynamic, short-term, and long-term conditions and evaluates the reinforcement effects of rockbolts and rockanchor of rock engineering structures under various rock mass conditions, both qualitatively and quantitatively. The mechanism and influencing factors of failure phenomena in rock engineering structures are the physical bases for why rock reinforcement and rock support are needed. Experimental studies on the mechanical behavior of the rockbolt and rockanchor system, as well as shotcrete, concrete linings, and steel ribs, provide actual data for developing appropriate constitutive laws for each member of a rock reinforcement and rock support system. Various analytical and numerical models provided tools for engineers to qualitatively and quantitatively evaluate the support and reinforcement effects of members of rock reinforcement and rock support systems for various rock engineering structures. Furthermore, it is vitally important to understand the behavior of members of rock support and rock reinforcement under dynamic loading conditions resulting from earthquakes, blasting, or rockburst. It is envisaged that the present book brings a clear understanding on rock support and the mechanics and reinforcement effects of rockbolts/ rockanchors, which has so far been poorly understood.

Chapter 2 is devoted to the mechanism and influencing factors of failure phenomena in rock engineering structures. It should be noted that the design of reinforcement and support members and the evaluation of the stability of structures are not possible unless one understands what rock mass really is. Most of the available approaches are either mechanically oriented without proper consideration of rock mass or geologically orientated without paying proper attention to the mechanics. In this respect, the present volume attempts to bridge the two approaches and bring a unified approach for the design of support and reinforcement systems for rock engineering structures, from not only the mechanical engineering but also the geological engineering point of view. The first part of this chapter is concerned with the geological description of rocks and of the formation and types of discontinuities in rocks and rock mass. Then the mechanical behavior of rock mass is discussed, considering the behaviors of intact rock, discontinuities, and the structure of the rock mass. In the second part of the chapter, the discussion of various modes of instability of rock engineering structures such as underground openings, slopes and foundations under both compressive and tensile stress fields, and the factors associated with the modes of instability are presented in relation with the structure of rock mass.

Chapter 3 is concerned with the present design philosophy of support and reinforcement for rock engineering structures. A brief description of available design approaches, such as empirical, analytical, and numerical methods, is given and discussed. Rock mass classification systems are extensively used for feasibility and pre-design studies, and often also for the final design. Nevertheless, it is pointed out that the considerations of global and local instability modes are essential for the design of rock support and rock reinforcement with a proper understanding of rock mass conditions. Several modern rock classification systems have been briefly explained, and quantitative assessments have been done based on RMQR. As it is required for an engineer to select the most appropriate method for determining design parameters in rock engineering, a brief overview of the kinds of rock loads and the procedures to determine their magnitude, empirical, analytical, and numerical techniques has been provided and discussed, with the objective of unifying the present methods of design. The presently available support members and their functions are briefly described and discussed, with an emphasis on rockbolts and rockbolting. Although various reinforcement effects of rockbolts are generally classified as internal pressure effect, improving the physical properties of ground, suspension effect, shear reinforcement, beam building effect, and arch formation effect, it is shown that they are, in a real sense, associated with their axial and shear responses and the interaction taking place between themselves and the surrounding medium under given loading and boundary conditions.

Chapter 4 described experimental studies undertaken on the mechanical behavior of the rockbolt system together with a reference to rockanchors, which are a special form of rockbolts. First, the mechanical behavior of the bolt material used in practice is given, then the experimental study undertaken for the anchorage performance of rockbolts in push-out and pull-out tests and subsequent shear tests on the mechanical behavior of interfaces within the system and grouting material are described. These experiments provide most appropriate procedures for mechanical characteristics of rock reinforcement elements.

A constitutive law for the bar is derived based on the classical incremental elasto-plasticity theory, as bar materials (such as steel) exhibit a non-dilatant plastic behavior. On the other hand, the constitutive law for the grout annulus and interfaces is derived based on the multi-response theory proposed by Ichikawa (1985) and Ichikawa *et al.* (1988), as the grout annulus and interfaces exhibit a dilatant plastic behavior. Then procedures to determine the parameters for the constitutive laws from the experimental data are described and several examples of applications to experiments are given.

Evaluation of the contribution of rockbolts/rockanchors for improving the properties of rock mass is described and the shear reinforcement effect of rockbolts on rock discontinuities is presented in view of some theoretical and experimental findings. These evaluations provide quantitative as well as qualitative data for the effect of rockbolts on rock mass and rock discontinuities.

A detailed presentation of estimation of pull-out capacity of rockbolts/rockanchors under various conditions are described. The evaluation of shear strength of interfaces in the boreholes of rockbolts/rockanchors, which are of great importance on the overall pull-out capacity, is described and pull-out experiments are simulated and compared with experimental results. Furthermore, a procedure for evaluating the effects of cyclic loading on the shear strength of interfaces is presented. The presented procedures can evaluate properly the pull-out capacity of rockbolts.

There is an increasing demand for underground openings for energy storage projects (i.e. CAES and SMES projects) and they are subjected to high internal pressures resulting in tangential tensile stresses around the cavities. The evaluation of reinforcement effect of mesh bolting on rock masses subjected to tensile stresses is presented and several practical examples are described. The procedures provide theoretical background for evaluating the tensile strengthening of rock mass by mesh-bolting.

Chapter 5 described the characteristics of various support elements, such as shotcrete, concrete liner, and steel ribs/sets. Support members fundamentally provide resistance against the movement of surrounding ground and/or intrusion of fluid into the open space, while reinforcement members such as rockbolts contribute to the mechanical properties of surrounding rock mass as well as restraining of ground movement into the open space. In some cases, support members may obstruct fluid flow into the surrounding medium. In other words, the support members are members separated from the surrounding ground, while the reinforcement members are an integral part of the rock mass on a macroscopic scale.

The support members are conventionally shotcrete (with/without fibers), concrete liners, steel ribs/sets, and steel liners in underground excavations. Regarding surface structures, shotcrete, steel, or reinforced concrete piles and retaining walls may be visualized as support members. Fundamentally, support members convey the loads to the stable ground if they are unclosed or internally resist the forces directly if they are closed. Experiments on the mechanical characteristics of shotcrete and concrete liners are presented and their constitutive modeling are described. The experiments on shotcrete and concrete are the most updated and extensive. Particularly, experimental studies on shotcrete cast in laboratory and in-situ covers uniaxial and triaxial compression tests and practical procedures for evaluating early-time properties of shotcrete are presented. Similar experiments for concrete liners are described. The appropriate constitutive models for shotcrete, concrete liners, and steel sets/liners are explained and the concepts for their mechanical modeling are also explained.

Chapter 6 described the models representing reinforcement and support systems in numerical analyses, particularly in finite element studies. Among all numerical analysis techniques, the finite element method is much superior to other numerical methods when the flexibility of mesh generation and size, ease of dealing with material nonlinearity, and formulation are taken into account. Some special elements such as rockbolt elements, shotcrete, and beam elements particularly for finite element analyses are explained.

First Aydan's rockbolt element, fully account for the mechanism involved in the rockbolt system (steel bar in axial and shear loading, interfaces and grout annulus in shear and normal loading), is described in detail. The mechanical modeling of the rockbolt system is presented, and the governing equations for the bar and grout annulus are derived in consideration of responses to the finite element model for the rockbolt system. In the two-dimensional case, the node number of this rockbolt element can be reduced to six. In some extreme cases, the node number may be reduced to four nodes. If the interactions between rockbolts and surrounding ground are neglected, the node number can be reduced to two, which would correspond to the bar element.

Shotcrete in numerical simulations can be modeled in various ways. As the thickness of shotcrete is so small compared to the size of structures, it can be modeled as shell elements under plane-stress condition. A two-noded element for shotcrete linings and named as shotcrete element is presented for two-dimensional analyses. The shotcrete element is similar to plate elements in three-dimensional and beam-like elements in two-dimensional. Steel ribs/sets are represented in finite element analysis using beam-elements with the consideration of Pasternak and Winkler type supports by replacing with the appropriate stiffness properties.

The essentials of nonlinear finite element analyses of rock engineering structures are briefly described. Furthermore, the discrete finite element method (DFEM) for assessing the response and stability of rock block systems, based on the finite element method together with the utilization of updated Lagrangian scheme, is briefly explained. This method is extended to incorporate the effect of rock reinforcement and rock support through the elements used for rock reinforcement and support systems in this chapter The original DFEM code developed by the author is extended to include the support and reinforcement members and re-named as DFEM-BOLT.

Chapter 7 was concerned with the analytical and numerical methods for evaluating support and reinforcement systems and their effects in underground excavations. Analytical solutions developed for a circular tunnel or spherical opening under hydrostatic stress condition and for tunnels subjected to non-hydrostatic stress state for evaluating the ground-response-support reaction, which incorporates various reinforcement and support members such as rockbolts, rockanchors, steel ribs/sets, shotcrete, concrete liners, and the face effect. Several examples of practical applications are given and compared with in-situ measurements on rockbolts installed in tunnels in squeezing rocks. In addition, the effect of long-term characteristics of rock mass is considered in analytical solutions. Furthermore, some slight modifications to the developed solutions are described to consider the effect of body forces in residual-plastic zones.

A semi-analytical method developed by Gerçek to evaluate the stress distribution around underground openings with various shapes under non-hydrostatic stress state based on complex variable method is used to obtain potential yield zones due to slippage of discontinuities in surrounding rock mass as well as of rock mass. This method is utilized to determine the length and number of rockbolts/rockanchors to prevent local instabilities. The method is expected to be a guideline for engineers dealing with discontinuous rock masses

A series of finite element simulations are presented to show the effects of various conditions for the effective utilization of reinforcement and support systems for underground structures. The effect of rockbolting with other support members is investigated in relation to some practical situations with the consideration of bolt spacing, the magnitude of the allowed displacement before the installation of the bolts, elastic modulus of the surrounding rock, and equipping rockbolts with bearing plates and bolting pattern. It is also shown that the concept of neutral point proposed to explain the reinforcement effect of grouted rockbolts is invalid and the so-called neutral point has nothing to do with the resistance offered by rockbolts. It just indicates the location where the highest resistance of bolts is mobilized. Furthermore, the response of reinforcement and support systems in rock mass with time-dependent characteristics is presented with a reference to an actual tunnel in Japan.

Analytical solutions are presented to evaluate the stress state around the internally pressurized circular cavities reinforced by rockbolts having a non-radial pattern, which is different from the conventional rockbolting pattern. Solutions are used for studying the effect of rockbolting, and implications of the parametric studies are discussed. Furthermore, a series of finite element studies was carried out and results are compared with analytical solutions.

Several examples are analyzed on the response of rockbolts in discontinuum, and their implications for interpreting field measurements of rockbolt performances are discussed. Furthermore, the presently available proposals on the suspension effect, the beam building effect, and the arch formation effect of rockbolts are re-examined and more generalized solutions are presented. In addition to covering the reinforcement effect of rockbolts against the sliding type of failure, solutions for the reinforcement effect of bolts against the flexural and columnar type of toppling failure are given. These applications provide engineers with what to expect from rockbolts and rockanchors in the reinforcement of discontinuum.

The essentials of concrete liner design for sub-sea tunnels and shafts subjected to high ground water pressure are explained and several practical examples are analyzed and discussed. Furthermore, the pressure development in concrete liner of a shaft excavated in frozen ground during thawing process is presented. It is expected that the practical examples would provide some essential guidelines to engineers.

In the final part, a special form of rock support, which is backfilling of abandoned room and pillar mines, is described. This is quite an important issue when abandoned mines or karstic caves exist beneath urbanized areas as well as transportation facilities. The outcomes of a series of experimental studies on the supporting effect of backfilling on the response and stability of abandoned mines, quarries, and karstic caves are presented and how to verify it with *in situ* monitoring is explained though some specific examples. In addition, several numerical analyses using DFEM are explained to evaluate the supporting effects of backfilling cavities. The procedures are expected to provide the ways to evaluate the supporting effects of backfilling.

Chapter 8 described the effect of support and reinforcement systems for the stabilization of rock slopes. Fundamentally, rock support for rock slopes may only be in the form of retaining walls, which have only a tiny effect on ordinary rock slopes. Although shotcrete with/without wire mesh or fibers is often used, the structural effect of shotcrete is fundamentally almost nil. However, shotcrete does prevent the deterioration of rock, the loosing of rock mass due to movements of small blocks, and internal erosion due to atmospheric agents and seepage in the long term, so its effect can be tremendous.

When piles of various types and rockbolts and rockanchors pass through the potential failure plane, they act as reinforcement. The emphasis in this chapter is given to reinforcement members, such as rockbolts and rockanchors. Most formulations are basically similar to those presented for underground structures. Nevertheless, reinforcement and support effects are described with the consideration of common modes of failure of rock slopes such as planar/wedge sliding, flexural or block toppling failure, and buckling failure. The methods are fundamentally based on limit equilibrium methods (LEM), and this chapter provides examples of analyses utilizing LEM, the finite element method, and the discrete finite element method (DFEM) incorporating Aydan's rockbolt element.

Some model experiments of jointed rock slopes with different reinforcement patterns are also presented to illustrate the effect of rockbolting against planar sliding and block-toppling modes. Furthermore, the experimental methods are compared with estimations from the limit equilibrium technique. This chapter provides engineers the most fundamental concepts and strategies to deal with the likely instability problems and to design appropriate reinforcement systems.

Chapter 9 was concerned with the stabilization of the foundations of bridges, pylons, and dams subjected to tension or compressive forces. Foundations of pylons and tunnel-type anchorage for suspension bridges are examples of foundations subjected to tension while sockets for stabilizing bridges and dam foundations would be under compression. Estimating the uplift (pull-out) capacity of anchors and anchorages embedded in rock is important in geotechnical engineering applications, such as the design of anchorages for suspension bridges, towers of any kind, gravity dams, and tsunami-protection walls for important civil engineering structures. Various approaches and their fundamental concepts to the design of anchor rocks with the consideration of anchor geometry and loading conditions are presented. Examples of applications include the potential use of rockanchors as foundations of pylons and of tunnel-type anchorage for suspension bridges. A special emphasis is given to the possible failure modes, which are quite important for the selection of appropriate limiting equilibrium method.

A unified design procedure is presented for the utilization of rockanchors as foundations of pylons with the use of short-term and long-term laboratory experiments, in-situ explorations, analytical solutions for pull-out capacity and the consideration of interface shear strength in boreholes, in which rock anchors are embedded. The design method for rockanchors foundations of pylons is applied to *in situ* pull-out tests at two locations, which are typical rock masses found in the Chubu district of Japan. The rock masses were jointed sandstone and granite. Further design nomograms are prepared together with comparisons of *in-situ* experiments. In addition, numerical analyses are carried out to investigate the stress and deformation states in the rockanchor systems under pull-out forces.

Suspension bridges, towers, and so on generally require large-scale anchorages. The design of anchorages varies depending on the structure type, as well as practices in each country. The details of four fundamental approaches for the design of anchorages (such as standards and empirical approaches, limiting equilibrium approaches, closed-form solutions, and numerical methods) are presented and their applications in practice are described. One of specific applications was a site where three *in situ* pull-out tests were done on cylindrical steel anchors. The second application was the examination of the design and to predict the post-construction performance of the tunnel-type anchorage of a 1570-m-long suspension bridge. The analyses are carried out to investigate the stability of the concrete anchorage body and surrounding rock on the suspension load of the bridge applied on the anchorage by considering two loading intensity conditions using the FEM.

When the bearing capacity of foundations under compression is likely to be influenced by structural weaknesses, such as bedding planes or major fracture zones, special counter measures are necessary to increase the shear resistance of discontinuities by active as well as passive rock anchors. In addition, steel-concrete piles, which may be viewed as composite cylindrical anchorages, may also be used. The analyses of the dam foundations are done through some limiting equilibrium approaches for simple conditions and the use of numerical analyses with the consideration of discontinuities in rock mass for complex conditions. Several examples of applications of limiting equilibrium method and numerical analyses are presented and discussed by considering various possible situations.

In the final part, rock socket with end-bearing was considered and closed form solutions were developed for evaluating the axial displacement, axial stress and shear stress along the interface between the socket and surrounding rock mass. This method was applied to an actual construction site in Okinawa Island. The results of computations showed that the actual amount of the end-bearing of rock sockets may be quite small despite the original design concept in view of their geometry and deformability characteristics of surrounding rock mass.

Chapter 10 dealt with dynamic issues such as rockburst, earthquakes, and blasting, which cause dynamic loads on rock support and rock reinforcement. As mentioned in previous chapters, rockbolts and rockanchors are commonly used as principal support members in underground and surface excavations. Some experiments on dynamic response of point-anchored rockbolt model under impulsive load were described in order to provide some physical evidences. The axial strain of the model rockanchor fluctuated and became asymptotic to the static strain level for the applied stress level. The experiments clearly showed that the loading of samples and structures, as well as the excavation of rock engineering structures, should be treated as a dynamic phenomenon. Then several numerical experiments on actual scale rockanchors were carried out using a dynamic finite element method. The numerical experiments also clearly illustrated that the rockbolts may experience loads under dynamic condition several times of those under static condition.

The yielding rockbolts are devised to investigate shock absorption resulting rockbursts. These rockbolts fundamentally utilize the displacement hardening frictional resistance while the tendon itself behaves elastically or elasto-strain hardening plastically under shock loads. The displacement capacity of yielding rockbolts of is at least 20 times that of typical grouted rebar. A formulation of the problem is given and its practical application was analysed and discussed.

The responses of rockbolts subjected to vibrations induced by turbines in an underground powerhouse were measured at several locations in relation to the assessment of the behavior of rockbolts and rockanchors during operation. The measurements were carried out at penstock, draft shaft, turbine top, and a rockbolt above the penstock and the maximum accelerations ranged between 120 gals and 250 gals. Furthermore, the vibrations were higher when the water flow was stopped or started in relation to the operation of the powerhouse turbines of this pumped-storage scheme.

Some model experiments were carried out on the development of axial forces in rockanchors and grouted rockbolts, stabilizing the potentially unstable blocks in sidewalls of the underground openings using shaking. The model rockbolts tested were rockanchors and fully grouted rockbolts. Experiments showed that the anchor force increased after each slip event in a step-like fashion and becomes asymptotic to a certain value thereafter. Furthermore, the block does not return to its original position. This implies that the axial force becomes higher than the applied pre-stress level following each slip event. Finally, the bolt force becomes sufficient to stop the sliding of the block, if it does not fail. The experiments clearly demonstrate that the bolts and anchors used as a part of the support system may experience greater load levels than that at the time of their installation after each passage of dynamic loads. This outcome has important implications in that reinforcement and support systems may fail during their service lives not simply due to their deterioration but also to dynamic loads resulting from different causes. The experimental situation is theoretically formulated and its applications to model experiments as well as the practical situations in large scale underground rock structures were done. Furthermore, some model experiments on fully grouted rockbolts passing through a discontinuity plane in a rock-pillar subjected to shaking loads were carried out. These experiments also showed that the axial strains of rockbolts were quite large at the close vicinity of discontinuity plane and some residual straining occurred.

An actual excavation of a circular tunnel supported rockbolts was considered as a dynamic problem under hydrostatic *in situ* stress condition. The comparisons of stresses induced for unbolted and bolted cases indicated that the radial and tangential stresses are less than those of the unsupported case, which indicates the reinforcement effect of rockbolts through the internal pressure as well as shear interaction between rockbolts and surrounding ground. The overall response of axial stress in rockbolts is quite similar to the changes of the stresses in rock mass. One important observation is that the axial stress in rockbolts would fluctuate and the rockbolts would experience larger axial stresses than those under static condition.

Finally blasting induced vibrations on the support and reinforcement system consisting of rockbolts and steel ribs were measured by using portable accelerometers QV3-OAM-SYC. The comparison of acceleration responses indicated that the vibration levels induced by blasting in steel ribs are much higher than those in rockbolts.

Chapter 11 described the mechanisms and techniques for evaluating corrosion in steel and iron materials in relation to the long-term performance and degradation of reinforcement and support systems and provides site examples. First the background of iron corrosion due to chemical or electrochemical action of water in the presence of oxygen and hydrogen was explained. Then factors affecting corrosion rate (such as environment, pH concentration of dissolved oxygen, and temperature) were presented. Some laboratory experiments and *in-situ* observations on the corrosion rate of iron and steel are described, and results of measurements and observations were tabulated. In addition, the possibility of evaluating corrosion state using non-destructive testing (such as electrical resistivity, p-wave velocity, and X-ray CT scanning technique) was discussed through some actual measurements. These techniques clearly showed some potential to determine the state of corrosion and its rate. Furthermore, a finite element modeling of corrosion problem is presented and applied to some practical situations. Some methods were also proposed to evaluate the failure time of rockanchors subjected to prestresses.

Observations and measurements in large-scale underground openings indicated that cable-type rockbolts or rockanchors are much more prone to corrosion in view of *in situ* pull-out experiments. Theoretical estimations for the service life of different types of rockbolts and rockanchors under pre-stress based on the experimentally measured corrosion rate confirm this experimental finding. A series of numerical analyses were carried out to investigate the effect of degradation of reinforcement and support system on the response and stability of a large scale underground cavern. As expected, the degradation of the support system increases the displacement field around the cavern. Continuum type analyses are valid for global stability of the caverns. However, rock mass always has discontinuities in different forms and these pre-existing discontinuities may constitute unstable rock blocks around the cavern. In such cases, the stabilization of such blocks by a support system may become the prime concern and emphasis to such conditions must be given.

The maintenance of existing rock engineering structures (such as underground power-houses, dams, and slopes) requires the reassessment of reinforcement and support members. Although *in situ* pull-out tests and lift-off tests would be the best options for assessing the soundness and acting stress of the rockbolts and rockanchors, performing such tests is costly and labor intensive. Furthermore, such tests may require the reinstallation of rockbolts and rockanchors if they are pulled out or broken during the tests. Therefore, the use of nondestructive tests to assess their soundness and/or acting stresses is desirable. Some theoretical models are developed for the axial and traverse dynamic nondestructive tests, and numerical simulations are carried out based on these fundamental equations by considering possible conditions *in situ*. Specifically, the bonding quality, the existence of corroded parts or couplers, and pre-stress are simulated through numerical experiments. The results of the numerical simulations can be used to interpret the dynamic responses to be measured in nondestructive dynamic tests *in situ*. Furthermore, a theoretical method is described for simulating lift-off experiments, which are classified as destructive testing. Several applications of this procedure to practice are given together with measurements obtained from some nondestructive tests under laboratory and *in situ* conditions.

Bibliography

Adali, S. & Rösel, R. (1980) Preliminary assessment of effect of grouted rockbolts in mine tunnels. *Trans. Inst. Min. & Metall., Sec.* pp. A190–198.

Akutagawa, S., Arimura, Y., Nakamura, E., Sakurai, S., Baba, S. & Mori, S. (2008) A new evaluation method of PS-anchor force by using magnetic sensor and its application in a large underground powerhouse cavern (in Japanese). *JSCE, Division F*, 64(4), 413–430.

Amberg, R. (1983) Design and construction of the Furka base tunnel. *Rock Mechanics and Rock Engineering*, 16, 215–231.

Arnold, A.B., Bisio, R.P., Haynes, D.J. & Wilson, A.O. (1972) Case history of three tunnel support failure, California Aqueduct. *Bulletin of Assoc. Eng. Geologist*, 9(3), 265–300.

Aydan, Ö. (1982) *Shaft Lining Design and Performance in North Selby Mine*. M.Sc. Thesis, Newcastle University, Newcastle upon Tyne, UK.

Aydan, Ö. (1985) Theoretical and observational approaches to the interaction among rock mass and rockbolts. Int. Rep., Dept. of Geotech. Engng. Nagoya University, July.

Aydan, Ö. (1989) *The Stabilisation of Rock Engineering Structures by Rockbolts*. Doctorate Thesis, Nagoya University, Nagoya, Japan.

Aydan, Ö. (1993) Reinforcement of underground openings by rock anchors (in Turkish). *Kaya Mekaniği Bülteni*, 9, 5–18.

Aydan, Ö. (1994) Chapter 7: Rock reinforcement and support. In: Vutukuri, V. & Katsuyama, K. (eds.) *Introduction to Rock Mechanics*. Industrial Pub. & Consul. Inc., Tokyo. pp. 193–247.

Aydan, Ö. (1994) The dynamic shear response of an infinitely long visco-elastic layer under gravitational loading. *Soil Dynamics and Earthquake Engineering*, Elsevier, 13, 181–186.

Aydan, Ö. (1995) Mechanical and numerical modeling of lateral spreading of liquefied soil. *The 1st Int. Conf. on Earthquake Geotechnical Engineering*, IS-TOKYO'95, Tokyo. pp. 881–886.

Aydan, Ö. (1997) Dynamic uniaxial response of rock specimens with rate-dependent characteristics. *SARES'97*. pp. 322–331.

Aydan, Ö. (1998) A simplified finite element approach for modeling the lateral spreading of liquefied ground. *The 2nd Japan-Turkey Workshop on Earthquake Engineering*, Istanbul.

Aydan, Ö. (2001) Comparison of suitability of submerged tunnel and shield tunnel for subsea passage of Bosphorus. *Geological Engineering Journal*, 25(1), 1–17.

Aydan, Ö. (2003) The moisture migration characteristics of clay-bearing geo-materials and the variations of their physical and mechanical properties with water content. *2nd Asian Conference on Saturated Soils*, UNSAT-ASIA 2003. pp. 383–388.

Aydan, Ö. (2004) Damage to abandoned lignite mines induced by 2003 Miyagi-Hokubu earthquakes and some considerations on its possible causes. *Journal of School of Marine Science and Technology*, 2(1), 1–17.

Aydan, Ö. (2007) The response of tunnels in liquefiable ground during earthquakes with special emphasis on Bosphorus Immersed tunnel. *Proceedings of 2nd Symposium on Underground Excavations for Transportation*. pp. 273–282.

Aydan, Ö. (2008) Investigation of the seismic damage to the cave of Gunung Sitoli (Tögi-Ndrawa) by the 2005 Great Nias earthquake. *Yerbilimleri*, 29(1), 1–16.

Aydan, Ö. (2008) New directions of rock mechanics and rock engineering: Geomechanics and geoengineering. *5th Asian Rock Mechanics Symposium*, ARMS5, Tehran. pp. 3–21.

Aydan, Ö. (2011) Some issues in tunneling through rock mass and their possible solutions. *Proc. First Asian Tunneling Conference*, ATS-15. pp. 33–44.

Aydan, Ö. (2012) The utilization of non-desctructive testing method for the evaluation of soundness of rockbolts and rock anchors used in rock engineering structures. *Elastic Wave Inspection Technology Association (EITAC)*, Invited Lecture, 110 pages.

Aydan, Ö. (2015) Some considerations on a large landslide at the left bank of the Aratozawa Dam caused by the 2008 Iwate-Miyagi intraplate earthquake. *Rock Mechanics and Rock Engineering*, Special Issue on Deep-Seated Landslides, 49(6), 2525–2539.

Aydan, Ö. (2016) *Time Dependency in Rock Mechanics and Rock Engineering*. CRC Press, Leiden, Leiden, ISRM Book Series, No. 2, 246 pages.

Aydan, Ö. (2016a) The state of art on large cavern design for underground powerhouses and long-term issues. In: *The Second Volume of Encyclopedia on Renewable Energy*. John Wiley & Sons, Hoboken, NJ. pp. 467–487.

Aydan, Ö. (2016b) Connecting Okinawa Prefecture to mainland. *Okinawa Construction Newspaper* (Okinawa Kensetsu Shinbun), No. 2823, Nov. 9, 2016. (In Japanese)

Aydan, Ö. (2016c) An integrated approach for the evaluation of measurements and inferences of *in situ* stresses. *RS2016 Symposium, 7th Int. Symp. on In-Situ Rock Stress*, Tampere. pp. 38–57.

Aydan, Ö. (2017) *Rock Dynamics*. CRC Press, Leiden, 463 pages.

Aydan, Ö., Akagi, T. & Ito, T. (1995b) Prediction of deformation behavior and design of support systems of tunnels in squeezing rocks. Development of Scientific Research of Monbusho, Research No. 05555139, 94 pages.

Aydan, Ö., Akagi, T., Ito, T., Ito, J. & Sato, J. (1995b) Prediction of deformation behavior of a tunnel in squeezing rock with time-dependent characteristics. *Numerical Models in Geomechanics*, NUMOG V, 463–469.

Aydan, Ö., Akagi, T. & Kawamoto, T. (1993) Squeezing potential of rocks around tunnels: Theory and prediction with examples taken from Japan. *Rock Mechanics and Rock Engineering*, 26(2), 137–163.

Aydan, Ö., Akagi, T. & Kawamoto, T. (1996) The squeezing potential of rock around tunnels: Theory and prediction with examples taken from Japan. *Rock Mechanics and Rock Engineering*, 29(3), 125–143.

Aydan, Ö., Akagi, T., Okuda, H. & Kawamoto, T. (1994) The cyclic shear behavior of interfaces of rock anchors and its effect on the long term behavior of rock anchors. *Int. Symp. on New Developments in Rock Mechanics and Rock Engineering*, Shenyang. pp. 15–22.

Aydan, Ö., Daido, M., Ito, T., Tano, H. & Kawamoto, T. (2006) Instability of abandoned lignite mines and the assessment of their stability in long term and during earthquakes. *4th Asian Rock Mechanics Symposium*, Singapore, Paper No. A0355 (on CD).

Aydan, Ö., Daido, M. & Ogura, Y. (2003) Dynamic characteristics and stability of abandoned mine sites. *Japan Backfilling Society, JUTEN*, (45), 11–17.

Aydan, Ö., Daido, M., Tano, H. & Ito, T. (2005c) The instability modes of abandoned lignite mines and the assessment of their long-term stability. *35th Japan Rock Mechanics Symposium*. pp. 57–62.

Aydan, Ö., Ebisu, S. & Kawamoto, T. (1995) The reinforcement of internally pressurized circular cavities by inclined mesh bolting. *The 8th Int. Congress on Rock Mechanics*, ISRM, Tokyo. pp. 517–520.

Aydan, Ö., Ebisu, S. & Komura, S. (1993) Pull-out tests of rock anchors and their failure modes. *Proc. Int. Symp. on Assessment and Prevention of Failure Phenomena in Rock Engineering*, Istanbul. pp. 285–294.

Aydan, Ö., Ebisu, S., Komura, S. & Kawamoto, T. (1995) A unified design method for rock anchor foundations of super-high pylons. *The Int. Workshop on Rock Foundation of Large-scaled Structures*, ISRM, Tokyo. pp. 279–284.

Aydan, Ö. & Ersen, A. (1983) Ground freezing method and its application (in Turkish). *Madencilik,* 22(2), 33–44.

Aydan, Ö. & Ersen, A. (1984) Stress, strain and temperature measurements in concrete liners of mine shafts and tunnels (in Turkish). *4th Coal Congress of Turkey,* Zonguldak. pp. 355–368.

Aydan, Ö. & ve Ersen, A. (1985) Determination of lining pressure on deep mine shafts (in Turkish). *Rock Mechanics Journal,* (3), 21–34.

Aydan, Ö. & Ersen, A. (1985) Estimation of pressures acting on the supports of mine shafts and road-ways (in Turkish). *Rock Mechanics Bulletin,* Turkish National Group for ISRM, 1(3), 21–34.

Aydan, Ö., Ersen, A., Ichikawa, Y. & Kawamoto, T. (1985a) A new analytic solution for stresses, strains and displacements in/along rockbolts (in Turkish). *Madencilik, Ankara,* 24(3), 27–36.

Aydan, Ö., Ersen, A., Ichikawa, Y. & Kawamoto, T. (1985b) Temperature and thermal stress distributions in mass concrete shaft and tunnel linings during the hydration of concrete (in Turkish). *Proc. of the 9th Mining Science and Technology Congress of Turkey,* Ankara. pp. 355–368.

Aydan, Ö. & Geniş, M. (2007) Assessment of dynamic stability of an abandoned room and pillar underground lignite mine. *Rock Mechanics Bulletin,* Turkish National Rock Mechanics Group, ISRM, (16), 23–44.

Aydan, Ö. & Geniş, M. (2010) Rockburst phenomena in underground openings and evaluation of its counter measures. *Journal of Rock Mechanics,* Turkish National Rock Mechanics Group, (Special Issue 17), 1–62.

Aydan, Ö. & Geniş, M. (2010) A unified analytical solution for stress and strain fields about radially symmetric openings in elasto-plastic rock with the consideration of support system and long-term properties of surrounding. *International Journal of Mining and Mineral Processing (IJMMP),* 1(1–32).

Aydan, Ö. & Geniş, M. (2014) A numerical study on the response and stability of abandoned lignite mines in relation to the excavation of a large underground opening below. In: Oka, F., Murakami, A., Uzuoka, R. & Kimoto, S. (eds.) *Proceedings of 14th Computer Methods and Recent Advances in Geomechanics.* 2015 Taylor & Francis Group, London. pp. 1031–1036.

Aydan, Ö., Ichikawa, Y., Ebisu, S., Komura, S. & Watanabe, A. (1990) Studies on interfaces and discontinuities and an incremental elasto-plastic constitutive law. *Int. Conf. Rock Joints,* ISRM. pp. 595–601.

Aydan, Ö., Ichikawa, Y. & Kawamoto, T. (1985) Load bearing capacity and stress distributions in/along rockbolts with inelastic behavior of interfaces. *The 5th Int. Conf. on Num. Meths. in Geomechanics,* Nagoya, 2. pp. 1281–1292.

Aydan, Ö., Ichikawa, Y. & Kawamoto, T. (1986) Reinforcement of geotechnical engineering structures by grouted rockbolts. *Proc. of Int. Symp. Engng. Complex Rock Formation,* Beijing. pp. 183–189.

Aydan, Ö., Ichikawa, Y. & Kawamoto, T. (1986c) A finite element for grouted rockbolts and their anchorage mechanism. *Proc. of the 1st Symp. on Num. Meths. in Geotechnical Engineering,* Tókyo. pp. 105–114.

Aydan, Ö., Ichikawa, Y. & Kawamoto, T. (1987c) Numerical analysis of reinforcement effects of rock-bolts. *Proc. of the 1st Symp. on Computational Methods,* Tokyo. pp. 235–242.

Aydan, Ö., Ichikawa, Y., Kawamoto, T. & Ersen, A. (1985a) Analytic solutions for axial stress, strain and deformation of rockbolts during pullout tests. *Madencilik,* 24(3), Sept., 27–36.

Aydan, Ö., Ichikawa, Y., Murata, K. & Shimizu, Y. (1991) An integrated systems for the stability of rock slopes. *Proc. Int. Conf. Computer Methods and Advances in Geomechanics,* Cairns, 1. pp. 469–474.

Aydan, Ö., Ito, T., Akagi, T. & Kawamoto, T. (1994) Theoretical and numerical modeling of swelling phenomenon of rocks in rock excavations. *Int Conf. on Computer Methods and Advances in Geomechanics,* IACMAG, Morgantown, 3. pp. 2215–2220.

Aydan, Ö. & Kawamoto, T. (1987) Toppling failure of discontinuous rock slopes and their stabilisation (in Japanese). *Journal of Japan Mining Society,* 103(1197), 763–770.

Aydan, Ö. & Kawamoto, T. (1991) A comparative numerical study on the reinforcement effects of rockbolts and shotcrete systems. *The 5th Int. Conf on Computer Methods and Advances in Geomechanics,* Cairns, 2. pp. 1443–1448.

Aydan, Ö. & Kawamoto, T. (1992) The stability of slopes and underground openings against flexural toppling and their stabilisation. *Rock Mechanics and Rock Engineering*, 25(3), 143–165.

Aydan, O. & Kawamoto, T. (1999) A proposal for the design of support system of large underground openings according to RMR rock classification system. *Engineering Geology Bulletin of Turkey*, 17, 103–110.

Aydan, Ö. & Kawamoto, T. (2000) Assessing mechanical properties of rock masses by RMR rock classification method. *Proc .of GeoEng 2000 Symposium*, Sydney, Paper No. OA0926.

Aydan, Ö. & Kawamoto, T. (2001) The stability assessment of a large underground opening at great depth. *17th Int. Min. Congress and Exhibition of Turkey*, IMCET 2001, Ankara, 1. pp. 277–288.

Aydan, Ö. & Kawamoto, T. (2004) The damage to abandoned lignite mines caused by the 2003 Miyagi-Hokubu earthquake and some considerations on its causes. *3rd Asian Rock Mechanics Symposium*, Kyoto, 1. pp. 525–530.

Aydan, Ö., Kawamoto, T., Ichikawa, Y. & Ersen, A. (1987b) The reinforcement of pillars with a discontinuity plane or discontinuity set by rockbolts (in Turkish). *Madencilik*, 26(2), 19–26.

Aydan, Ö., Kinbara, T., Uehara, F. & Kawamoto, T. (2005) A numerical analysis of non-destructive tests for the maintenance and assessment of corrosion of rockbolts and rock anchors. *35th Japan Rock Mechanics Symposium*, Tokyo. pp. 371–376.

Aydan, Ö. & Kumsar, H. (2010) An experimental and theoretical approach on the modeling of sliding response of rock wedges under dynamic loading. *Rock Mechanics and Rock Engineering*, 43(6), 821–830.

Aydan, Ö., Kumsar, H., Toprak, S. & Barla, G. (2009) Characteristics of 2009 L'Aquila earthquake with an emphasis on earthquake prediction and geotechnical damage. *Journal of Marine Science and Technology*, Tokai University, 7(3), 23–51.

Aydan, Ö., Kyoya, T., Ichikawa, Y. & Kawamoto, T. (1987) Anchorage performance and reinforcement effect of fully grouted rockbolts on rock excavations. *The 6th Int. Congress on Rock Mechanics*, ISRM, Montreal, 2. pp. 757–760.

Aydan, Ö., Kyoya, T., Ichikawa, Y. & Kawamoto, T. (1987a) Reinforcement effects of rockbolts and their analysis. *Proc. of the 22 Japan National Conf. on Soil Mechs. and Foundation Engineering*, JSSMFE, Niigata, 1. pp. 923–926.

Aydan, Ö., Kyoya, T., Ichikawa, Y. & Kawamoto, T. (1988c) Flexural toppling failures in slopes and underground openings. *Proc. of the 9th Western Japan Rock Mechs. Symp.*, Kumamoto. pp. 72–80.

Aydan, Ö., Kyoya, T., Ichikawa, Y., Kawamoto, T., Ito, T. & Shimizu, Y. (1988) Three-dimensional simulation of an advancing tunnel supported with forepoles, shotcrete, steel ribs and rockbolts. *The 6th Int. Conf. on Num. Meths. in Geomechanics*, Innsbruck, 2. pp. 1481–1486.

Aydan, Ö., Kyoya, T., Ichikawa, Y., Kawamoto, T. & Shimizu, Y. (1988b) A model study on failure modes and mechanism of slopes in discontinuous rock mass. *Proc. of the 23 National Conf. on Soil Mechs. and Foundation Engng.*, JSSMFE, Miyazaki, 1. pp. 1089–1092.

Aydan, Ö., Kyoya, T., Ichikawa, Y., Kawamoto, T. & Shimizu, Y. (1989) A model study on failure modes and mechanisms of slopes in discontinuous rock mass. *The 24th Annual Meetings of Soil Mechanics and Foundation Eng. of Japan*, Miyazaki, 415. pp. 1089–1093.

Aydan, Ö., Mamaghani, I.H.P. & Kawamoto, T. (1996b) Application of discrete finite element method (DFEM) to rock engineering. *North American Rock Mechanics Symp.* Montreal, 2. pp. 2039–2046.

Aydan, Ö. & Nawrocki, P. (1998) Rate-dependent deformability and strength characteristics of rocks. *Int. Symp. On the Geotechnics of Hard Soils-Soft Rocks*, Napoli, 1. pp. 403–411.

Aydan, Ö., Ohta, Y., Daido, M., Kumsar, H., Genis, M., Tokashiki, N., Ito, T. & Amini, M. (2011a) Chapter 15: Earthquakes as a rock dynamic problem and their effects on rock engineering structures. In: Zhou, Y. & Zhao, J. (eds.) *Advances in Rock Dynamics and Applications*. CRC Press, Taylor and Francis Group, Leiden. pp. 341–422.

Aydan, Ö., Ohta, Y., Geniş, M., Tokashiki, N. & Ohkubo, K. (2010) Response and earthquake induced damage of underground structures in rock mass. *Journal of Rock Mechanics and Tunneling Technology*, 16(1), 19–45.

Aydan, Ö., Ohta, Y., Geniş, M., Tokashiki, N. & Ohkubo K. (2010) Response and stability of underground structures in rock mass during earthquakes. *Rock Mechanics and Rock Engineering*, 43(6), 857–875.

Aydan, Ö. & Paşamehmetoğlu, G. (1994) In-situ measurements and lateral stress coefficients in various parts of the earth. *Rock Mechanics Bulletin, TNGRM*, 10, 1–17.

Aydan, Ö., Rassouli, F. & Ito, T. (2011b) Multi-parameter responses of Oya tuff during experiments on its time-dependent characteristics. *Proc. 45th US Rock Mechanics/Geomechanics Symp.*, ARMA, San Francisco. pp. 11–294.

Aydan, Ö., Sakamoto, A., Yamada, N., Sugiura, K. & Kawamoto, T. (2005a) The characteristics of soft rocks and their effects on the long-term stability of abandoned room and pillar lignite mines. *Post Mining 2005*, Nancy, France.

Aydan, Ö., Sakamoto, A., Yamada, N., Sugiura, K. & Kawamoto, T. (2005b) A real time monitoring system for the assessment of stability and performance of abandoned room and pillar lignite mines. *Post Mining 2005*, Nancy, France.

Aydan, Ö., Sezaki, M. & Kawamoto, T. (1992) Mechanical and numerical modeling of shotcrete. *Int. Symp. on Numerical Models in Geomechanics*, NUMOG IV, Swansea. pp. 757–764.

Aydan, Ö., Sezaki, M. & Kawamoto, T. (1992) Mechanical and numerical modeling of shotcrete. In: Pande, G. & Pietruszczak, S. (eds.) *Proceeding of the 4th International Symposium on Numerical Models in Geomechanics*. Swansea, Great Britain, 1. pp. 757–765.

Aydan, Ö., Shimizu, Y. & Kawamoto, T. (1992) The stability of rock slopes against combined shearing and sliding failures and their stabilisation. *Int. Symp. on Rock Slopes*, New Delhi. pp. 203–210.

Aydan, Ö. & Tano, H. (2012a) The damage to abandoned quarries and mines by the M9.0 east Japan Mega Earthquake on March 11, 2011. *The 41st Rock Mechanics Symposium of Japan, Paper No. 410011*, Tokyo. pp. 129–134.

Aydan, Ö. & Tano, H. (2012b) The damage to abandoned mines and quarries by the Great East Japan Earthquake on March 11, 2011. *Proceedings of the International Symposium on Engineering Lessons Learned from the 2011 Great East Japan Earthquake*, March 1–4, 2012, Tokyo, Japan. pp. 981–992.

Aydan, Ö. & Tano, H. (2012c) The observations on abandoned mines and quarries by the Great East Japan Earthquake on March 11, 2011 and their implications. *Journal of Japan Association on Earthquake Engineering*, 12(4), 229–248.

Aydan, Ö., Tano, H. & Geniş, M. (2007) Assessment of long-term stability of an abandoned room and pillar underground lignite mine. *Rock Mechanics Bulletin*, Turkish National Rock Mechanics Group, ISRM, No. 16, 1–22.

Aydan, Ö., Tano, H., Ideura, H., Asano, A., Takaoka, H., Soya, M. & Imazu, M. (2016) Monitoring of the dynamic response of the surrounding rock mass at the excavation face of Tarutoge Tunnel, Japan. *EUROCK2016*, Ürgüp. pp. 1261–1266.

Aydan, Ö. & Tokashiki, N. (2007) Some damage observations in Ryukyu Limestone Caves of Ishigaki and Miyako islands and their possible relations to the 1771 Meiwa Earthquake. *Journal of The School of Marine Science and Technology*, 5(1), 23–39.

Aydan, Ö. & Tokashiki, N. (2011) A comparative study on the applicability of analytical stability assessment methods with numerical methods for shallow natural underground openings. *The 13th International Conference of the International Association for Computer Methods and Advances in Geomechanics*, Melbourne, Australia. pp. 964–969.

Aydan, Ö. & Tokashiki, N. (2013) An experimental study on the supporting effect of backfilling on abandoned room and pillar mines and quarries and karstic caves. *The 47th US Rock Mechanics/Geomechanics Symposium*, San Francisco, ARMA 13–379, 10 pages.

Aydan, Ö., Tokashiki, N. & Geniş, M. (2012a) Some Considerations on Yield (Failure) Criteria in Rock Mechanics ARMA 12–640. *46th US Rock Mechanics/Geomechanics Symposium*, Chicago, Paper No. 640, 10 pages (on CD).

Aydan, Ö., Tokashiki, N. & Seiki, T. (1996) Micro-structure models for porous rocks to jointed rock mass. *APCOM'96*, 3, 2235–2242.

Aydan, Ö., Tokashiki, N. & Tano, H. (2011c) Report on the monitoring of rock mass behavior before and after backfilling at Gushikawa castle Remains. *Institute of Technology and Science of Nagoya*, 38 pages.

Aydan, Ö., Tsuchiyama, S., Kinbara, T., Uehara, F. & Kawamoto, T. (2008) A numerical study on the long-term performance of an underground powerhouse subjected to varying initial stress state, cyclic water head and temperature variations. *The 12th International Conference of International Association for Computer Methods and Advances in Geomechanics (IACMAG)*, Goa, India. pp. 3875–3882.

Aydan, Ö., Tsuchiyama, S., Kinbara, T., Uehara, F., Tokashiki, N. & Kawamoto, T. (2008) A numerical analysis of non-destructive tests for the maintenance and assessment of corrosion of rockbolts and rock anchors. *The 12th International Conference of International Association for Computer Methods and Advances in Geomechanics (IACMAG)*, Goa, India. pp. 40–45.

Aydan, Ö., Uehara, F. & Kawamoto, T. (2012) Numerical study of the long-term performance of an underground powerhouse subjected to varying initial stress states, cyclic water heads, and temperature variations. *International Journal of Geomechanics, ASCE*, 12(1), 14–26.

Aydan, Ö. & Ulusay, R. (2013) Application of RMQR classification system to rock-support design for underground caverns and tunnels. *Proc. of the 3rd Int. Symp. on Underground Excavations for Transportation*, İstanbul. pp. 387–398.

Aydan, Ö. & Ulusay, R. (2014) Rock Mass Quality Rating (RMQR) System: Its application to estimation of geomechanical characteristics of rock masses and to rock support selection for underground caverns and tunnels. *Proceedings of the 8th Asian Rock Mechanics Symposium*, Sapporo, pp. 2075–2084.

Aydan, Ö., Ulusay, R., Kumsar, H. & Ersen, A. (1996) Buckling failure at an open-pit coal mine. *EUROCK'96*. pp. 641–648.

Aydan, Ö., Ulusay, R. & Tokashiki, N. (2013) A new Rock Mass Quality Rating System: Rock Mass Quality Rating (RMQR) and its application to the estimation of geomechanical characteristics of rock masses, *Rock Mech. and Rock Eng.*, 47, 1255–1276.

Aydan, Ö., Watanabe, S. & Tokashiki, N. (2008) The inference of mechanical properties of rocks from penetration tests. *5th Asian Rock Mechanics Symposium (ARMS5)*, Tehran. pp. 213–220.

Barla, G. & Cravero, M. (1972) Analysis of stress around underground openings reinforced with rockbolts. *Procs. Int. Symp. für untertagbau*, Lüzern.

Barla, G. & Cravero, M. (1972) The distribution of stress around underground openings reinforced with rock bolts. *International Symposium on Underground Openings*, September.

Barton, N. (1995) The influence of joint properties in modeling jointed rock masses. *8th Int. Rock Mech. Congress*, Tokyo. pp. 1023–1032.

Barton, N. & Bandis, S.C. (1987) Rock joint model for analyses of geological discontinua. *Proc. Int. Symp. on Constitutive Laws for Engineering Materials: Theory and Applications*, Elsevier, Amsterdam. pp. 993–1002.

Barton, N. & Grimstad, E. (1994) The Q-system following twenty years of application in NATM support selection. *Felsbau*, 12(6), 428–436.

Barton, N., Harvik, L., Christianson, M. & Vik, G. (1986) Estimation of joint deformations, potential leakage and lining stresses for a planned urban road tunnel. *Int. Symp. on Large Rock Caverns*. pp. 1171–1182.

Barton, N., Lien, R. & Lunde, I. (1974) Engineering classification of rock masses for the design of tunnel supports. *Rock Mech.*, 6(4), 189–239.

Barton, N., Løset, F., Lien, R. & Lunde, J. (1980) Application of the Q system in design decisions concerning dimensions and appropriate support for underground installations. *Int. Conf. on Sub Surface Space, Rockstore, Stockholm, Sub Surface Space*, 2. pp. 553–561.

Baudendistel, M., Malina, H. & Müller, L. (1970) Einfluss von Discontinuitaten auf die Spannungen und Deformationen in der Umgebung einer Tunnelröhre. *Rock Mechanics*, 2, 17–40.

Beard, M.D. & Lowe, M.J.S. (2003) Non-destructive testing of rock bolts using guided ultrasonic waves. *International Journal of Rock Mechanics & Mining Sciences*, 40, 527–536.

Bernaix, J. (1966) *Contribution à l'étude de la stabilité des appuis de barrages*. Thèse, Paris.

Beyl, Z.S. (1945–1946) Rock pressure and roof support. *Colliery Engineering*, Sept. 1945–Oct. 1946.

Bieniawski, Z.T. (1970) Time-dependent behavior of fractured rock. *Rock Mechanics*, 2, 123–137.

Bieniawski, Z.T. (1973) Engineering classification of jointed rock masses. *Trans. of the South African Inst. of Civil Engineers*, 15, 335–344.

Bieniawski, Z.T. (1976) Rock mass classification in rock engineering. In: Bieniawski, Z.T. (ed.) *Exploration for Rock Engineering, Proc. of the Symp.* 1, Balkema, Cape Town. pp. 97–106.

Bieniawski, Z.T. (1989) *Engineering Rock Mass Classifications*. John Wiley & Sons, New York.

Bieniawski, Z.T. & van Tonder, C.P.G. (1969) A photoelastic-model study of stress distribution and rock ure around mining excavations. *Experimental Mechanics*, 9(2), 75–81.

Birön, C. & Arıoğlu, E. (1983) *Design of Supports in Mines*. John Wiley & Sons, New York.

Bishop, A.W. (1955) The use of slip circle in the stability analysis of slopes. *Geotechnique*, 5, 7–17.

Bjürström, S. (1974) Shear resistance of hard rocks reinforced by grouted untensioned bolts. *Proc. of the 3rd Int. Congr. on Rock Mechs.*, ISRM, Denver, 2B. pp. 1194–1199.

Bobet, A. (2001) Analytical solutions for shallow tunnels in saturated ground. *Journal of Engineering Mechanics*, 127(12), 1258–1266.

Bowden, F.P. & Tabor, D. (1964) *The Friction and Lubrication of Solids*. Clarendon Press, Oxford.

Brady, B.G. & Brown, E.T. (2005) *Rock Mechanics for Underground Mining*. Kluwer Academic Publishers, New York.

Brawner, C.O., Pentz, D.L. & Sharp, J.C. (1971) Stability studies of a footwall slopes in layered coal deposit. *Proc. of the 13th US Symp. on Rock Mechanics.* pp. 329–365.

Bray, J.W. (1967) A study of jointed and fractured rock. *Rock Mech. and Eng. Geol.*, 5(2, 3, 4), 117–136, 197–216.

Brown, E.T. (1981) Rock characterization, testing and monitoring. In: Brown, E.T. (ed.) *Pub. Commis. on Testing Methods, ISRM*. Pergamon Press, Oxford.

Brown, E.T., Bray, J.W., Hoek, E. & Ladanyi, B. (1983) Ground response curves for rock tunnels. *J. Geotech. Div., ASCE*, 109(1), 15–39.

Brown, E.T., Bray, J.W., Ladanyi, B. & Hoek, E. (1983) Characteristic line calculations for rock tunnels. *J. Geotech. Engineering*, ASCE, 109, 15–89.

Buys, B.J., Heyns, P.S. & Loveday, P.W. (2009) Rock bolt condition monitoring using ultrasonic guided waves. *The Journal of the Southern African Institute of Mining and Metallurgy*, 108, 95–105.

Cavers, D.S. (1981) Simple methods to analyse buckling of rock slopes. *Rock Mechanics*, 14, 87–104.

Chajes, A. (1974) *Principles of Structural Stability Theory*. Civil Eng. and Eng. Mech. Series. Prentice Hall, Englewood Cliffs, NJ.

Chaki, S. & Bourse, G. (2009) Guided ultrasonic waves for non-destructive monitoring of the stress levels in prestressed steel strands. *Ultrasonics*, 49, 162–171.

Chesson, E., Faustino, N.L. & Mounze, W.H. (1965) High strength bolt subjected to tension and shear. *Proc. of ASCE, J. Struct. Div.*, 91(ST5), 155–180.

Coates, D.F. (1981) Rock Mechanics Principles. Canadian Government Pub Centre, 410 pages.

Cording, E.J. (1973) Geological considerations in shotcrete design: Uses of shotcrete for underground structural support. *Proceedings of the Engineering Foundation Conference*, Maine, pp. 175–199.

Cording, E.J., Hendron, A.J. & Deere, D.U. (1972) Rock engineering for underground caverns. *ASCE, Symp. on Underground Chambers*, 567–600.

Cox, R.M. (1974) Why some bolted mine roofs fail. *Trans. Soc. Min. Engrs.*, AIME, 256, 167–171.

Crotty, J.M. & Wardle, L.J. (1985) Boundary integral analysis of piece-wise homogenous media with structural discontinuities. *Int. J. Rock Mech. Min. Sci.*, 22(6), 419–427.

Crouch, S.L. & Starfield, A.M. (1983) *Boundary Element Methods in Solid Mechanics*. Allen & Unwin, London.

Cundall, P.A. (1971) A computer model for simulating progressive, large-scale movements in blocky rock systems. *Int. Symp. on Rock Fracture, II-8*, Nancy, France.

Cundall, P.A. (1971) *The Measurement and Analysis of Acceleration in Rock Slopes*. Ph.D. Thesis, University of London (Imperial College), London.

Daemen, J.J.K. (1983) Slip zones for discontinuities parallel to circular tunnels or shafts. *Int. J. Rock Mech. Min. Sci.*, 20(3), 135–148.

Deere, D.U., Peck, R.B., Schmidt, B. & Monsees, J.E. (1969) *Design of Tunnel Liners and Support Systems*. US Dept. of Transportation, Washington, DC.

Desai, C.S., Zaman, M.M., Lightner, J.G. & Siriwardane, H.J. (1984) Thin layer element for interfaces and joints. *Int. J. Num. Anal. Meth. Geomech.*, 8, 19–43.

Detournay, E. (1983) *Two-Dimensional Elastoplastic Analysis of a Deep Cylindrical Tunnel Under Non-Hydrostatic Loading*. PhD thesis, University of Minnesota, Minneapolis, MN. pp. 1–133.

Detzlhofer, H. (1968) Rockfalls in pressure galleries. *Felsmechanik und Ingenieurgeologie*, (Suppl. 4), 158–180.

Detzlhofer, H. (1970) Experience in securing gallery deformations in headings in weak and yielding rock. *Rock Mechanics*, (Suppl. 1), 69–86.

Dezhen, G. & Sijing, W. (1982) Fundamentals of geomechanics for rock engineering in China. *Rock Mechs.*, (Suppl. 12), 75–87.

Döring, T. (1965) The equilibrium of jointed rock masses with a parallel set of joints. *Rock Mechs. and Eng. Geology*, 3(2), 19–26.

Drucker, D.C. & Prager, W. (1952) Soil mechanics and plastic analysis for limit design. *Quarterly of Applied Mathematics*, 10(2), 157–165.

Ebisu, S., Aydan, Ö. & Kawamoto, T. (1994) Reinforcement of highly pressurised gas storage caverns by inclined rockbolting (in Japanese). *The 9th National Rock Mechanics Symposium of Japan*, Tokyo. pp. 659–665.

Ebisu, S., Aydan, Ö. & Komura, S. (1993) Mechanism of interface failure of rock anchors. *Proc. Int. Symp. on Assessment and Prevention of Failure Phenomena in Rock Engineering*, Istanbul. pp. 677–686.

Ebisu, S., Aydan, Ö., Komura, S. & Kawamoto, T. (1992) Characterization of jointed rock masses for rock anchor foundations. *Int. Conf. on Fractured and Jointed Rock Masses*, Lake Tahoe. pp. 150–157.

Ebisu, S., Aydan, Ö., Komura, S. & Kawamoto, T. (1994a) Assessing the mechanical properties of rock masses for rock anchors (in Japanese). *Journal of Japan Society of Civil Engineers, Geotechnical Division*, 505(III-29), 267–276.

Ebisu, S., Aydan, Ö., Komura, S. & Kawamoto, T. (1994b) Failure modes of rock anchors during pull-out tests (in Japanese). *Journal of Japan Society of Civil Engineers, Construction Division*, 504(VI-25), 157–166.

Ebisu, S., Komura, S. & Aydan, Ö. (1991) The behavior of rock anchors *in situ* pull-out tests (in Japanese). *Proc. of the 12th W. Japan Symposium on Rock Engineering*, Kumamoto. pp. 91–96.

Ebisu, S., Komura, S., Aydan, Ö. & Kawamoto, T. (1992) A comparative study on various rock mass characterization methods for rock surface structures. *Proc. of the Int. Symp. Rock Characterization, ISRM*, Chester.

Ebisu, S., Komura, S., Aydan, Ö. & Kawamoto, T. (1992) Characterization of rock masses for rock anchor foundations. *Proc. of the Int. Symp. Fractured and Jointed Rock Masses, ISRM*, Lake Tahoe.

Egger, P. (1973a) Einfluss des Post-Failure-Verhaltens von Fels auf den Tunnelausbau unter besanderer Berücksichtigung des Ankerausbau. Veröffentlichungen des Institutes für Bodenmechanik und Felsmechanik, Univ. of Karlsruhe.

Egger, P. (1973b) Influence of post-failure behavior of rocks in tunneling. *Institutes für Bodenmechanik und felsmechanik, der Universität Fridericiana*, Karslruhe.

Egger, P. (1973c) Rock stabilization. In: L. Müller (ed.) *Rock Mechanics, Courses and Lectures*, Springer, Udine, 241–297.

Egger, P. & Fernandes, H. (1983) A novel triaxial press – Study of anchored jointed models (in French). *Proc. of the 5th Int. Congr. on Rock Mechanics, ISRM*.

Elfman, S. (1969) Design of a tunnel in a rock burst zone. *Proceedings of the Symposium on Large Permanent Underground Openings*, Oslo, pp. 47–53.

Elsworth, D. (1986) Wedge stability in the roof of a circular tunnel: Plain strain condition. *Int. J. Rock Mech. Min. Sci.*, 23(2), 177–181.

Erguvanlı, K. (1973) *Engineering Geology* (in Turkish). Istanbul Teknik University, Istanbul.

Erguvanlı, K. & Goodman, R.E. (1972) Applications of models to engineering geology for rock excavations. *Bulletin of the Assoc. of Eng. Geologist*, 9(1).

Eringen, A.C. (1961) Propagations of elastic waves generated by dynamical loads on a circular cavity. *J. Applied Mechanics, ASME*, 28(2), 218–212.

Eringen, A.C. (1980) *Mechanics of Continua*. John Wiley & Sons Ltd (1st), 2nd edition. Robert E. Krieger Publ. Co., Huntington, New York, 520 pages.

Ersen, A. (1983) *Design of Concrete Shaft Linings in Frozen Strata at Whitemoor Mine*. M.Sc. Thesis, Newcastle University, Newcastle upon Tyne, UK.

Ersen, A. & Aydan, Ö. (1984) *Stress, Strain and Temperature Measurements at Concrete Lined Mine Shafts By Wibrating Wire Gauges*, vol. 4. Coal Congress of Turkey, Zonguldak, 289–303.

Evans, I. (1964) An expanding-bolt seam-tester: A theory of tensile breakage. *Int. J. Rock Mech. Min. Sci.*, 1(4), 459–474.

Evans, W.H. (1960) Roof bolting and the stabilisation of natural arches on roadways. *Colliery Engineering*, 293–296.

Everling, G. (1964) Model tests concerning the interaction of ground and roof support in gate roads. *Int. J. of Rock Mechs. & Min. Sci.*, 1, 319–326.

Ewy, R.T., Cook, N.G.W. & Myer, L.R. (1988) Hollow cylinder tests for studying fracture around underground openings. *Proc. of the 19th US Symp. on Rock Mechs*. pp. 67–74.

Fabbri, D. (2004) The Gotthard base tunnel: Fire/Life safety system. *The 6th annual Tunneling Conference*, Sydney, 30–31 August 2004, 10 pages.

Feder, G. (1976) The reinforcement effect of systematic bolting on underground cavities in an isotropic medium (in German). *Berg-und Hüttenmännische Monatshefte*, (6).

Fellenius, W. (1936) Calculation of the stability of earth dams. *Trans. 2nd Congr. on Large Dams*, Washington, 4, 445–462.

Fenner, R. (1938) Researches on the notion of ground stress (in German). *Gluckauf*, 74, 681–695.

Foote, P. (1964) An expanding-bolt seam-tester. *Int. J. Rock Mech. Min. Sci.*, 1(2), 255–275.

Freeman, T.J. (1978) The behavior of fully-bonded rock bolts in the Kielder experimental tunnel. *Tunnels and Tunneling*, 37–40.

Fujita, K. (1981) Use of slurry and earth-pressure-balance shield in Japan. *Int. Congress on Tunneling*, Tunnel 81, Bd. 1,383–406. Dtisseldorf. (Ed.: Dtisseldorfer Messegesellschaft mbH – NOWEA – in Zusammenarbeit mit der Deutschen Gesellschaft fiir Erd-und Grundbau e.V.).

Gaudreau, D., Aubertin, M. & Simon, R. (2004) Performance assessment of tendon support systems submitted to dynamic loading. *Proc. of 5th Int. Symp. on Rock Support*, Taylor and Francis, Abingdon. pp. 299–312.

Geisler, H., Wagner, H., Zieger, O., Mertz, W. & Swoboda, G. (1985) Practical and theoretical aspects of the three-dimensonal analysis of finally lined intersections. *Proc. of the 5th Int. Conf. on Num. Meths. in Geomechanics*, Nagoya, Japan. pp. 1175–1192.

Geniş, M. & Aydan, Ö. (2006) Assessment of dynamic stability of an abandoned room and pillar underground lignite mine. *8th National Rock Mechanics Symposium of Turkey*, 29–38.

Gerçek, H. (1988) Calculation of elastic boundary stresses for rectangular underground openings. *Mining Science and Technology*, 7, 173–182.

Gerçek, H. (1993) Qualitative prediction of failures around non-circular openings. In: Paşamehmetoğlu, A.G. et al. (eds.) *Proc. Int. Symp. on Assessment and Prevention of Failure Phenomena in Rock Engineering*. A.A. Balkema, Rotterdam. pp. 727–732.

Gerçek, H. (1996) Special elastic solutions for underground openings. In: *Milestones in Rock Engineering*, The Bieniawski Jubilee Collection, Balkema, Rotterdam. pp. 275–290.

Gerçek, H. (1997) An elastic solution for stresses around tunnels with conventional shapes. *Int. J. Rock Mech. Min. Sci.*, 34(3–4), paper No. 096.

Gerçek, H. & Geniş, M. (1999) Effect of anisotropic *in situ* stresses on the stability of underground openings. *Proc. of the Ninth Intl. Congress on Rock Mechanics, ISRM*, Balkema, Rotterdam, 1, 367–370.

Ghaboussi, J. (1988) Fully deformable discrete analysis using a finite element approach. *Comput. Geotech*, 5, 175–195.

Ghaboussi, J., Wilson, E.L. & Isenberg, J. (1973) Finite element for rock joints and interfaces. *J. Soil Mechs. and Found. Eng. Div., ASCE, SM10*, 99(10), 833–848.

Goodman, R.E. (1976) *Methods of Geological Engineering in Discontinuous Rocks*. West Publishing, San Francisco.

Goodman, R.E. (1977) Analysis of jointed rocks. In: *Finite Elements in Geomechanics*, Wiley & Sons, New York. pp. 351–375.

Goodman, R.E. & Bray, J.W. (1976) Toppling of rock slopes. *Proc. of Specialty Conference on Rock Engineering for Foundations and Slopes, ASCE*, 2, 201–234.

Goodman, R.E., Taylor, R. & Brekke, T.L. (1968) A model for the mechanics of jointed rock. *J. Soil Mechs. and Found. Eng. Div., ASCE, SM3*, 94(3), 637–659.

Greenspan, M. (1944) Effect of small hole on the stresses in a uniformly loaded plate. *Quarterly J. of Appl. Mathematics*, 2(1), 60–71.

Grimstad, E. & Barton, N. (1993) Updating the Q-system for NMT. *Proc. Int. Symp. on Sprayed Concrete–Modern Use of Wet Mix Sprayed Concrete for Underground Support*, Fagernes. Norwegian Concrete Assn., Oslo. pp. 46–66.

Haas, C.J. (1976) Shear resistance of rockbolts. *Trans. of Am. Soc. Min. Engrs., AIME*, 260, 32–41.

Hamming, R.W. (1973) *Numerical Methods for Scientists and Engineers*. Int. Student ed. McGraw-Hill, New York.

Hayashi, M. & Fujiwara, Y. (1963) A mechanism of anisotropic dilatancy and shear failure of laminately jointed rock masses (in Japanese). *Central Res. Inst. of Electric Power Industry, CRIEPI*, Res. Rep. No.67096.

Henke, A. (2005) Tunneling in Switzerland: From long tradition to the longest tunnel in the world. *Proceedings of Int. Symp. on World Long Tunnels*. pp. 57–70.

Heuze, F.E. & Goodman, R.E. (1973) Finite element and physical model studies of tunnel reinforcement in rock. *Proc. of the 15th US Symp. on Rock Mechs.* pp. 37–67.

Hibino, S. & Motojima, M. (1981) Effects of rockbolting in jointy rocks. *Proc. of Int. Symp. on Weak Rocks*, Tokyo, 2. pp. 1057–1062.

Hill, J.L. & Bauer, E.R. (1984) An investigation of the causes of cutter roof failure in a central Pennsylvania coal mine: A case study. *Proc. of the 25th US Symp. on Rock Mechanics.* pp. 603–614.

Hoek, E. & Bray, E.T. (1977) *Rock Slope Engineering*, Revised 2nd ed. The Institution of Mining and Metallurgy, London.

Hoek, E. & Brown, E.T. (1980) *Underground Excavations in Rock*. Inst. Min. & Metall., London.

Hoffmann, H. (1974) Zum Verformungs-und Bruchverhalten regelmassig geklufteter Felsboschungen. *Rock Mech.*, (Suppl. 3), 31–34.

Hollingshead, G.W. (1971) Stress distribution in rock anchors. *Can. Geotech. J.*, 8(4), 588–592.

Horino, F.C., W.F. Duval & B.T. Brady (1971) The use of rockbolts or wire rope to increase the strength of fractured model pillars. *U.S. Dept. Interior, BM*, IN23.U7, No.7568, 622.06173.

Hutchinson, J.N. (1971) Field and laboratory studies of a fall in upper chalk cliffs at Joss Bay, Isle of Thanet. *Proc. of the Roscoe Memorial Symp.*, Cambridge Univ. pp. 692–706.

Ichikawa, Y. (1985) *Fundamentals of the Incremental Elasto-plastic Theory for Rock-like Materials* (in Japanese), Dr. Thesis, Faculty of Engineering, Nagoya University, Nagoya, Japan. 132 pages.

Ichikawa, Y., Kyoya, T., Aydan, Ö., Yoshikawa, K., Kawamoto, T. & Tokashiki, N. (1988) Deformation and failure of rocks under weak cyclic loading and incremental elasto-plasticity theory. *Memoirs of Faculty of Engineering, Nagoya University*, 40(2), 273–326.

Ikeda, K. (1970) A classification of rock conditions for tunnelling. *First Internationak Congress on Engineering Geology*, IAEG, Paris, 1258–1265.

Imazu, M., Ideura, H. & Aydan, Ö. (2014) A monitoring system for blasting-induced vibrations in tunneling and its possible uses for the assessment of rock mass properties and *in situ* stress inferences. *Proc. of the 8th Asian Rock Mechanics Symposium*, Sapporo. pp. 881–890.

Inglis, C.E. (1913) Stresses in plates due to the presence of cracks and sharp corners. *Transactions of the Institute of Naval Architects*, 55, 219–241.

Isaac, I.D. & Bubb, C. (1981) Geology at dinorwic. *Tunnels & Tunneling*, 20–25.

Ivanović, A. & Neilson, R.D. (2008) Influence of geometry and material properties on the axial vibration of a rock bolt. *International Journal of Rock Mechanics and Mining Sciences*, 45, 941–951.

Jaeger, J.C. (1962) *Elasticity, Fracture and Flow*, 2nd ed. Methuen, London.

Jaeger, J.C. & Cook, N.G.W. (1979) *Fundamentals of Rock Mechanics*, 3rd ed. Chapman & Hall, London.

John, C.M. & van Dillen, D.F. (1983) Rockbolts: A new numerical representation and its application in tunnel design. *Proc. of the 24th US Symp. on Rock Mechs.*, 13–25.

Joule, J. (1842) On a new class of magnetic forces. *Annals of Electricity, Magnetism, and Chemistry*, 8, 219–224.

Kaiser, P.K. (1979) Tunnel design: Conclusions from long-term model tests. *Proc. of the 4th Int. Congr. on Rock Mechs., ISRM*, Montreux, 3. pp. 236–238.

Kamemura, K., Husni, S., Harada, T. & Shishido, Y. (1986) Construction of underground powerhouse cavern: CIRATA hydroelectric power project. *Proc. of the Int. Congr. on Large Underground Space*, Florence, 2, 390–399.

Kastner, H. (1961) *Statik des Tunnel- and Stollenbaues*, ("Design of. Tunnels"), 2nd ed. Springer-Verlag, Berlin.

Kastner, H. (1971) *Statik des Tunnel und Stollenbaus auf der Grundlage Geomechanischer Erkenntnisse*. Springer-Verlag, Berlin.

Kawamoto, T. & Aydan, Ö. (1988) Reinforcement effects of rockbolts (in Japanese). *Electric Power Civil Engineering*, (214), 3–13.

Kawamoto, T. & Aydan, O. (1999) A review of numerical analysis of tunnels in discontinuous rock masses. *International Journal of Numerical and Analytical Methods in Geomechanics*, 23, 1377–1391.

Kawamoto, T., Aydan, Ö. & Tsuchiyama, S. (1991) Analysis and review of design for large underground cavern (in Japanese). *Tunnels and Underground*, 23(3), 31–37.

Kawamoto, T., Aydan, Ö. & Tsuchiyama, S. (1991) A consideration on the local instability of large underground openings. *Proc. of Int. Conf., Geomechanics '91*, Hradec. pp. 33–41.

Kawamoto, T., Ichikawa, Y. & Kyoya, T. (1988) Deformation and fracturing behavior of discontinuous rock mass and damage mechanics theory. *Int. J. Num. Anal. Meth. Geomech.*, 12(1), 1–30.

Kawamoto, T., Kyoya, T. & Aydan, Ö. (1994) Numerical models for rock reinforcement by rockbolts. *Int. Conf. on Computer Methods and Advances in Geomechanics, IACMAG*, Morgantown, 1. 33–45.

Kawamoto, T., Obara, Y. & Ichikawa, Y. (1983) A base friction apparatus and mechanical properties of a model material (in Japanese). *J. of Min. & Metall. Inst. of Japan*, Tokyo, 99, 1–6.

Kirsch, G. (1898) Die theorie der elastizitat und die bedürfnisse der festigkeitslehre. *Veit Ver. Deut. Ing.*, 42, 797–807.

Kitatsu, T. & Nishimura, M. (1982) Fundamental research into the interaction mechanism between rockbolts and rockmass (in Japanese). *Annual Research Report, Okumura-Gumi*, (8), 53–76.

Komura, S., Aydan, Ö., Ebisu, S. & Kawamoto, T. (1992) Design methodology for rock anchor foundations. *Proc. Int. Symp. Rock Support*. pp. 491–497.

Kondo, Y. & Saka, S. (1965) *Concrete Handbook* (in Japanese), Asakura Books, Tokyo.

Kovari, K. (1977) The elasto-plastic analysis in the design practice of underground openings. In: Gudehus, G. (ed.) *Finite Elements in Geomechanics*. John Wiley & Sons, New York. pp. 377–442.

Koyama, Y. (1997) Railway construction in Japan. *Japan Railway Transport and Review*, 36–41.

Kreyszig, E. (1983) *Advanced Engineering Mathematics*. John Wiley & Sons, New York.

Krsmanovic, D., Tufo, M. & Langof, Z. (1965) Some aspects of the rupture of a rock mass. *Fels-mechanik und Ingenieurgeologie*, 3, 143–155.

Kumsar, H., Aydan, Ö. & Ulusay, R. (2000) Dynamic and static stability of rock slopes against wedge failures. *Rock Mechanics and Rock Engineering*, 33(1), 31–51.

Kutter, H.K. (1974) Mechanisms of slope failure other than pure sliding. In: Müller, L. (ed.) *Rock Mechanics*, International Center for Mechanical Sciences, Courses and Lectures No. 165. Springer, New York.

Ladanyi, B. (1974) Use of the long-term strength concept in the determination of ground pressure on tunnel linings. *3rd Congr. Int. Soc. Rock Mech.*, Denver, Vol. 2B. pp. 1150–1165.

Ladanyi, B. (1993) Time-dependent response of rock around tunnels. In: *Comprehensive Rock Engineering*, Vol. 2, Elsevier, Amsterdam. pp. 77–112.

Laguerre, L., Aime, J.C. & Brissaud, M. (2002) Magnetostrictive pulse-echo device for nondestructive evaluation of cylindrical steel materials using longitudinal guided waves. *Ultrasonics*, 39, 503–514.

Laguerre, L., Christian, J. & Brissaud, M. (2000) Generation and detection of elastic waves with magnetoelastic device for the nondestructive evaluation of steel cables and bars. *Proceedings, 15th World Conference on Nondestructive Testing*, 15–21 October, Roma, Italy.

Lama, R.D. & Vutukuri, V.S. (1978) *Handbook on Mechanical Properties of Rocks*. Trans Tech Publications, Clausthal, Germany.

Lancaster, P. & Salkauskas, K. (1986) *Surface and Curve Fitting*. Academic Press, London, 280.

Lang, T.A. (1961) Theory and practice of rock bolting. *Trans. Soc. Min. Engrs., Am. Inst. Min. Metall. Petrolm Engrs.*, 220, 338–348.

Lang, T.A. & Bischoff, J.A. (1982) Stabilisation of rock excavations using rock reinforcement. *Proc. of the 23rd US Symp. on Rock Mechs.* pp. 935–943.

Li, C.C. (2010) A new energy-absorbing bolt for rock support in high stress rock masses. *International Journal of Rock Mechanics & Mining Sciences*, 47, 396–404.

Lombardi, G. (1970) The influence of rock characteristics on the stability of rock cavities. *Tunnels and Tunneling*, Jan. and March.

Lorig, L.J. (1985) A simple representation of fully bonded passive rock reinforcement for hard rocks. *Computers and Geotechnics*, 1, 79–97.

Løset, F. (1997) Engineering geology: Practical use of the Q-method. NGI report 592046–4.

Ludvig, B. (1983) Shear tests on rock bolts. *Proc. of the Int. Symp. on Rock Bolting*, Abisko. pp. 113–123.

Mamaghani, I.H.P., Baba, S., Aydan, Ö. & Shimizu, S. (1994) Discrete finite element method for blocky systems. *Computer Methods and Advances in Geomechanics*, IACMAG, Morgantown, 1, 843–850.

Marence, M. & Swoboda, G. (1995) Numerical model of rock bolts under consideration of rock joint movements. *Rock Mechanic and Rock Engineering*, 28(3), 145–165.

Miki, K. (1986) *Introduction to Rock Mechanics* (in Japanese). Kajima Publishing Association, Tokyo, 315.

Mindlin, R.D. (1940) Stress distribution around a tunnel. *Trans. ASCE*, 105, 1117–1153.

Mindlin, R.D. (1949) Compliance of elastic bodies in contact. *J. Appl. Mech.*, 16, 259–268.

Moosavi, M. (1997) *Load Distribution along Fully Grouted Cable Bolts Based on Constitutive Models Obtained from Modified Hoek Cells*. Ph.D. Thesis, Queen's University, Canada.

Müller, L. (1978) Removing mis-conceptions on the new Austrian tunneling method. *Tunnels and Tunneling*, Oct., 29–32.

Muskhelishvili, N.I. (1953) *Some Basic Problems of the Mathematical Theory of Elasticity*. Noordhoff, Groningen.

Ngo, D. & Scordelis, A.C. (1967) Finite element analysis of reinforced concrete beams. *J. ACI*, 152–163.

Niimi, Y., Komura, S., Aydan, Ö. & Ebisu, S. (1990) Studies on the mechanical behavior of a rock anchor (in Japanese). *Proc. 8th Domestic Rock Mechs. Symp.*, Tokyo. pp. 37–42.

Nilsen, B., Palmstrøm, A. & Stille, H. (1999) Quality control of a sub-sea tunnel project in complex ground conditions. *Challenges for the 21st Century, Proc. World Tunnel Congress '99*, Balkema, Oslo. pp. 137–145.

Obert, L. & Duvall, W.I. (1967) *Rock Mechanics and the Design of Structures in Rock*. John Wiley & Sons, New York.

Oda, M., Yamabe, T., Ishizuka, Y., Kumasaka, H., Tada, H. & Kimura, K. (1993) Elastic stress and strain in jointed rock masses by means of crack tensor analysis. *Rock Mech. Rock Engng.*, 26(2), 89–112.

Ohnishi, Y., Sasaki, T. & Tanaka, M. (1995) Modification of the DDA for elasto-plastic analysis with illustrative generic problems. *35th US Rock Mechanics Symp.*, Lake Tahoe. pp. 45–50.

Ohta, Y., Nakatani, E., Daido, M. & Aydan, Ö. (2004) Multi-parameter response of rockmass around abandoned mines during deformation and fracturing. *34th Japan Rock Mechanics Symposium.* pp. 483–488.

Okagbue, C.O. & Abam, T.K.S. (1986) An analysis of stratigraphic control on river bank failure. *Engineering Geology*, 22, 231–245.

Ortlepp, W.D. (1992) Implosive-load testing of tunnel support. In: *Rock Support in Mining and Underground Construction*, Balkema, Rotterdam. pp. 675–682.

Owada, Y. & Aydan, Ö. (2005) Dynamic response of fully grouted rockbolts during shaking (in Japanese). *The School of Marine Science and Technology, Tokai University*, 3(1), 9–18.

Owada, Y., Daido, M. & Aydan, Ö. (2004) Mechanical response of rock anchors and rockbolts during earthquakes (in Japanese). *34th Japan Rock Mechanics Symposium.* pp. 519–524.

Owen, D.R.J. & Hinton, E. (1980) *Finite Element in Plasticity: Theory and Practice*. Pineridge Press Ltd., Swansea.

Panek, L.A. (1956) Theory of model testing as applied to roof bolting. *U.S. Bureau of Mines*, R.I. 5154.

Panek, L.A. (1962a) The effect of suspension in bolted bedded mine roof. *U.S. Bureau of Mines*, R.I. 6138.

Panek, L.A. (1962b) The combined effect of friction and suspension in bolting bedded mine roof. *U.S. Bureau of Mines*, R.I. 6139.

Panek, L.A. (1965a) Design of bolting systems to reinforce bedded mine roof. *U.S. Bureau of Mines*, R.I. 5155.

Panek, L.A. (1965b) Principles of reinforcing bedded roof with bolts. *U.S. Bureau of Mines*, R.I. 5156.

Panet, M. (1969) Several rock mechanics problems observed in the Mont Blanc tunnel (in French). *Bull. Liasion Labo. Routiers P. et Ch.*, Dec.(42), 115–145.

Pellet, F. (1994) *Strength and Deformability of Jointed Rock Masses Reinforced by Rock Bolts*. Ph.D. Thesis, No. 1169, Ecole Polytechnique Federale de Lausanne, 205 pages.

Perzyna, P. (1966) Fundamental problems in viscoplasticity. *Advances in Applied Mechanics*, 9(2), 244–368.

Pietruszczak, S. & Mroz, Z. (1981) Finite element analysis of deformation of strain-softening materials. *Int. J. Num. Meth. Eng.*, 17, 327–334.

Pistone, P.E.S. & del Rio, J.C. (1982) Excavation and treatment of the principal faults in the tailrace tunnel of the Rio Grande I hydroelectric complex, Argentina. *Proc. of the IV Congr., Int. Assoc. of Eng. Geol.*, New Delhi, 4, 139–152.

Rabcewicz, L. (1955) Bolted support for tunnels. *Mine and Quarry Engineering*, 153–159.

Rabcewicz, L. (1957a) Model tests with anchors in cohesionless material (in German). *Die Bautechnik*, 34(5), 171–173.

Rabcewicz, L. (1957b) Bolting in tunneling to replace the conventional methods (in German). *Schweizerische Bauzeitung*, 75(9), 123–131.

Rabcewicz, L. (1964) New Austrian tunneling method. *Water Power*, 453–457 (Nov. 1964), 511–515 (Dec. 1964) and 19–24 (Jan. 1965).

Rabcewicz, L. (1969) Stability of tunnels under rock load. *Water Power*, 225–229 (June), 266–273 (July) and 297–302 (Aug.).

Rabcewicz, L. & Golser, J. (1973) Principles of dimensioning the supporting system for the new Austrian tunneling method. *Water Power*, 88–93 (March).

Rabcewicz, L. & Gölser, I. (1979) Principles of dimensioning the supporting system for the new Austrian tunneling method. *Water Power*, 88–93 (March).

Ramsay, J.G. & Huber, M.I. (1987) *The Techniques of Modern Structural Geology.* Academic Press, London.

Rashid, Y.R. (1968) Ultimate strength analysis of prestresses concrete pressure vessels. *Nuclear Engineering and Design,* 7, 334–344.

Reik, G. & Soetomo, S. (1986) Influence of geological conditions on design and construction of Cirata powerhouse cavern. *Proc. of the Int. Symp. on Large Rock Caverns,* Helsinki, 1. pp. 195–208.

Rescher, O.J. (1968) Erfahrungen beim ausbau des Kavernzentrale Veytaux mit spritzbeton und Felsankern. *Felsmechanik und Ingenieur Geol.,* (Suppl. IV), 216–253.

Roberts, M.K.C. & Brummer, R.K. (1988) Support requirements in rockburst conditions. *Journal of the South African Institute of Mining and Metallurgy,* 88(3), 97–104.

Roko, R.O. & Daemen, J.J.K. (1983) A laboratory study of bolt reinforcement influence on beam building, beam failure and arching in bedded mine roof. *Proc. of the Int. Symp. on Rock Bolting,* Abisko. pp. 205–218.

Romana, M. (1985) New adjustment ratings for application of Bieniawski classification to slopes. *Proceedings of the International Symposium on the Role of Rock Mechanics in Excavations for Mining and Civil Works,* International Society of Rock Mechanics, Zacatecas. pp. 49–53.

Ross, C.T.F. (1988) The analysis of thin-walled membrane structures using finite elements: Finite element analysis of thin-walled structures. Editor Bull, J.W., Elsevier. *App. Sci. Pub.,* 93–132.

Sakamoto, A., Yamada, N., Sugiura, K., Aydan, Ö., Tano, H. & Hamada, M. (2006) Failure modes and responses of abandoned lignite mines induced by earthquakes and the evaluation of their stability. *4th Asian Rock Mechanics Symposium,* Singapore, Paper No. A0493 (on CD).

Sakurai, S. (1993) Back analysis in rock engineering. *Comprehensive Rock Engineering,* 4, 543–569, Balkema.

Savin, G.N. (1961) *Stress Concentrations around Holes.* Pergamon, Oxford.

Semenza, E. & Ghirotti, M. (2000) History of 1963 Vaiont slide: The importance of geological factors. *Bulletin of Engineering Geology and the Environment,* 59, 87–97.

Sezaki, M. (1990) *A Research on the Utilisation of Tunneling Data-Base and the Design and Construction of Shotcrete with the Consideration of Its Hardening Process.* Doctorate Thesis, Nagoya University, Nagoya, Japan. (In Japanese)

Sezaki, M., Aydan, Ö. & Kawamoto, T. (1994) A consideration on ground response curve (in Japanese). *Journal of Japan Society of Civil Engineers, Geotechnical Division,* 499(III-28), 77–85.

Sezaki, M., Aydan, Ö., Kawata, T., Moussa, A. & Swoboda, G. (1992) Numerical modeling for the representation of shotcrete hardening and face advance of tunnels excavated by bench excavation method. *Int. Symp. on Numerical Models in Geomechanics,* NUMOG IV, Swansea, 707–716.

Sezaki, M., Aydan, Ö. & Yokota, H. (1994) Non-destructive testing of shotcrete for tunnels. *Int. Conf. on Inspection, Appraisal, Repairs & Maintenance of Buildings & Structures,* Bangkok, 209–215.

Sezaki, M., Kawata, T., Swoboda, G., Aydan, Ö. & Moussa, A. (1992) Numerical modeling for the representation of shotcrete hardening and face advance of tunnels excavated by bench excavation method. In: Pande, G.N. & Pietruszczak, S. (eds.) *Num. Models in Geomech.* Balkema, Rotterdam. pp. 707–717.

Sezaki, M., Kibe, T., Ichikawa, Y. & Kawamoto, T. (1989) An experimental study on the mechanical properties of shotcrete (in Japanese). *J. Soc. Materials Sience, Japan,* 38(434), 1336–1340.

Shi, G.H. (1988) *Discontinuous Deformation Analysis: A New Numerical Model for the Statics and Dynamics of Block Systems.* Ph.D. Thesis, Department of Civil Engineering, University of California, Berkeley, 378 pages.

Shimizu, Y. (1992) *A Fundamental Research on the Assessment of the Stability of Slopes in Discontinuous Rock Masses.* Doctorate Thesis, Nagoya University, Nagoya, Japan.

Shimizu, Y., Aydan, Ö., Ichikawa, Y. & Kawamoto, T. (1988) An experimental study on the seismic behavior of discontinuous rock slopes (in Japanese). *The 42nd Annual Meeting of Japan Society of Civil Engineers,* III-180. pp. 386–387.

Shimizu, Y., Aydan, Ö., Ichikawa, Y. & Kawamoto, T. (1988) Dynamic stability and failure modes of slopes in discontinuous rock mass (in Japanese). *Geotechnical Journal. of JSCE*, 406(III-11), 189–198.

Shimizu, Y., Aydan, Ö. & Kawamoto, T. (1992) The stabilisation of rock slopes by rockbolting and shotcreting. *Int. Symp. on Rock Slides*, New Delhi. pp. 345–352.

Singh, B. (1973) Continuum characterization of jointed rock masses, part I: The constitutive equations'. *Int. J. Rock Mech.Min Sci.*, 10, 311–335.

Skudrzyk, F.J., Bandopadhyay, S. & Rybachek, S.C. (1986) Mining-induced landslide in permafrost. *Proc. of Int. Symp. on Geotechnical Stability in Surface Mining*, Calgary. pp. 169–177.

Snyder, V.W. (1983) Analysis of beam building using fully grouted rockbolts. *Proc. of the Int. Symp. on Rock Bolting*, Abisko. pp. 187–194.

Snyder, V.W. & Krohn, R.L. (1982) An experimental study of beam building mechanisms using fully grouted bolts in bedded mine roof. *Proc. of the Symp. on Strata Mechanics*, Newcastle upon Tyne. pp. 234–236.

Sokolnikoff, I.S. (1956) *Mathematical Theory of Elasticity*, 2nd ed. McGraw-Hill, New York.

Spencer, E.E. (1967) A method of the analysis of the stability of embankments assuming parallel inter-slice forces. *Géotechnique (London)*, 17, 11–26.

Sperry, P.E. & Heuer, R.E. (1979) Excavation and support of Navajo tunnel No.3. *Proc. of the Rapid Excavations and Tunneling Conf.*, 1, 539–571.

Stacey, T.R. & Ortlepp, W.D. (1994) Rockburst mechanisms and tunnel support in rockburst conditions. In: Rakowski, Z. (ed.) *Geomechanics 93*, Balkema, Rotterdam. pp. 39–46.

Stille, H. (1983) Theoretical aspects on the difference between prestressed anchor bolt and grouted rockbolt in squeezing rock. *Proc. of the Int. Symp. on Rockbolting*, Abisko. pp. 65–73.

Talobre, J. (1957) *The Mechanics of Rocks* (in French) Dunod, Paris.

Tang, C. (1997) Numerical simulation of progressive rock failure and associated seismicity. *International Journal of Rock Mechanics and Mining Sciences*, 34(2), 249–261.

Tannant, D.D., Brummer, R.K. & Yi, X. (1995) Rockbolt behavior under dynamic loading: Field tests and modeling. *Int. J. of Rock Mechanics & Mining Sci. & Geomech. Abstracts*, 32(6), 537–550.

Tannant, D.D. & Buss, B.W. (1994) Yielding rockbolt anchors for high convergence or rockburst conditions. *Proc. 47th Canadian Geotechnical Conference*, Halifax.

Terzaghi, K. (1946) *Introduction to Tunnel Geology in Rock Tunneling with Steel Supports*. Commercial Shearing and Stamping Co., Youngstown, OH.

Terzaghi, K. (1946) Rock defects and loads on tunnel supports. In: Proctor, R.V. & White, T.L. (eds.) *Rock Tunneling with Steel Supports*. Youngstown, OH, 1. pp. 17–99.

Terzaghi, K. (1946) *Rock Tunneling with Steel Supports*, Commercial Shearing and Stamping Co., Youngstown, OH.

Terzaghi, K. (1960) Stability of steep slopes on hard, unweathered rock. *Geotechnique*, 12, 251–270.

Terzaghi, K. & Richart, F.E. (1952) Stresses in rock about cavities. *Geotechnique*, 3, 57–90.

Tharp, T.M. (1983) Mechanics of failure for rock masses subjected to long-term tensile loading – analysis of large naturally occurring cantilevers. *Proc. of the 24th US Symp. on Rock Mechanics*. pp. 309–318.

Timoshenko, S.P. & Gere, J.M. (1961) *Theory of Elastic Stability*, 2nd ed. McGraw-Hill, New York.

Timoshenko, S.P. & Goodier, J.N. (1951) *Theory of Elasticity*, 2nd ed. McGraw-Hill, New York.

Timoshenko, S.P. & Goodier, J.N. (1970) *Theory of Elasticity*, 3rd ed. Wiley, New York.

Tokashiki, N. & Aydan, Ö. (2010) The stability assessment of overhanging Ryukyu limestone cliffs with an emphasis on the evaluation of tensile strength of Rock Mass. *Journal of Geotechnical Engineering, JSCE*, 66(2), 397–406.

Tsuchiya, T. (1981) Research for design of system rock-bolt tunnelling method. *Proceedings of the International Symposium on Weak Rocks*, Tokyo, pp. 1063–1068.

Tsuchiya, Y., Kurokawa, T., Matsunaga, T. & Kudo, T. (2009) Research on the long term behavior and evaluation of concrete lining of the Seikan tunnel. *Soils and Foundations, JGS*, 49(6), 969–980.

Tsuneyoshi, F., Yamada, N., Izumi, Y. & Miki, K. (1998) Construction of Trans-Tokyo Bay Highway. *IABSE Reports*, 78, 43–48.

Uchida, Y., Harada, T. & Urayama, T. (1993) Behavior of discontinuous rock during large underground cavern excavation. *International Symposium on Assessment and Prevention of Failure Phenomena in Rock Engineering*, Istanbul. pp. 807–816.

Ulusay, R., A. Ersen & Ö. Aydan (1995) Buckling failure at an open-pit coal mine and its back analysis. *The 8th International Congress on Rock Mechanics, ISRM*, Tokyo, pp. 451–454.

Ünal, E. (1983) *Design Guidelines and Roof Control Standards for Coal Mine Roofs*. Ph.D. Thesis, Pennsylvania State University, University Park, 355 pages.

Ünal, E. (1992) Rock reinforcement design and its application m mining. *Proceedings of the International Symposium on Rock Support*, Canada. pp. 541–547.

US Army Corps of Engineers (1994) *Engineering and Design – Rock Foundations*. EM1110–1–2908, 121 pages.

Verruijt, A. (1997) A complex variable solution for a deforming circular tunnel in an elastic half plane. *Int. J. Numer. Anal. Methods Geomech.*, 21(2), 77–89.

Vesic, A.S. (1971) Breakout resistance of objects embedded in ocean bottom, *Journal of the Soil Mechanics and Foundation Division*, ASCE, Vol. 97, No. SM9, Paper 8372, pp. 1183–1205.

Villari, E. (1865) Change of magnetization by tension and by electric current. *Annual Review of Physical Chemistry*, 126, 87–122.

von Karman, T. (1911) Testigkeitsversuche unter allseitigem Druck. *Z. Ver. Dtsch. Ingenieure*, 55, 1749–1757.

Walton, G. & Coates, H. (1980) Some footwall failure modes in South Wales opencast workings. *Proc. of the 2nd Int. Conf. on Ground Movements and Structures*, Cardiff. pp. 435–451.

Watanabe, A., Komura, S., Aydan, Ö. & Ebisu, S. (1990a) Mechanical properties of interfaces governing the bearing capacity of rock anchors (in Japanese). *Proc. 22nd Annual Rock Mechs. Symp.*, Tokyo. pp. 291–295.

Watanabe, A., Komura, S., Aydan, Ö. & Ebisu, S. (1990b) *In situ* pull-out tests on failure modes of rock anchors and their mechanism (in Japanese). *Tsuchi to Kiso.*, 38(5), 27–32.

Weiss-Malik, R.F. & Kuhn, A.K. (1979) Roof stabilization in a smooth-bored tunnel, Chicago T.A.R.P. *Proc. of the 20th US Symp. on Rock Mechanics.* pp. 225–232.

Whittaker, B.N. & Reddish, D.J. (1989) Subsidence: Occurrence, prediction and control. *Developments in Geotechnical Engineering*, 56, Elsevier, Amsterdam.

Wickham, G.E., Tiedemann, H.R. & Skinner, E.H. (1972) Support determination based on geologic predictions. In: Lane, K.S. & Garfield, L.A. (eds.) *Proc. North American Rapid Excav. Tunneling Conf.* Soc. Min. Engrs, Am. Inst. Min. Metall. Petrolm Engrs., Chicago, New York. pp. 43–64.

Wittke, W. (1964) A numerical method of calculating the stability of slopes in rocks with systems of plane joints (in German). *Felsmechanik und Ingenieurgeologie*, (Suppl. 1), 103–129.

Wittke, W. (1972) Influence of the shear strength of the joints on the design of prestressed anchors to stabilize a rock slope. *Proc. of the Geotech. Conf.*, 1.

Wullschläger, D. and O. Natau (1983) Studies on the composite system of rock mass and non-prestressed grouted rock bolts. *Proceddings of the International Symposium on Rock Bolting*, Abisko, pp. 75–85.

Yap, L.P. & Rodger, A.A. (1984) A study of the behavior of vertical rock anchors using the finite element method. *Int. J. Rock Mech. and Min. Sci.*, 22(2), 47–61.

Yoshinaka, R., Shimizu, T., Sakaguchi, S., Arai, H. & Kato, E. (1986) Reinforcing effect of rockbolt in rock joint model. *Proc. of the Int. Symp. on Engineering in Complex Rock Formations*, Beijing. pp. 922–928.

Yüzer, E. & Vardar, M. (1983) *Rock Mechanics* (in Turkish). Mine Faculty, Istanbul Teknik University, Istanbul.

Zachmanoglou, E.C. & Thoe, D.W. (1986) *Introduction to Partial Differential Equations with Applications*. Dover Pub. Inc., New York.

Zienkiewicz, O.C. (1977) *The Finite Element Method*. McGraw-Hill, London, 787 pages.

Zienkiewicz, O.C. & Pande, G.N. (1977) Time-dependent multi-laminate model of rocks – a numerical study of deformation and failure of rock masses. *Int. J. Num. Anal. Meth. Geomech.*, 1, 219–247.

Zienkiewicz, O.C., Valliappan, S. & King, I.P. (1969) Elasto-plastic solutions of engineering problems: Initial stress finite element approach. *Int. J. Num. Meths. Engng.*, 1, 75–100.

Zoback, M.D., Tsukahara, H. & Hickman, S.H. (1980) Stress measurements at depth in the vicinity of the San Andreas Fault: Implications for the magnitude of shear stress at depth. *Journal of Geophysical Research*, 85(B11), 6157–6173.

Subject index

action 10, 32, 44, 50, 67, 96, 137, 138, 223, 256, 264, 266, 289, 301, 316, 409, 410
  arch 137, 223, 228, 248, 256, 264, 266
  cathode 410
  chemical 32
  electrochemical 409, 410
  Initial shearing 10
  Internal pressure 67
  load 50, 248
  reinforcement 289
  shotcrete 228
  splitting 10
  stitching 316
  support 301
  wedging 138
Adali 68, 463
airflow 39
Akutagawa 463, 452
analytical solution 40, 41, 177, 178, 191–208, 235–240, 432, 458–469
atmospheric 56, 279, 289, 459
axisymmetric 45, 156, 160, 163, 191, 200, 235, 237, 330, 332, 349, 350
  Body 163, 350
  cavity 235
  Coordinate system 156
  cylindrical 45
  Finite element 330, 332
  object 156
  state 160
  structure 191

bar 5, 20, 43, 62, 65, 67, 70–80, 91–166, 187, 213, 222, 223, 240–252, 309, 321–335, 375–378
  Axial stiffness 155, 240
  cylindrical 67, 69
  element 155, 159, 240
  Glass-fiber 70, 71
  grout 62
  material 5

radius 187
Shear resistance 251
steel 20, 40, 62, 67, 70–80, 156, 222, 223, 242–252, 321–335, 375–378
Barla 155, 466, 468
behavior 32, 40–45, 114–123, 134, 150–162, 170–178, 190, 227, 279, 326–330, 368, 380
  elastic 150, 162, 170, 327–330, 368
  elasto-plastic 40–45, 62, 114–123, 178, 227, 330, 357
  elasto-visco-plastic 32, 380
  Long-term 32, 279, 326
  visco-elastic 190
blasting 404–408, 428, 452–461
Bieniawski 16, 33–37, 53–56, 189, 469, 471
  Prismatic block 10
body force 42, 171, 189, 458
borehole 41, 56, 64–83, 104–187, 208, 282, 322–332, 456, 460
breakout 41, 208
Brekke 472
Brown 16, 41, 68, 97, 178–179, 189, 203, 270, 469, 472

cavern 33, 39, 48, 49, 69, 150, 395, 396, 428, 462
cavity 32, 40, 212, 235, 238, 241, 351, 404
  circular 235, 404
  spherical 40
chemical reaction 409
concrete Lining 39–42, 148, 178, 191, 238, 268–271, 455
condition 172, 181–238, 246, 309, 330–340, 371, 378, 405, 432, 456
  boundary 181–238, 246, 309, 330–340, 371, 456
  initial 172, 378, 405, 432
Cook 41, 208, 471–272
corrosion 6, 62, 65, 320, 326, 409–426
creep 189, 190, 198, 279, 426, 435
  device 279
  failure 189, 198

creep (*continued*)
   secondary 190
   strength 190
   tertiary 190
   test 189

dam 5–6, 14, 56, 65–66, 69, 317, 354, 366, 429,
   459, 460, 462
   Aratozawa 65
   arch 66, 366
   foundation 14, 317, 354, 366, 460
   gravity 354, 366, 367, 459
   Malpasset 65, 366
damping 44, 174, 378–379, 430–432,
   437–438
   coefficient 378–379, 430
   matrix 174
   ratio 430
degradation 6, 56, 65, 198–199, 279–282,
   326–327, 409–453, 462
discontinuity 7, 13–27, 43–59, 69, 101–106, 13,
   69, 101–106, 209–212, 242–266, 290–310,
   366–385, 461
   formation 7, 13–27, 43–59
   natural 47–49
   plane 13, 69, 101–106, 209–212, 242–266,
      290–310, 366–385, 461
   set 13–27, 43–59
   spacing 23
displacement transducer 78
drilling 64, 322, 404

earthquake 6, 34, 49, 255, 271–272, 455,
   354–356, 375, 396, 455–460
   2003 Miyagi-Hokubu 271
   2005 Nias 272
   2008 Iwate-Miyagi 271, 375
   2009 L'Aquila 271
   2011 Great East Japan 271
effect 5, 69, 101, 105, 137, 177, 188, 242, 243,
   256, 456, 458
   arch formation 256, 456, 458
   Beam building 69
   dowel 101, 105
   Internal pressure 69, 137, 188, 456
   suspension 5, 69, 177, 242, 243, 456, 458
Egger 41, 68, 69, 105, 106, 178, 201, 241, 246, 470
Electrical resistivity 273, 412–414, 462
element 5, 42–44, 155, 159, 171–174, 196, 233,
   289, 334–335, 457–459
   contact 42–43, 171–174
   interface 44, 159
   rockbolt 5, 155, 159, 166–167, 196, 233, 289,
      334–335, 457–459
   shotcrete 166–167, 233, 457
   solid 44

equation of motion 40, 42, 171–172, 177,
   378–379, 430
Eringen 42, 405, 471
Ersen 150, 183, 202–203, 466, 471, 478
experiment 33, 35, 138, 271, 272, 274, 279,
   292, 301
   Long-term 279
   Physical model 292, 301
   shotcrete 138
   Short-term 271, 272, 274
   Trapdoor 33, 35

fault 9, 53, 59
   columnar 61, 248, 252, 300–302
   flexural 16, 23, 46, 49–51, 248–253, 296–299,
      365
   Toppling 16, 23, 46, 49–51, 61, 248–253,
      296–299, 300–302, 365
flow rule 73
forepole 40, 224–227
foundation 6, 69, 317, 365, 366, 367, 459, 460
   bridge 69, 365, 370, 374
   dam 6, 317, 366, 367, 459, 460
   pylons 317
friction angle 27–33, 37–39, 51, 97–105,
   181, 202, 210–213, 227, 250, 262, 290,
   306–312, 340, 351, 368, 385, 387, 395

Gauss divergence theorem 172
geoengineering 45
Goodier 41, 208, 373, 477
Grout annulus 82–93, 115–119, 127, 155–164,
   199, 204, 220, 222, 330, 332, 450, 456, 457

Hamming 471
Hinton 42, 170, 171, 475
Hoek 41, 68, 73, 178–179, 189, 469, 472

Ichikawa 5, 89, 456, 465–477
Ikeda 356, 472
Imazu 467, 473
Impact wave 430–453
Inglis 41, 208, 473
Iteration Scheme 43, 170, 171, 263
   Direct 170
   Newton-Raphson 42, 170, 210, 263, 368
   Regula Falsi 210
   Secant 42, 170, 368
   Tangential 42, 170
Ito 464

Jaeger 13, 41, 208, 473

Kirsch 41, 207, 211, 473
Kreyszig 380, 473
Kumsar 24, 466, 468, 474

Ladanyi 190, 469, 474
law 377, 404, 423
    mass conservation 423
    Voigt-Kelvin 377, 404
Lift-off testing 452
Lining 178, 191, 238, 268–271, 455
    Concrete 178, 191, 238, 268–271, 455
    design 268–269

measurement 5, 59, 139, 177, 196, 357, 381,
        383, 386, 458
    Displacement response 383, 386
    field 5, 59, 177, 458
    Ground stress 357
    In-situ 196, 447, 458
    vibration 381
    Wave velocity 139, 425
method 5, 42–45, 155, 171–175, 191–196, 232,
        289, 297–298, 302, 331–333, 356, 367,
        377, 380, 423, 457–461
    discrete finite element 5, 42, 43, 171, 289,
        293, 298, 302, 367, 457–459
    finite difference 42, 380
    finite element 42–45, 155, 171–175, 191–196,
        232, 289, 297, 331–333, 356, 367, 377,
        423, 457–461
    limiting equilibrium 113, 169
    secant 42, 368
    tangential stiffness 42
mine 244, 271–272, 281–288, 411, 421, 459
    abandoned 271–272, 281–288, 411,
        421, 459
    room and pillar 244, 271–272, 285, 459
Mitake 279, 411, 414
modulus 148–162, 178–179, 187, 200, 205,
        217–248, 325–374, 425–437
    deformation 178, 428
    elastic 148–162, 179, 187, 200, 205, 217–248,
        325–374, 425–437, 458
    shear 187, 200
mudstone 16, 137, 448–450
Müller 468–474

Nawrocki 172, 190, 426, 466
Neutral point 220, 458
Nondestructive testing 409–452
North Selby Mine shaft 268, 270
nuclear waste 268
numerical analysis 45, 53, 155, 161, 445, 457

overburden 40, 183, 227, 232, 283,
        285, 421
Owen 42, 170, 171, 475
Oya tuff 272–276

Phase angle 443

ratio 39, 53, 111, 139, 145, 148, 150, 179, 181,
        200, 203, 227–236, 268, 356, 374, 435
    Poisson's 39, 111, 139, 145, 148, 150, 179,
        181, 200, 203, 227–236, 268, 356, 374, 435
    stress 53
relaxation 9, 137, 189
Rock 7–16, 59, 137, 270
    Igneous 7–16, 59
    Sedimentary 7–16, 59, 137, 270
Rock Classifications 33, 35–40, 64,
        351–352, 456
    DENKEN (CRIEP) 351–352, 352, 356
    NEXCO 33
    Q-Value System 33, 37–39, 64, 351–352
    Rock mass quality rating (RMQR) 33, 39–40,
        64, 351–352, 456
    Rock mass rating (RMR) 33, 35–37, 64,
        351–352
    Rock quality designation (RQD) 33, 35, 64,
        351–352
rockanchor 51, 62–66, 67, 69, 111–133,
        166, 177, 199–208, 251, 289, 292, 296,
        301–302, 317–336, 365–374, 375–387,
        394–396, 410, 424–453, 455–462
rockbolt 3–6, 35–42, 48–51, 61–66, 67–135,
        151, 155–170, 177–178, 185–214, 217–270,
        288–304, 309, 312–316, 322–335, 365,
        370, 375–393, 397–403, 406–408, 411–412,
        423–453, 455–462
Rock reinforcement 6, 31, 374, 375, 377, 455,
        456, 458, 460
Rock salt 16
Rock support 6, 31, 271, 374, 375, 377, 455
Ryukyu limestone 272–273

Sandstone 20, 65, 270, 322, 327, 460
sensor 282, 383, 386, 451–453, 444
    acceleration 444
    AE 282, 383
    displacement 383, 386
    magnetic 451–453
Shale 20, 137, 225
shape function 160, 166, 173
shear 70, 84, 89, 91–96, 102–106, 116–130, 158,
        162, 187, 199, 200, 209–220, 242–248,
        290, 297, 325, 328, 330, 334, 341, 345,
        370, 372–375, 430, 431, 436–437, 460
    modulus 94, 117, 118, 158, 162, 187, 200,
        325, 370, 431, 436–437
    strain 89, 116, 117, 158, 159, 375
    stress 70, 84, 91–96, 102–106, 116–130,
        158, 199, 209–220, 242–248, 290, 297,
        328, 330, 334, 341, 345, 372–375, 430,
        431, 460
simultaneous equation system 309
sliding plane 264, 306, 308, 383–389

softening 115–130, 179, 194, 275–279, 326,
    332–334
strength 13, 16, 39, 43, 50, 51, 98–101, 105,
    134, 135, 139–149, 179, 190–193, 202,
    210, 236, 240, 243, 248, 250, 252, 347,
    275, 350, 351, 356, 359, 365, 368,
    425–427, 435
  tensile 16, 39, 43, 50, 51, 98–101, 105, 134,
    135, 210, 236, 240, 243, 248, 250, 252, 347,
    350, 351, 356, 359, 365, 368, 425–427, 435
  triaxial 13
  uniaxial compressive 139–149, 179, 190–193,
    202, 275
support ring 39, 233

Taylor expansion 173
Terzaghi 16, 33–41, 208, 268, 477
test 65, 89, 111, 310, 385, 387, 396
  direct shear 65, 89, 111, 310
  tilting 385, 387, 396
timoshenko 2, 20, 41, 208, 373, 477
tunnel 2, 5, 20, 66, 178, 232, 267, 317–357,
    407–408, 425, 427, 448–450, 459, 461
  anchorage 5, 66, 317–357
  circular 178, 191, 461
  Japan 425, 427
  Navajo irrigation 20

Seikan 267
subsea 267, 459
Tarutoge 407–408
Tawarazaka 197–198, 232
type 5, 45, 452
  hyperbolic 45
  sliding 5
  toppling 452

Ulusay 40, 53, 316, 468, 474, 478
Underground power station 425, 453
  mazegawa 425
updated Lagrangian scheme 43, 171, 458

Vardar 9, 478

Wittke 24, 61, 478

X-Ray CT Scanning Technique 421, 422, 462

yield criterion 41, 43, 96, 111, 178, 179, 190,
    202–210, 261, 306, 308, 341, 343, 347,
    349
  Hoek-Brown 41, 178, 179
  Mohr-Coulomb 41, 43, 96, 111, 178, 179, 190,
    202–210, 261, 306, 308, 341, 343, 347, 349
yield zone 41, 57, 207, 208, 458

ISRM Book Series

Book Series Editor: Xia-Ting Feng

ISSN: 2326-6872

Publisher: CRC Press/Balkema, Taylor & Francis Group

1. Rock Engineering Risk
   Authors: John A. Hudson & Xia-Ting Feng
   2015
   ISBN: 978-1-138-02701-5 (Hbk)

2. Time-Dependency in Rock Mechanics and Rock Engineering
   Author: Ömer Aydan
   2016
   ISBN: 978-1-138-02863-0 (Hbk)

3. Rock Dynamics
   Author: Ömer Aydan
   2017
   ISBN: 978-1-138-03228-6 (Hbk)

4. Back Analysis in Rock Engineering
   Author: Shunsuke Sakurai
   2017
   ISBN: 978-1-138-02862-3 (Hbk)

5. Discontinuous Deformation Analysis in Rock Mechanics Practice
   Author: Yossef H. Hatzor, Guowei Ma & Gen-hua Shi
   2017
   ISBN: 978-1-138-02768-8 (Hbk)

6. Rock Reinforcement and Rock Support
   Author: Ömer Aydan
   2018
   ISBN 978-1-138-09583-0 (Hbk)

Printed in the United States
by Baker & Taylor Publisher Services